Governability of Fisheries and Aquaculture: Theory and Applications

MARE Publication Series

Volume 7

Series Editors

Svein Jentoft, *University of Tromsø, Norway*
Svein.jentoft@uit.no

Maarten Bavinck, *University of Amsterdam, The Netherlands*
J.M.Bavinck@uva.nl

The MARE Publication Series is an initiative of the Centre for Maritime Research (MARE). MARE is an interdisciplinary social-science network devoted to studying the use and management of marine resources. It is based jointly at the University of Amsterdam and Wageningen University (www.marecentre.nl).

The MARE Publication Series addresses topics of contemporary relevance in the wide field of 'people and the sea'. It has a global scope and includes contributions from a wide range of social science disciplines as well as from applied sciences. Topics range from fisheries, to integrated management, coastal tourism, and environmental conservation. The series was previously hosted by Amsterdam University Press and joined Springer in 2011.

The MARE Publication Series is complemented by the Journal of Maritime Studies (MAST) and the biennial People and the Sea Conferences in Amsterdam.

For further volumes:
http://www.springer.com/series/10413

Maarten Bavinck • Ratana Chuenpagdee
Svein Jentoft • Jan Kooiman
Editors

Governability of Fisheries and Aquaculture: Theory and Applications

Springer

Editors
Maarten Bavinck
Department of Human Geography
Planning and International Development Studies
University of Amsterdam
Amsterdam, The Netherlands

Ratana Chuenpagdee
Department of Geography
Memorial University of Newfoundland
St. John's, NL, Canada

Svein Jentoft
Faculty of Biosciences, Fisheries and Economics
University of Tromsø
Tromsø, Norway

Jan Kooiman
Centre for Maritime Research
University of Amsterdam
Amsterdam, The Netherlands

ISSN 2212-6260 ISSN 2212-6279 (electronic)
ISBN 978-94-007-6106-3 ISBN 978-94-007-6107-0 (eBook)
DOI 10.1007/978-94-007-6107-0
Springer Dordrecht Heidelberg New York London

Library of Congress Control Number: 2013934113

Printed on acid-free paper

Springer is part of Springer Science+Business Media (www.springer.com)

Foreword

Around four decades ago professionals working on fisheries problems generally agreed there was a crisis in the world's fisheries. The world's fish catch had leveled off, many formerly productive fisheries were badly depleted, and a few had utterly collapsed. Since then some progress has been made, but the overall trend worldwide has continued downward. True, the world's total fish supply increased over this time, but that was mainly a result of increased targeting of smaller species further down the marine food web, and also the phenomenal growth of aquaculture. Indeed, aquaculture contributed only around 3% of the world's fish supply four decades ago, but now contributes almost 40% – and now raises environmental concerns that were only dimly anticipated before its phenomenal surge.

Over the past years several theoretical approaches have been proposed for analyzing and alleviating this crisis, but perhaps none have proposed a theoretical framework as holistic and promising as the one in this ambitious book, which urges that fisheries and aquaculture systems be viewed through the lens of *governability*, or their capacity for governance.

The present volume is a follow-up to the authors' earlier book, *Fish for Life* (2005), which only raised the topic of governability in its concluding chapters. It precedes a planned third volume that will explore methods for applying governability theory in real-world situations.

The core authors here have been collaborating on the theoretical development of the governability concept for more than a decade, and this volume represents the fruition of their many formal discussions, meetings, workshops, research, and scholarly work. And now joined by more than a dozen additional authors, here presented is a rich diversity of intellectual, philosophical, theoretical, and methodological insights, as well as case studies that explore these insights in six different culture regions.

For assessing a particular system's capacity for governance, the authors urge beginning by conceptually deconstructing the system into three sub-systems: *the system to be governed, the governing system,* and *the governing interactions* between these two. Then, with each part's components and dynamics richly illuminated, they urge synthesizing an "image" of the system's *governance capacity* by focusing on its inherent components and dynamics, which either promote or discourage that.

Furthermore, they suggest the *system to be governed* should be analyzed with regard to four cardinal features of all social systems: *diversity, complexity, dynamics*, and *differences in scale*. And they also suggest that it should be investigated from the perspective of *a set of concerns* that impact governability – here emphasizing such contemporary concerns as *social justice, livelihoods, food security*, and *ecosystem health.*

For analyzing the *governing system* a number of foci are likewise suggested, including the *governance realms* that pertain to *the state, markets, civil society*, and *hybrids* of these. And to analyze the dynamics of the *governing interactions*, they urge looking for interactional factors that ultimately promote or inhibit effective governance.

Obviously, by taking such a broadly holistic approach, it is unlikely that a panacea will emerge, much less a formulaic set of rules for assessing the governability of a fisheries or aquaculture system. Each system will still have to be studied with regard to its own particular history, organization, dynamics, and contemporary circumstances – whatever these may be.

The authors also acknowledge that the governance of some social systems may pose a "wicked problem," that is, a problem that is so complex that it is highly resistant to resolution or improvement. In such cases no solution may be possible, and there may be no foreseeable end point at which the problem could ever be considered to be resolved. But into these seemingly hopeless situations the authors now beam a new ray of light, stressing that by viewing them through the lens of governability their internally contradictory concerns, disjointed components, confounded interactional dynamics, and other obstacles to their governability can be identified. Indeed, the governability approach's emphasis on discovering governing interactions is what distinguishes it from most of the other major theoretical approaches that have been offered heretofore.

It now seems clear that the crisis in the world's fisheries was a much larger and more complex problem than many had imagined. Yet, examining it through the lens of governability may offer the best hope for alleviating it – as well as alleviating similar crises in other social systems.

James R. McGoodwin

James R. McGoodwin is author of Crisis in the World's Fisheries: People, Problems, and Policies (Stanford University Press, 1990). A former Fellow in the Marine Policy and Ocean Management Program of the Woods Hole Oceanographic Institution, he is Professor Emeritus in the Department of Anthropology, and Research Affiliate in the Institute of Arctic and Alpine Research at the University of Colorado, Boulder, Colorado

Preface

Coasts and seas are special environments that give rise to a set of unique activities, cultures and governance issues. This volume highlights the situation of capture fisheries and aquaculture from a global perspective, and takes an interactive governance approach. It is a sequel to an earlier publication in the series entitled, *Fish for Life – Interactive Governance for Fisheries* (2004). The present volume continues where the earlier book left off, and explores the concept of governability and its application to fisheries.

The MARE Publication Series commenced in 2004 under the auspices of Amsterdam University Press and produced five volumes on coastal and marine topics. This is the first volume to be produced in collaboration with Springer Academic Publishers. We are particularly grateful to Martine van Bezooijen(†) and Fritz Schmuhl for facilitating the transition and taking us on.

As we have contributed to this volume, Dr. Mirjam Ros-Tonen (University of Amsterdam) has stepped in as guest series editor. We are grateful for her help.

University of Tromsø — Svein Jentoft
University of Amsterdam — Maarten Bavinck

He began to pity the great fish that he had hooked. He is wonderful and strange and who knows how old he is, he thought. Never have I had such a strong fish nor one who acted so strangely. Perhaps he is too wise to jump. He could ruin me by jumping or by a wild rush. But perhaps he has been hooked many times before and he knows that this is how he should make his fight.

Ernest Hemingway
(*The Old Man and the Sea*)

Here is the Sea of Indifference, glazed with salt [...]
this is the sea town of myth and story   when the fishing fleets
went bankrupt   here is where the jobs were   on the pier
processing frozen fishsticks   hourly wages and no shares [...]
then yes let it be   these are small distinctions
where do we see it from is the question

Adrienne Rich
(*An Atlas of the Difficult World*)

Acknowledgements

This book has been some years in the making. The idea originated back in 2005 with the publication of a preceding volume – *Fish for Life* – that launched the concept of governability in its final chapters. The members of the Fisheries Governance Network (FISHGOVNET), which had been established through project funding by the European Commission, agreed that this topic was critical for the development of the interactive governance approach and decided to take it forward, first through a special issue for the *Journal of Transdisciplinary Environmental Studies*. The Stockholm resilience conference of 2008 then provided the editors with an occasion to decide on the outline and schedule of this volume. Other meetings – some with a fuller group of authors, others with the editors alone – took place at regular intervals in Ottawa, Gothenburg, St John's, Bangkok and Amsterdam.

The present volume could not have come about without the encouragement and support of a number of people and organizations. The editors would first of all like to mention the backing of the University of Amsterdam, Memorial University, and the University of Tromsø, that provided them with the opportunity to pursue their interests in this field. Ratana Chuenpagdee also acknowledges support from Canada Research Chairs program and Social Sciences and Humanities Research Council of Canada. Svein Jentoft acknowledges the support from the POVFISH project and the POVPEACE program of the Norwegian Research Council.

We would like to thank Miriam Ros for coordinating the review process on behalf of the editors of the MARE Publication Series. The two reviewers provided important, and in some respects, contradictory views on how to improve the manuscript – we have taken their suggestions to heart and integrated them in our own way. We are more than grateful to Philip van der Krogt, who did a wonderful job language-editing the manuscript. Our thanks are also due to Fritz Schmul and others at Springer Publishers, who have guided us skillfully through the publishing process. Russ McGoodwin was willing to go through the manuscript and write a scintillating foreword, for which we extend our gratitude.

Finally we would like to take the opportunity to thank the contributors to this book, who have put up with a stream of comments and requests from our side. They have been remarkably patient with us, and we hope they feel that the manuscript has improved in the process.

The Editors

Contents

Contributors

Sharon Almerigi People Dynamics Associates, St. James, Barbados

Maarten Bavinck Department of Human Geography, Planning and International Development Studies, University of Amsterdam, Amsterdam, The Netherlands

Ratana Chuenpagdee Department of Geography, Memorial University of Newfoundland, St. John's, NL, Canada

Raquel De la Cruz Modino Institute of Social and Political Sciences, University of La Laguna, Tenerife, Spain

Lucia Fanning Marine Affairs Program, Dalhousie University, Halifax, NS, Canada

Katia Frangoudes Center for the Law and Economics of the Sea, University of Brest, Brest, France

Svein Jentoft Faculty of Biosciences, Fisheries and Economics, University of Tromsø, Tromsø, Norway

Derek Johnson Department of Anthropology, University of Manitoba, Winnipeg, MB, Canada

Jan Kooiman Centre for Maritime Research, University of Amsterdam, Amsterdam, The Netherlands

Yajie Liu Department of Economics, Norwegian University of Science and Technology, Trondheim, Norway

Robin Mahon Centre for Resource Management and Environmental Studies (CERMES), University of the West Indies, St. Michael, Barbados

Begoña Marugán-Pintos Department of Sociology, University Carlos III of Madrid, Getafe, Madrid, Spain

Patrick McConney Centre for Resource Management and Environmental Studies (CERMES), University of the West Indies, St. Michael, Barbados

Paul Onyango Department of Aquatic Sciences and Fisheries, University of Dar es Salaam, Dar es Salaam, Tanzania

José J. Pascual-Fernández Institute of Social and Political Sciences, University of La Laguna, Tenerife, Spain

Roger S.V. Pullin Independent Consultant Aquatic Biologist, Makati City, Philippines

Joeri Scholtens Department of Human Geography, Planning and International Development Studies, University of Amsterdam, Amsterdam, The Netherlands

Andrew Song Department of Geography, Memorial University of Newfoundland, St. John's, NL, Canada

U. Rashid Sumaila Fisheries Economics Research Unit, Fisheries Centre, University of British Columbia, Vancouver, BC, Canada

Part I
Introducing Governability

Chapter 1
Governability – New Directions in Fisheries Governance

Ratana Chuenpagdee, Svein Jentoft, Maarten Bavinck, and Jan Kooiman

Abstract The chapter explains the origin and conception of the book, provides the rationale for its contents and describes its goals. It also includes a brief description of each chapter, how the chapters are linked and presented to illustrate the complex and wicked reality of the fisheries and aquaculture governance, and the utility of applying a systematic analytical lens like governability in addressing them. While the book focuses on the four main concerns identified in the earlier publication, *Fish for Life* (ecosystem health, social justice, livelihood, and food security), it argues that the framework can be applied to address concerns and governability challenges faced in other spheres.

Keywords Governability • Interactive governance • Fisheries • Aquaculture • Wicked problems

R. Chuenpagdee (✉)
Department of Geography, Memorial University of Newfoundland, St. John's, NL, Canada
e-mail: ratanac@mun.ca

S. Jentoft
Faculty of Biosciences, Fisheries and Economics, University of Tromsø, Tromsø, Norway
e-mail: svein.jentoft@uit.no

M. Bavinck
Department of Human Geography, Planning and International Development Studies, University of Amsterdam, Amsterdam, The Netherlands
e-mail: j.m.bavinck@uva.nl

J. Kooiman
Centre for Maritime Research, University of Amsterdam, Amsterdam, The Netherlands
e-mail: jkooiman@xs4all.nl

M. Bavinck et al. (eds.), *Governability of Fisheries and Aquaculture: Theory and Applications*, MARE Publication Series 7, DOI 10.1007/978-94-007-6107-0_1,

Why are fisheries in crisis worldwide (McGoodwin 1990; Clark 2006)? What stops us from making progress and halting the over-exploitation of inland and marine fisheries, as well as unsustainable coastal and ocean development (Pauly et al. 2005; Roberts 2007)? Why do we have a hard time defining whether we need an ecosystem-based, wealth-based or livelihood approach to fisheries (see for instance, Berkes 2011)? Could it be that we still do not agree on what the real concerns are or fully understand the extent of the problems and challenges they raise for governance? We posit that part of what is missing is a holistic perspective to examine and address these problems, which may seem obvious and simple, but are complex and wicked in reality. This book offers a lens and a systematic approach for analyzing the nature of problems and challenges concerning the governance of fisheries and aquaculture. It explores where these problems are situated, and where potential solutions may be found. The key concept of the current focus is governability.

This book follows the line of investigation offered in the first product of the Fisheries Governance Network (www.fishgovnet.org), *Fish for Life – Interactive Governance for Fisheries* (Kooiman et al. 2005). In *Fish for Life*, which succeeded *Creative Governance* (Kooiman et al. 1999), we applied the theory of interactive governance and governability, as developed by Jan Kooiman in *Modern Governance* (1993) and *Governing as Governance* (2003), to fisheries and aquaculture. These theoretical groundings have since been further elaborated and explored in various settings around the world, and resulted in a series of publications, including special issues in the *Journal of Transdisciplinary Environmental Studies* (Kooiman et al. 2008) and *Fish and Fisheries* (Chuenpagdee and Sumaila 2010). The conceptual framework that this approach offers has generated a methodology to examine the characteristics of the aquatic environment, the natural as well as the human dimensions, including institutions governing the interactions between them. It has also stimulated the phrasing of research questions essential to addressing the complexity of the challenges that a better governance of fisheries and aquatic system would require.

Encouraged by the richness of the interactive governance perspective and the governability concept, as well as the opportunities that they bring to the fisheries discourse, in this volume we continue to pursue what this approach may reveal in settings where governors and stakeholders face real issues and challenges. Our deliberation is based on the premise that these challenges are varied and can be felt among governments and communities globally. The diagnosis of the problem must therefore come from their interactions and align with their ideas of what the solutions may be. Such emphasis on interactions distinguishes the interactive governance approach from others, which tend to presume that the tasks of governing lie either with the state or with communities.

Fish for Life contains a systematic analysis of the major, generalized and overall features of fishery systems that are seen as parts of a fishing chain: from the ecosystem to the consumer. It ends with a discussion of the governability of such systems, introduced as a diagnostic tool to understand the extent to which fisheries systems "are governable, i.e., have characteristics that facilitate or hamper governance" (Kooiman et al. 2005, 351). The assumption was, and still is, that governability of fisheries systems varies from one to the next, depending on the constellations of specific features.

The present volume continues where *Fish for Life* left off, highlighting governability as "the overall capacity for governance" of a societal system (see Chap. 2 by Kooiman and Bavinck this volume). However, rather than venturing straight away into the outline of an assessment framework, we begin by providing theoretical foundations for investigating major concerns in fisheries and aquaculture, followed by applying the governability concept in real world situations.

Part I contains this introductory chapter and Chap. 2, which revisits the concept and theory behind the interactive governance approach and governability. Part II investigates governability from the viewpoint of the four societal concerns that are highlighted in *Fish for Life* – ecosystem health, social justice, livelihood, and food security. These concerns are examined for the role they play in enhancing or inhibiting governability. The diversity, complexity, dynamics and scale associated with the systems-to-be-governed, the governing system and their interactions are portrayed in light of these concerns. Part III applies a governability perspective to specific themes and topics in fisheries and aquaculture. These include issues related to trawl fisheries in India, marine protected areas in Spain, salmon farming in Canada and Norway, gender relations in Galician shellfish gathering, and poverty in small-scale fisheries in Lake Victoria. In Part IV, we provide an overview of approaches and methods that can be utilized to examine various aspects of fisheries systems that impact on governability. Chapters in this part illustrate some of these methods as they are applied in the real world situations, including the Caribbean and Malawi. Part V is about looking forward. It examines the next steps necessary for a full and systematic assessment of governability that can be applied to fisheries, aquaculture and other contexts. By drawing on the lessons provided in earlier chapters, the final two chapters offer both an analytical framework and some general reflections.

Like *Fish for Life*, this volume "stands somewhere between an academic monograph and a multi-author, edited volume" (Kooiman et al. 2005, 8). The nucleus of the author group has been working together on interactive governance for over a decade. Added to the list of authors are younger scholars who have been attracted by the potentialities of the interactive governance approach for fisheries and aquaculture. While each author (set of authors) is of course responsible for the content of the chapter to which their name is attached, and the volume allows for a diversity of insights, the editors have made deliberate efforts to compose an integrated perspective.

Although originating in the social sciences, the book's ambit is inter-, if not transdisciplinary. The author group consists of representatives from a range of social science disciplines, as well as economists and marine biologists. All of us have made efforts to master the interactive governance perspective and to infuse it with our own insights from our experiences on the ground. In so doing, we have crossed many disciplinary boundaries and hopefully contributed to a more holistic approach.

While the present chapter explains why we believe this book is needed, the next chapter by Kooiman and Bavinck (Chap. 2) provides the theoretical foundation underlying the interactive governance and governability approach. It places an emphasis on the governing roles and interactions between state, market and civil society, by examining two inter-related systems, i.e., a system-to-be-governed, a governing system, and the governing interactions mediating between them. Together, these two chapters of Part I serve as an introduction to the theory and concept underlying the applications that follow in the rest of the book.

Part II consists of five chapters, the first of which (Chap. 3 by Jentoft and Chuenpagdee) offers an overview of issues and concerns in fisheries and aquaculture, and the problems they pose to governance. They argue that improving governability requires, first and foremost, recognition of the wickedness of these problems and the need for an interactive governance approach to examine and address them. Chapters 4, 5, 6 and 7 take these concerns and explore them in greater detail. Jentoft (Chap. 4) focuses on social justice implications. While arguing that justice is an important value in itself, he also explores the functional aspects of justice in providing more legitimacy and compliance among affected stakeholders. Livelihood concerns are addressed in Chap. 5 by Johnson. His chapter links employment, livelihood strategies and wellbeing as major targets for fisheries governance, while recognizing their inherent wickedness. In so doing, references are made to key principles for improving livelihood governability, including inclusiveness, reflexivity, adaptive capacity, precautionarity and social justice. The next concern is related to food security. Here, Pullin (Chap. 6) makes a strong case for fish constituting an essential contribution to food security at all levels; the continuity of its provision, however, is threatened by inter-sectoral and environmental problems in fisheries and aquaculture. The chapter also reviews potential interventions for improving governability along the fish chain. The final chapter in this part (Chap. 7), by Pascual-Fernández and Chuenpagdee, elevates the concern to another level, where the overall health of the marine and ocean ecosystems is of key importance. The authors employ the governability concept to examine stressors affecting the ability of the governing system to address ecosystem challenges and disturbance.

In Part III, the book turns to a number of thematic applications of the governability concept in fisheries and aquaculture. It starts with a chapter that further explores the concept and ends with another that illustrates how major social transformation may enhance governability. In between are chapters that aim to understand, assess, appraise and improve governability as they deal with various issues and concerns in different parts of the world. The stage is set in Chap. 8, by Bavinck and Kooiman, as they explore the variations in governability that occur in fisheries systems. Making use of fieldwork material from South Asian fisheries, as well as perspectives from legal pluralism and institutional studies, the chapter offers a sense of the range of governability issues that may prevail in any given region. Chapter 9 by Onyango and Jentoft brings the analysis to Lake Victoria in Tanzania to demonstrate the importance of understanding governability when confronting difficult challenges like poverty. They illustrate how interactive governance and the governability framework can be applied in the study of poverty in small-scale fisheries at various scales. By addressing the wickedness of poverty, they emphasize the opportunities for local communities to elevate their livelihoods and wellbeing.

Scholtens and Bavinck follow in Chap. 10 with an assessment of a specific case of Palk Bay trawl fisheries in the South Asian region, illustrating that better adjustment and match between the fishery's system-to-be-governed and the governing system may go a long way in increasing governability. In Chap. 11, Liu, Chuenpagdee and Sumaila shift the focus to aquaculture. Given that it is one of the world's fastest growing food producing industries, providing many benefits for society, governance challenges are expected to occur. They appraise what these are, drawing on lessons from salmon farming in Norway and Canada. Their analysis shows that governability of salmon aquaculture industry can be elevated through technological improvement, economic-based instruments, and innovative governance solutions.

In Chap. 12, De la Cruz Modino and Pascual-Fernández apply the concept to marine protected areas (MPAs) in Spain, arguing for the roles of local communities in enhancing the implementation and effectiveness of marine conservation and thus the sustainability of small-scale fisheries in the area. They also emphasize the importance of partnership and the step zero for MPA governance, which in turn helps improve governability. The final chapter in this part addresses another key issue in fisheries governance, gender inequality. Frangoudes, Marugán-Pintos and Pascual-Fernández exemplify this in Chap. 13 with the struggles that women who gather shellfish in Galicia face in their attempts to gain control over the exploitation of the resources. The case study demonstrates how local organization can be transformed to empower women and eliminate inequality, thereby enhancing the overall governability of a fishery system.

Recognizing that numerous methods and approaches may be employed to examine the various aspects of fisheries systems that affect governability, the chapters of Part IV present an eclectic suite of what these may be in empirical settings. In Chap. 14, Chuenpagdee and Mahon offer a broad overview of existing tools and techniques for investigating the characteristics of the system-to-be-governed, the capacity of the governing system and the quality of their interactions. Their examples include both natural science approaches (life histories, trophodynamics and food web studies) and social science methods (oral histories, stakeholder surveys and discourse analyses). Song and Chuenpagdee (Chap. 15) follow with an illustration of a specific tool called the 'damage schedule,' a paired-comparison method aimed at capturing the relative importance of issues concerning fisheries and aquaculture that might result in a prioritization of goals. The application to illegal fishing in Lake Malawi reveals people's underlying values and highlights what it takes to enhance governability in a way that benefits the ecosystems and the communities. The next approach, presented by Mahon and McConney (Chap. 16), is commonly used in social network analysis. The chapter demonstrates the utility of a network perspective in revealing system characteristics, such as where institutional links are deficient and how power is distributed within the fisheries governance system. This information is essential, they argue, for determining where interventions for improving governability are required. Networks and cooperation across scales are also important in fisheries governance, as illustrated by Almerigi, Fanning, Mahon and McConney (Chap. 17) with the example of the marine ecosystem-based management initiative for the Wider Caribbean Region. The authors employ a facilitation process to explore multi-level functionality, shared vision, guiding principles and priority actions of institutions and actors in the region in addressing key issues and identifying strategies for collaboration.

The final part (Part V) contains two chapters, one by Chuenpagdee and Jentoft and the other one by Kooiman. In the former (Chap. 18), the authors draw on the contributions of this book to summarize key elements of the governability assessment framework. Their view focuses on the steps that follow after wicked problems in a fishery system are identified and the properties that affect governability have been examined. They propose two further enquiries: one into the capacity of the governing system to produce desirable outcomes, and the other into factors affecting and being affected by governing interactions. In a similar vein, Kooiman reflects on the fruitfulness of the interactive governance perspective and what it offers for the transdisciplinary study of fisheries and aquaculture systems in the final chapter (Chap. 19). Based on a scrutiny of basic concepts, he presents his views on achieving a full understanding of the complexities involved in securing healthy social and ecological systems through a governability lens.

Although this book emphasizes global challenges in fisheries and aquaculture governance, the interactive governance theory and the analytical governability perspective that we are offering have relevance for other societal sectors. Many of the concerns and problems discussed in this book are certainly not unique to fisheries and aquaculture, but may be found elsewhere. Other concerns, such as climate change and globalization, may also be examined in the same fashion. It is therefore important that this book be read not just as another contribution to the field of fisheries. With many publications having been generated over a period of more than a decade, interactive governance has moved beyond the initial formative and combative stage. It is now time to bring the perspective to other sectors that are confronted with similar concerns and governability challenges. This will be the aim of the upcoming compilations by the Fisheries Governance Network.

References

Berkes, F. (2011). Implementing ecosystem-based management: Evolution or revolution? Fish and Fisheries. doi:10.1111/j.1467-2979.2011.00452.x.

Chuenpagdee, R., & Sumaila, R. (Eds.). (2010). Fisheries governance. Fish and Fisheries, 11(Special Issue), 233–314.

Clark, W.C. (2006). The worldwide crisis in fisheries: Economic models and human behaviour. Cambridge: Cambridge University Press.

Kooiman, J. (1993). Modern governance: New government-society interactions. London: Sage Publications.

Kooiman, J. (2003). Governing as governance. London: Sage.

Kooiman, J., van Vliet, M., Jentoft, S. (1999). Creative governance: Opportunities for fisheries in Europe. Aldershot: Ashgate.

Kooiman, J., Bavinck, M., Jentoft, S., Pullin, R. (Eds.). (2005). Fish for life: Interactive governance for fisheries. Amsterdam: Amsterdam University Press.

Kooiman, J., Bavinck, M., Chuenpagdee, R., Mahon, R., Pullin, R., Salagrama, V. (2008). Governability of aquatic resources: Conceptual exploration and applications. Journal of Transdisciplinary Environmental Studies, 7(Special Issue), 1.

McGoodwin, J.R. (1990). Crisis in the world's fisheries: People, problem, and politics. Stanford: Stanford California Press.

Pauly, D., Watson, R., Alder, J. (2005). Global trends in world fisheries: Impacts on marine ecosystem and food security. Philosophical Transactions of the Royal Society B, 360, 5–12.

Roberts, C. (2007). The unnatural history of the sea. Washington, DC: Island Press.

Chapter 2
Theorizing Governability – The Interactive Governance Perspective

Jan Kooiman and Maarten Bavinck

Abstract This chapter presents the conceptual foundations of governability and interactive governance upon which it is based. Interactive governance is a theoretical perspective that emphasizes the governing roles of state, market and civil society. Interactions between these realms are argued to be an important factor in the success or failure of whatever governance takes place. Governability refers to the quality of governance in a societal field, such as fisheries. Diversity, complexity, dynamics and scale are argued to be major variables influencing the governability of societal systems and their three components: a system-to-be-governed, a governing system and a system of governing interactions mediating between the two.

Keywords Complexity • Diversity • Dynamics • Governability • Governance • Interaction • System

Introduction

This chapter introduces two concepts. The first, 'interactive governance', emphasizes solving societal problems and creating opportunities through interaction between civil, public and private persons and organizations. Testing its feasibility has begun with work on capture fisheries and aquaculture. The second concept,

J. Kooiman
Centre for Maritime Research, University of Amsterdam,
Amsterdam, The Netherlands
e-mail: jkooiman@xs4all.nl

M. Bavinck (✉)
Department of Human Geography, Planning and International
Development Studies, University of Amsterdam, Amsterdam, The Netherlands
e-mail: j.m.bavinck@uva.nl

M. Bavinck et al. (eds.), *Governability of Fisheries and Aquaculture: Theory and Applications*, MARE Publication Series 7, DOI 10.1007/978-94-007-6107-0_2,

'governability', provides a conceptual basis for assessing and improving the governance of societal systems, such as fisheries. There is a close relationship between the two terms. An attempt to improve governance inevitably results in the need to explore and assess governability. Vice versa, the governability of societal systems can only be understood in/with reference to their basic qualities. Building upon the earlier *Fish for Life* volume (Kooiman et al. 2005), this chapter first discusses the main characteristics of the interactive governance approach in the context of other perspectives on governance. This is followed by an overview and elaboration of the governability concept.

The Range of Governance Theory

Ideas about governance have been around for a long time and few are therefore entirely unfamiliar with its significance. Traditionally governance has been related to governments and what governments do. In more recent social science interpretations, however, governance has acquired a broader meaning. Here, government is not the only institution acting as governor. Private enterprises, civic organizations, communities, political parties, universities, the media, and the general public, among others, are all in one way or another involved in governance.

As is the case with other concepts in the popular vocabulary, the term 'governance' has different meanings for those who use it (for overviews see Pierre 2000; Kjær 2004; Ezzamel and Reed 2008; Lockwood et al. 2010; Osborne 2010; Torfing et al. 2012). These differences often revolve around the perceived role of the state, viewed in a normative or in an analytical sense. In the more normative approaches, such as those offered by the World Bank (1989, 2004) and the oft-quoted book, 'Reinventing Government' (Osborne and Gaebler 1992), governments are often seen as failing to live up to the expectations of those whom they govern. This is shown in particular by analyses of weak, unstable, collapsing or failed states. Where the state is unable to govern effectively, other actors from market and civil society move into prominent governing positions. The recent financial crisis may, however, reverse this trend, as governments have been recognized as playing a crucial role in maintaining public services and preventing disorder.

Governance theory comes in different versions and schools, all of which share the view that governance is beyond government. This implies that it is possible to have more governance but less government, and that the solution to many present day challenges must involve other sectors of society in some form of a public-private partnership arrangement. This state of affairs is caused by societal realities of diversity, dynamics and complexity, which preclude the state from acting as a sole governor.

But there are also more analytically based conceptions of governance to which we add the one developed in this book. Among such approaches are those who view governance as networks (Rhodes 1997; Sørensen and Torfing 2007), and perspectives that distinguish governance according to the scale-level at which it takes place. See for example the journals *Governance, Global Governance*, and *Public Management Review.*

Interactive Governance

In line with other approaches, the interactive governance perspective applied here proceeds from the assumption that societies are governed by a combination of efforts (see Torfing et al. 2012 for a related conception of interactive governance). These governing mixes respond to ever growing diversity, dynamics and complexity, as well as major concerns such as poverty and climate change. The main sources for discussions of 'governance', as conceptualized in this book, are, 'Governing as Governance' (Kooiman 2003), and its application in fisheries (Kooiman et al. 1999; Bavinck et al. 2005; Kooiman et al. 2005).

Governance is the aggregate of governing activities carried out by societal actors in response to public needs and visions. It is generally organized and routine, rarely harmonious but typically interactive. Kooiman and Bavinck (2005, 17) thus define interactive governance as:

> The whole of interactions taken to solve societal problems and to create societal opportunities, including the formulation and application of principles guiding those interactions and care for institutions that enable them.

The emphasis on interactions constitutes the main innovation of this approach. Governing interactions are exchanges between actors that contribute to the tackling of societal problems and opportunities. The adjective 'societal' refers to everything that has a common or public dimension; it stands opposed to 'private' activities. The definition includes a reference to institutions and principles, considered vital for any governance interaction. Our supposition is also that governance arrangements lacking a normative basis ultimately suffer from ineffectiveness and illegitimacy.

The interactive perspective on governance proposes that society is comprised of a large number of governing actors, who are constrained or enabled by their surroundings. Actors are any social unit possessing agency or power of action. This includes individuals, associations, firms, governmental agencies and international bodies. The surroundings are captured by the concept of structure. Structure refers to the social frameworks within which actors operate, including culture, law, politics and economics, but also natural conditions such as geography and ecosystems. According to sociological reasoning, actors are continuously making changes to these structures while at the same time being subjected to their influence (Berger and Luckmann 1966; Giddens 1984). The analysis of governance therefore requires that we pay attention to both of these dimensions.

Governance is and can be based on a broad range of principles, none of which have an inherent primacy over another. Our approach to governance does, however, contain normative elements, chief among them the notion that 'interacting' is often a more effective way of governance than 'going it alone'. Broad societal participation in governance is an expression of democracy and a desirable state of affairs.

Interactive governance therefore advocates wide participation in governance from a normative as well as from a practical point of view.

Rather than being fixed, interactive governance suggests that values, principles and goals are simultaneously crafted and expressed as actors engage in social-political exchange. Goals are then not given but negotiated, and are not stable but vary according to the relative strength of the participants that come and go. Furthermore, governance rarely starts from a grand plan. To the extent that such plans do exist, they are an outcome that is typically arrived at through an incremental process of interactive, experience-based learning. As a research tool, governance theory insists that goals are treated as an empirical question rather than assumed. What are the goals? How do they come about? Whose are they? What do they mean? Similar questions are asked of the various groups of governors or stakeholders active in the governance process: Who are they? What exactly is it that they have at stake? Who defines who they are? How do they come forward and make themselves heard (Jentoft et al. 2011)?

Governability

Within the interactive governance perspective, governability is defined as "the overall capacity for governance of any societal entity or system" (Kooiman et al. 2008, 3), whereby a societal system is understood to consist of a combination of human and natural characteristics. This definition draws attention to governors' differential ability to solve societal problems and to create societal opportunities. What these problems and opportunities actually are depends on societal perceptions and is therefore subjective. But problems and opportunities can also be determined on the basis of scientific analysis.

The definition of governability is built on the notion that societies, or parts thereof termed societal systems, are made up of three related components: a system-to-be-governed, a governing system, and governing interactions (See Fig. 2.1). Our approach suggests that governability depends on qualities of the object of governance (the system-to-be-governed), its subject (the governing system) and the relationship (governing interactions) between the two (Kooiman et al. 2008, 2010). Governors, the governed, and their interactions all contribute to the available governability.

This perspective has important consequences for an assessment of governing capacities. Rather than attributing failure to a shortcoming of only the governing system, it urges for a holistic assessment of the situation at hand. Governability depends on the ability of a governing system to deliver on the challenges that the system-to-be-governed raises. The latter, however, may well exceed the governors' capacities. There are thus limitations to governability (Jentoft 2007). In other words, not every situation is equally governable.

Governability also refers less to a state than to a variable process – it is not a capacity that is given once and for all. Rather, governability changes in response to internal as well as external conditions. Fisheries, for example, are often regarded as

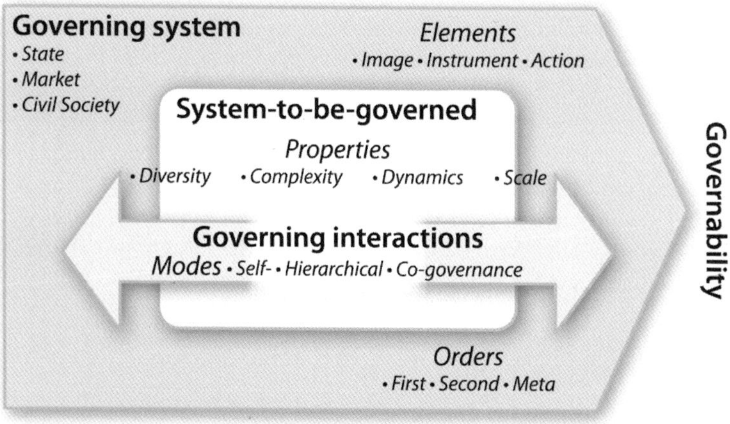

Fig. 2.1 Interactive governance perspective of a societal system

complex adaptive systems in which unpredictability is a key factor (Mahon et al. 2008). External factors are also beyond the control of governors. A Marine Protected Area (MPA) may therefore be well-governed, but still lack sufficient governability due to the interference of outside disturbances (Jentoft et al. 2007, 2012). Governability thus comes not only with limitations but also with vacillating potentialities. It is therefore argued that the act of governing necessitates flexibility and learning, and must account for the possibility of disappointment and failure.

Societal Systems

Systems theory has a rich pedigree, both in the natural and in the social sciences. Recent attention for issues of complexity, chaos and resilience has contributed to a resurgence of interest in this academic stream. In the field of fisheries, for example, scholars are now making fruitful use of complex adapted systems theory (Rammel 2007).

In the following, we use the systems concept as a heuristic tool, without teleological or functional connotations. A general and, for our purposes, workable definition of a societal system is the whole of interrelations among a given number of entities belonging to the natural and social worlds (Kooiman et al. 2008). This formulation contains the notion that systems can be circumscribed in different ways and at different scale levels. Moreover, it contends that no one approach is intrinsically better than another. Instead we suggest that the designation of a system's contours depends on the nature of the research enquiry.

Irrespective of the way we define systems, they are always part of larger events and structures. It is therefore useful to conceive of systems as divisions of other entities rather than solitary units. A particular capture fishery is nested within a larger fishery, as well as within larger coastal dynamics. Its governing system too is nested in larger administrative units that scale up to the international level. What happens in any one system at a particular level has consequences for other levels.

Societal systems are 'rich' in the sense that their parts and their interrelations have many facets and histories. Systems also tally with the adage that the whole is more than the sum of its parts – they possess qualities that go beyond the features of the individual units. In keeping with this understanding, interactive governance considers governability as a composite property. Phrased differently, the governability of any societal system depends on the nature of the system-to-be-governed, the governing system, and the governing interactions taken together.

Diversity, Complexity, Dynamics and Scale

Interactive governance theory argues that societal systems are inherently diverse, complex and dynamic, and that these traits pose fundamental challenges to their governability at different scales. Although diversity, complexity and dynamics are crucial, their intensity varies from one system to the next. One societal system may therefore be less diverse, complex and dynamic than another, with important consequences for their overall governability. That this is true also for fisheries is illustrated by Bavinck and Kooiman (Chap. 8, this volume).

The diversity, complexity and dynamics of the relations between parts ensure uncertainty and unpredictability in system behavior. There is nothing unusual about this. Uncertainty and unpredictability are simply the consequence of actors or units acting and interacting without anyone having the chance to make out what the impact of these actions and interactions are on the system as a whole. In many cases the outcomes of actions or interactions are indirect and delayed because of system effects. In other cases there may be all kinds of unexpected side effects (Jervis 1997).

Mechanisms like these have important consequences for those taking part in a system, but also for their study. Reductionist approaches, which focus on specific actors or interactions, fail because they lack a larger (system) picture. But holistic approaches, in which the system as a whole is the unit of analysis, are also fruitless. Combinations need to be made. In the meantime it is necessary to keep in mind that, citing Cilliers, "we never can know complex, dynamic and diverse things completely" (2002, 79). Uncertainty therefore always qualifies our assessments of governability, thereby suggesting a precautionary approach.

We noted above that systems rarely exist in isolation, and that it is useful to view them as part of bigger wholes. This directs us to further examine the relevance of 'scale' for governance and governability. Fisheries and coastal zones, as natural systems and the social and governance systems related to them, function on varying spatial and temporal scales. The matching of scale in the operation of these systems is therefore an important aspect affecting governability.

In previous publications we pointed to scale as an essential characteristic of societal systems, next to diversity, complexity and dynamics. Scale was defined as referring "to time and space dimensions of systems to be governed as well as to governing systems" (Kooiman and Bavinck 2005, 14). The spatial dimensions of fisheries are easily illustrated. Some fish species only occur in limited geographical areas, while others span the globe. Additionally, some fishers – like the ecosystem people referred to by McGoodwin (1990) – make a living from resources in their immediate environments. 'Biosphere people', however – such as the 'roving bandits' referred to by Berkes et al. (2006) – operate on a far larger geographical scale, seeking out the target species where it is to be found. Governance too takes place at various geographical scale levels, from the village councils or *panchayats* of southern India (Bavinck 2001) and the *cofradias* of Spain (Pascual-Fernandez et al. 2005) to global institutions such as the Food and Agriculture Organization (FAO) and the Law of the Sea Tribunal (Suarez de Vivero et al. 2005). Each set of governors and governing arrangements thus matches a particular geography.

Time scales are also relevant to the governability of societal systems. Time plays a role in ecology (for example, the life cycle of a fish species or the time needed to destroy or rebuild an ecosystem), as well as in practices of capture, trade, and processing. Such scales also include the time perspectives of the human actors involved – the periods over which they assess, judge, plan and expect things to happen. In fisheries, for example, seasonality is an important phenomenon – fishers adapt their gear and fishing practices according to the species that prevail in certain seasons. The rhythm of fishing seasons thereby forms a unique time scale, unfamiliar to outsiders.

System-to-Be-Governed

In this volume we are mainly interested in societal systems belonging to the category of 'primary processes'. That is, those activities that meet basic human needs. In addition, we focus specifically on systems that connect natural and social phenomena. Capture fisheries – which include harvesting, processing and marketing – is thus a social process that is integrated with, and relies on, specific sets of ecological and geophysical conditions. The same is true for aquaculture. In line with this understanding, the system-to-be-governed of capture fisheries and aquaculture has frequently been depicted as a fish chain leading from the ecosystem to the consumers' plates (Johnson et al. 2005; Thorpe et al. 2005; Bush and Oosterveer 2007)

Fish chains run from marine ecology, through harvesting, processing, marketing and distribution, to the end consumer and, in so doing, generally cross multiple scale levels. It follows that the shrimp extracted from a particular ecosystem in, for example, South India, proceeds – through transactions at local, national and international levels – to the consumers' plates in Europe, North America, or East Asia. Another species, one of little international demand, is destined more for the local market, and follows a different chain route with its own configuration of actors. As Johnson et al. (2005) point out, fish chains have tremendous range of variation and complexity. Their drivers may reside within the market or, conversely, within the local dynamics of a fishery.

There are ways other than a fish chain to conceptualize the system-to-be-governed, and in the course of this book we will highlight at least two of these other perspectives. The first is the Social Ecological Systems approach, which, focusing on a territorial entity, distinguishes between two parts – a socio-economic and a natural subsystem (Berkes et al. 2003; Walker et al. 2004; Armitage and Johnson 2006). The Complex Adaptive Systems approach, which emphasizes the capacity of complex systems to self-organize or adapt, is a variation hereof (Wilson 2006; Mahon et al. 2008). The second approach informs Part II of this book: the respective chapters focus on societal concerns such as ecosystem health, social justice, livelihood and employment, and food security. Concerns differ from principles in that they emerge not from analysis but from political and social debate (Chuenpagdee et al. 2005). This vantage point therefore possesses the advantage of poignancy. We consider the discussion of such concerns – which could even be considered as sub-systems of system-to-be-governed – worthwhile for determining governability issues. In addition, such norm-driven approaches to systems-to-be-governed bring some order to the otherwise unlimited number of potential factors relevant for asking governability questions.

Governing System

A governing system includes the total set of mechanisms and processes that are available for guidance, control and steerage of the system-to-be-governed in question. Interactive governance theory distinguishes three governing realms, state, market and civil society, each with specific governing characteristics and features.

Realms of Governing

In almost all parts of the world, *states* are still the most central governing entity. Governments intervene in society all the time, and try to influence, steer and control from the local level to the international. Their governing activities vary from sector to sector, and are in constant flux. Recent tendencies have been described under headings as ‘regulatory’, ‘enabling’, ‘supermarket’, ‘corporate’ or ‘bargaining’ states (Kooiman 2003). A major division in state performance is the one between South and North. Not only do the life stories of states in these two regions vary substantially, the challenges they face differ enormously as well. Politics is obviously a major part of the governing system. Without a political dimension, in which the goals of governance are negotiated and established, the governing system will remain an empty shell. This aspect is often ignored or taken for granted in studies of fisheries governance.

Governance cannot exist without bureaucracy either. All major organizations, public or private, make use of bureaucratic styles of management in one way or another. Public bureaucracies seldom have a good reputation, although recently there is a revived interest in their positive contributions (Olsen 2005). Bureaucracies are here to stay and certainly deserve a place in our thinking about governability. This also applies to phenomena such as corruption and non-compliance. These phenomena are not temporary, incidental or limited to specific parts of the world, as was thought for a long time (Williams 2000). Of course they also occur in fisheries (Hauck 2008; Hanich and Tsamenyi 2009).

How can we conceptualize the *market* as part of governing system and assess its contribution to governability? Answers to this question depend heavily on one's disciplinary vantage point. Shipman (1999), for example, views the market as productive and efficient in thc allocation of resources; it makes full use of capacity as well as employment, and aims at optimum growth. Lindblom (2001), on the other hand, considers the market to be a system of society-wide coordination with mutual interactions in the form of transactions. It is not competition that coordinates the market, but instead a combination of competition and social cooperation. Fligstein (2001) views markets through the lens of politics. In his opinion, firms desire stable markets; they defend and control their positions not only through exchange transactions but also by using power (also Swedberg 2005). All three perspectives (and there are many more) emphasize the part played by markets in the governing process. The first highlights markets' capacity to self-organize competition, the second stresses the role of competition and cooperation, and the last one draws attention to competition and power. But markets also have their limitations, as is brought out by the term 'market failure', and the economic crises of the past decade. Market approaches prevail in contemporary fisheries management, such in the promotion of transferable quotas and seafood labelling.

While much debated, civil society is not a very clearly defined entity. Non-profit organizations such as religious bodies, professional associations, social movements and non-governmental organizations (NGOs) are generally considered to make up its core; academic institutions and the media are boundary cases (Müller 1996). Observers frequently consider civil society to make a positive contribution to governance and governability, for example by providing a moral foundation (Wolfe 1989). But there is criticism as well (Lewis 2002; Lewis and Opoku-Mensah 2006). Changing normative criteria on effectiveness and even legitimacy, as well as predilections to 'bring the state back in', play a role in this re-evaluation. From the interactive governance viewpoint, civil society's main contribution to governability is the channeling of societal activities for governance purposes. However the central role of user-groups as part of civil society in the governance of fisheries must also be critically reviewed (Jentoft and McCay 2003; Mikalsen et al. 2007).

The distinction between three realms of governing we have made here is of course an analytical one. Moreover, it is a simplification of societal reality. The boundaries between state, market and civil society are permeable and constantly changing. For a long time the state has been considered to interpenetrate the other two realms, but recently the market has regained some of its lost territory. There are also many so-called hybrid institutions, and their role in governance appears to be growing (Van Tulder and Van der Zwart 2006).

Elements of Governing

In order to execute their governing tasks or activities, governors have three resources available, which we have called governing elements: images, instruments and action. Images are sets of governing ideas, instruments give these ideas substance, and action puts these instruments to work.

Anyone involved in governance, in whatever capacity or authority, makes use of *images* (see Kooiman 2003, 29–44; Boulding 1956; Jervis 1989). Such images can consist of visions, knowledge, facts, judgments, presuppositions, wishes, goals, hypotheses, theories and convictions. The form of an image thereby helps to identify the governance challenges and tasks ahead (Jentoft et al. 2010). In contemporary society, the production and utilization of knowledge for governance is a vast social process in which scientists, politicians, opinion leaders, and others all play a role. These processes are of course open to various forms of manipulation, and the end result may either be enlightenment or confusion.

Governors wishing to move from one state of affairs to another need *instruments* to do so. A wide array of instruments – or tool boxes (Hood 1983) – is available to public and private governing organizations alike (for an overview of the tools of fisheries see Charles 2001 and Degnbol et al. 2006). Traditional instruments consist of rules and regulations, taxes, fines and subsidies, while more recent ones include covenants and certification. A worrisome aspect is the piling of instruments, which sometimes contradict one another. For this reason, one can notice the law of diminishing returns at work in many areas of governance.

The *action* element of governance conveys the fact that without sufficient will or support, images and instruments – however convincing and applicable – remain up in the air. Governance action and reaction are best seen as chain processes – one governing entity takes the initiative and is followed by others. The chaotic nature of societal processes ensures that small incidents can have major consequences (the butterfly effect), and even non-action may have certain unintended ramifications. Due to the interconnectedness of modern societies, it is more appropriate to speak of complexes of societal action than of collective action (Kooiman 2003). Moreover, the diversity of modern societies means that the action potential for governance is not only a societal attribute, but a political one as well. The two are related because in the long run no political action can be taken without societal support; vice versa societal activism cannot flourish without some kind of political expression.

Orders of Governing

In the first order of governance, governing actors try to tackle *problems* or create *opportunities* as a day-to-day exercise. Globally, two perspectives of problem-solving can be distinguished. One takes a linear path and reasons from a problem to its solution. The advantage of this model is its simplicity. Its limitation on the other hand is that one particular characteristic is taken as dominant for problem definition as well as for the solution offered. In the non-linear alternative, problem-definition and solution-finding are seen as an iterative and dynamic process with scope for identifying different forces at work. The disadvantage is a perceived lack of clarity and available stopping rules.

Many efforts have been made to design typologies of problems, such as between highly, moderate or ill-structured problems; distributive, regulatory and re-distributive problems; or problems with a few or many dimensions (Kooiman 2003). Jentoft and Chuenpagdee (2009) have noted that governance problems are often 'wicked', meaning for instance that the idea of the solution tend to inform the definition of the problem (Rittel and Webber 1973; Jentoft and Chuenpagdee 2009).

Problem-solving and opportunity-creating activities are embedded in a second order of institutional settings, which enable governance, sustain it and give it focus. Although many scholars and policymakers consider governance to be synonymous with institutions (many definitions of governance are in institutional terms), the systematic care for institutions as a category of governance activities is neglected. Institutions are considered to consist of relatively enduring sets of rules stimulating, enabling, regulating or controlling human behaviors. They can do this in formal or informal ways. They will surely be changed over time, although the faster they change the less influence they may have. But institutions are also made up of the organizations that decide, effectuate and monitor the implementation of rules.

Meta governance is the third order of governance. It is where decisions on the values and principles of the governing of governance are made (Kooiman 2003; Kooiman and Jentoft 2009; Peters 2010). Such principles and values often remain in the curtains and unvoiced. When they are made explicit and deliberated upon, however, principles are rarely absolute – they guide rather than prescribe. Applying them in governance means making choices. Choices that can be difficult because the normative notions confined in them are often in conflict. As governance choices can be differentiated, so too can governance norms, principles and maybe even values. Some values, such as those pertaining to human rights, are universal and part of substantive governance everywhere. Other values are specific to certain situations, settings, and cultures. The *Fish for Life* volume (Kooiman et al. 2005) sets forth a series of meta governance principles applicable to the elements, orders and modes of governance.

Governing Interactions

Governing interactions are mutually influencing relations between two or more actors or entities in a governance setting. Theoretically, we distinguish between two levels, an intentional (actor) level and a structural level of governing interactions, as well as a set of interaction modes. These are discussed below.

Governing Interactions at the Actor Level

The minimal condition for governance interactions is the willingness or ability of actors to participate. The opportunity for participation is not at all obvious, however, a fact that clearly emerges against the background of empowerment movements – for the poor and women in some parts of the world, and for consumer empowerment in other regions. Nor is it difficult to understand that governance also involves the need for people involved in a system-to-be-governed to pro-act or re-act to activities by governing institutions – public as well as private – and thus make governing a process of interaction.

Citizen involvement or public participation has always been part of the theory and practice of democratic institutions, particularly at local levels (see for an overview Special Issue *Public Administration Review* 2005, 5). The classical example is the 'participation ladder', where the essence is located in power-sharing (Arnstein 1969). Under the influence of new thinking about direct, participatory and deliberative democratic thinking, as well as the ideas and practices of empowerment and inclusion, the analysis of participation has been given a new impulse (Parkins and Mitchell 2005; Papadopoulos and Warin 2007), also in fisheries (Gray 2005).

Scholars generally recognize the advantages of citizens participating in public affairs. They identify positive effects such as development, education, learning, integration, improvement of results, better understanding and control, legitimacy, and accountability. But disadvantages are noted as well: participation is thus argued to be a false notion, inefficient, politically naïve, unrealistic, disruptive and dangerous. It frequently results in excess time consumption, costs, hostility, worse outcomes, loss of control and complacency (Roberts 2004). A paradox may arise when more participation results in less influence as seems to be the case in European fisheries governance under the influence of devolution (Suarez de Vivero et al. 2008)

Governing Interactions at the Structural Level

Interactive governance theory suggests that the central features of system-to-be-governed and governing system are normally reflected in governing interactions. A well-organized and powerful societal sector will thus find these qualities represented in governing interactions. Highly developed connections between state, market and civil society are mirrored in a multitude of participatory interactions (Lovan et al. 2004). However, two concepts deserve special attention for their mediating and structuring contribution: public sphere and social capital.

The idea of the public sphere, as conceptualized by Habermas (1989), has become an important part of social and political theorizing about the nature of modern society. In Habermas' view, the public sphere is the social realm in which the public organizes itself as the bearer of public opinion by conferring and communicating in an unrestricted and rational fashion about matters of general interest. Only a few scholars in the field of fisheries have made systematic use of Habermasian thought

(Van der Schans 2001). The notion that there is a public interest in fisheries that reaches beyond the role of stakeholders deserves more attention, however (Mahony et al. 2010).

We lean on Bourdieu (1986) and Coleman (1990) in conceptualizing social capital (see Kooiman 2003). Both authors consider social capital to be a structural resource that is unevenly distributed and expressed at different levels of societal aggregation (Foley and Edwards 1999). This perspective enables social capital to be seen as sets of interactions, which take place within a societal context having historical antecedents and showing broad stratification patterns (such as societal sectors or 'social fields' as Bourdieu calls them). Halpern elaborates it as "societies not being composed of atomized individuals, but people connected with one another through the social fabric of intermediate social structures affecting greatly with whom, and how, we interact and co-operate" (2005, 3). The concept of social capital is widely applied in fisheries discourse today for explaining governance outcomes (Grafton 2005; Gutiérrez et al. 2011).

Modes of Governing

In the reality of societal governance an enormous variety of interactions can be observed. To come to grips with them we distinguish between three types: a self-governing mode, a hierarchical mode, and a co-governing mode of interactions.

Self-Governing Interactions

Self-governance refers to the capacity of social entities to govern themselves. While in modern societies this is seldom true in an absolute sense, it is useful to realize that without sustaining a certain self-governing capacity, societal governance is an altogether impossible task (as the history of many totalitarian regimes has shown). Forms of self-governance are found in all societies and to a much greater extent than is often realized. In fisheries, self-governance is frequently described under the labels of 'customary management' or 'sea tenure' (Ruddle et al. 1992; Bavinck 2005).

But beware, some of what is for example sold as self-regulation is actually a form of re-regulation, or changing centralized forms of public control into 'steering at a distance'. Self-organized governance varies from spontaneous types embodied by social action groups and social movements to institutionalized types representing interest groups. Such initiatives differ in organizational form, strategies applied, and styles of interaction between system-to-be-government and governing system.

What interest groups are is a matter of much debate and although insight into the phenomenon has grown considerably, there is still little accumulation of knowledge

(Coen 2007; Beyers et al. 2008). 'Stakeholder' is a related concept that is applied to private actors or entities involved in a public matter or issue (Mitchell et al. 1997; Bryson 2004; Buanes et al. 2004). In the earlier stages of debate, stakeholders were mainly seen in the context of commercial firms. In contemporary development discourse, as in other disciplinary approaches, the concept has now become widely applied.

Participation in social movements is far from commonplace, and even large movements mobilize only relatively small proportions of the population (Meyer et al. 2002; Nash 2005). The collective character of social action is not self-evident either. The manner in which social discontent is transformed into organized action has always been a key issue in social movement literature. It has been given some attention in fisheries (Sundar 2012).

Hierarchical Governing Interactions

Hierarchical governance is the usual style in which governments interact with their citizens, either as individuals or collected in groups or organizations. This governance mode, however, is also practiced in the private sector as many private organizations have hierarchical structures. The terms 'policy' and 'management' subsume much of what hierarchical governance is about. Both imply a flow moving from governing system to system-to-be-governed. It is interesting to note that fisheries management is often criticized for relying too much on a hierarchical model of decision-making (e.g. Raakjær 2009).

'Policy' has been theorized in many different ways. An overview counts at least ten policy process theories, some more popular and elaborated than others (Sabatier 1999). For a long time, actor-oriented notions of policy process were the usual mode of analysis. More recently, however, an interest has grown in the relation of policy to broader societal and governance processes (e.g. Hill and Hupe 2009). Along the way, positivist, neo-positivist, modernist and post-modernist analytical and constructivist approaches to policy studies have waxed and waned. The most classical approach is still to distinguish between stages or phases within a policy process (Kooiman 2003). Although not undisputed, this approach is still useful to "help disaggregate an otherwise seamless web of public policy transactions [...] and transitions distinguished by differentiated actions and purposes" (DeLeon and Resnick-Terry 1999, 24).

'Management' is relatively new concept in the debate on public governance. In recent decades, the fields of Public Management and later, New Public Management have expanded enormously, with major attention being paid to matters such as efficiency, effectiveness, value for money, excellence, and performance (Ferlie et al. 2005). Although the main focus is on 'how to run a service as a business', external relations have also been highlighted. The attention given to the client or customer of services is an indication of the interest available for the field of governing interactions.

Co-governing Interactions

Collaborative and cooperative governance interactions are growing in number and in importance. This raises an important question: why are groups, organizations and authorities interested in sharing their governance responsibilities and conducting activities together instead of alone? Mutual interdependencies are often mentioned as the main reason for such interactions (Huxham 1996). In the field of social-political governing, parties may collaborate, co-operate, co-ordinate and communicate 'sideways' without any one actor playing a central or dominating role. In earlier publications we made a conceptual distinction between collaboration and cooperation, collaboration being the day-to-day interaction where actors or entities work together, and cooperation a more formal attunement of activities. Two forms of co-governing stand out: networks and co-management (see Kooiman 2003).

One of today's catchwords is 'networks'. Its definition varies in meaning from an overall theory of society (Castells 1996) to very detailed, precise and quantitative analyses of particular types of interactions between people or entities in various societal domains. Functionalist explanations for societal networks emphasize the need for resources, combating common environmental uncertainties and strategic considerations. Interactions are mainly described as being of a horizontal nature, although minor hierarchical elements can also develop by, for example, linking-pin organizations. Approaches to networks that explicitly conceptualize them as modes of governance merit special attention. Some authors in this field even consider networks almost exclusively in terms of governance (Rhodes 1997; Sørensen and Torfing 2007). See Mahon and McConney (Chap. 15, this volume) for an application of network analysis to fisheries.

Co-management differs from networks by identifying specific tasks to be carried out. Authors in this field expect that by involving stakeholders directly in the governance process, a positive feedback loop will develop. This is because: governing knowledge becomes more adequate, resulting in more satisfactory governing measures, which in turn lead to higher management legitimacy and compliance, "accepting the regulations as appropriate and consistent with [...] persisting values and world views" (Jentoft 1989, 139; cf. Wilson et al. 2003).

Working with Governability

We noted above that the governability of societal systems is a function of its three components: system-to-be-governed, governing system and governing interactions. Each component possesses its own governability aspects, some of which add to overall governability, while others detract from it. In the following pages we discuss earlier contributions to the discussion, which have highlighted the relationship between system-to-be-governed and governing system. We call attention here to two different approaches: one that makes use of criteria, the other emphasizing 'match'.

In one of the concluding chapters of the *Fish for Life* volume, Kooiman and Chuenpagdee (2005) postulate an evaluative framework for governability that corresponds gracefully with the four pillars of interactive governance theory. The 'features' (diversity, complexity, dynamics and scale), which permeate the system-to-be-governed, governing system and governing interactions, are first of all to be assessed according to the criterion of representation: "the manner and degree to which the features of a fisheries system correspond with those in its governing system" (Kooiman and Chuenpagdee 2005, 347–8). The utility of this criterion has been further investigated by Bavinck and Salagrama (2008). The second pillar, 'elements' (images, instruments and action), is gauged by the criterion of rationality: are the elements in tune with or supportive of each other? For 'modes' (hierarchical, co-governance, self-governance) Kooiman and Chuenpagdee suggest using 'responsiveness' as a measuring stick: "does the mix of governing modes respond to the varying governing needs of [the variety of fishing] types?" (2005, 347). The final theoretical pillar, orders (first, second and meta), is to be evaluated with reference to the norm of 'performance'.

The framework is not elaborated further in this first writing. The authors do, however, provide a pointer for how an evaluation exercise of this kind might actually be conducted. The 'simplified version' of an evaluative framework, which they present, consists of a matrix with different fishing métiers on the vertical axis and the four evaluative criteria on the horizontal axis. Each fishery is given one of three scores (high, medium and low) for each criterion, after which the scores are totaled to form a composite quality of governability.

This assessment approach, which is based on the application of evaluative criteria deduced from the theory of interactive governance, is expanded upon in a special issue for the Journal of Transdisciplinary Environmental Studies published in 2008. After a general introduction to the concept of governability (Kooiman et al. 2008) and its application to the realms of fisheries, aquaculture and coastal zones (Chuenpagdee et al. 2008), the special issue presents two case studies on the governability of fisheries, one in the Caribbean (Mahon et al. 2008) and the other in the Bay of Bengal (Bavinck and Salagrama 2008). Jentoft and Chuenpagdee (2009) continue in the same vein, replacing the scoring of items in the matrix mentioned above with the formulation of questions that guide assessment.

The *Fish for Life* volume also stands at the root of another, more practical, bottom-up approach to assessing governability. Discussing the governability of fish chains and the ubiquity of diversity, complexity and dynamics, Mahon et al. emphasize that "effective fisheries governance will as fully as possible reflect its operating context" (2005, 353). This formulation, which applies the verb 'to reflect', comes close to what Kooiman and Chuenpagdee (2005) label, 'representation'. What the first authors mean by 'reflection' emerges more clearly in the remainder of the cited chapter, which deals with the dynamics and uncertainties of the fish chain and the resultant need for governing system to boost its adaptive capacity.

Mahon et al. (2005) pinpoint mismatches of scale – spatial, temporal and organizational – as an important factor impacting on governability, and voice agreement with Costanza et al. (1998), who include the implementation of governance at the appropriate scale as one of the important ocean governance principles. Echoing Kooiman and Chuenpagdee's (2005) reference to 'performance' as a criterion for evaluating governance orders, these authors emphasize upward and downward linkages and the integration of the overall governance system. 'Match' and 'mismatch' are important descriptors for the relationship between system-to-be-governed and governing system. Less complex fisheries chains thus require significantly different governing arrangements than do more complex chains. For example, a large commercial fishery that uses a few large vessels to exploit a few relatively stable resources with outputs that are processed and sold in supermarkets may be inherently more governable than a widely dispersed, small-scale fishery from which products are distributed freshly by a large number of middlemen with little organization of either fishers or distributors (Mahon et al. 2005, 351)

From the contrast drawn between a large-scale and a small-scale fishery one can readily imagine that a governing system may be in difficulty if it does not adapt itself. The idea ventured in this quote suggests that, because of their relative simplicity, large commercial fisheries are 'inherently more governable', is corrected some time later by another set of authors (Mahon et al. 2008) who suggest that fisheries chains of different complexities require different 'governance mixes'. More complex fisheries would thus necessitate "a large component of 'letting go' of past control systems to allow space for self-organization" (Mahon et al. 2008, 110).

Continuing along a similar train of thought, Jentoft (2007) formulates the requirement of 'match' in terms of 'compatibility'. He argues that the governing system and the system-to-be-governed should be 'isomorphic' and 'mutually responsive'. According to this author, compatibility is not a consequence of natural adaptation but instead of deliberate intervention, planning and institutional design by societal actors such as legislative bodies, planning agencies and civic organizations. This can happen alone or, preferably, in concert as governing capacity and interaction is enhanced through collaboration.

The perspective of assessing and enhancing match is explored further in a special section of the journal, *Fish and Fisheries* (volume 11, 2010). This section includes two research papers on the adjustment process between governing needs and governing capacities in Malawi and Lake Victoria (Song and Chuenpagdee 2010; Onyango and Jentoft 2010), and two applications of the governability concept from the vantage points of anthropology and economics (Johnson 2010; Sumaila 2010). One of the aspects elaborated is the determination of appropriate 'images' for governance (also see Jentoft et al. 2010).

The study of match has finally resulted in a number of papers on the limitations of governability and the chance that governability in particular fisheries situations may be restricted for reasons beyond immediate control (Jentoft 2007; Jentoft et al. 2007). This perspective highlights governance dilemmas (Jentoft 2007), hard choices (Kooiman and Jentoft 2009) and wicked problems (Jentoft and Chuenpagdee 2009), and suggests a precautionary approach on the basis of 'sensible foolishness' (Jentoft 2007).

We have noted two approaches to the study of governability in fisheries in previous writings, based on the application of criteria, and alternatively on the study of match. These investigations have not resulted, however, in a clear choice for one approach over the other. Instead, the tendency has been to let a hundred flowers blossom. The following chapters build upon these variegated efforts at developing an assessment framework for governability.

Conclusion

This chapter had a theoretical purpose: to introduce readers to the interactive governance approach and the concept of governability, and indicate the relevance hereof for governing primary processes such as fisheries and aquaculture. It has hopefully become clear that, from the interactive governance perspective, governance is not just about the selection of appropriate tools or instruments from a standard toolbox. With every fishery representing a unique constellation of factors – located in the system-to-be-governed, the governing system as well as in the governing interactions – there is no single type of governability situation available. Instead, the governabilities of fisheries vary along many different axes, with some systems-to-be-governed being more prone to limitations than others. Stakeholders and others responsible for governance in different countries and historical time periods have dissimilar priorities – the weight they attach to concerns such as environmental health, social justice or economic progress may be quite different. Then, fisheries systems also vary greatly in the importance of and experience with varieties of governing interactions. The dissimilarities in factors influencing governability will naturally impact on the way governability is assessed, as well as on the strategies employed toward improvement.

Another reason is the fact that fisheries vary substantially as to their diversity, complexity, dynamics and scale. Interactive governance posits that these features have important implications for governability, with some fisheries facing an intrinsically 'easier' governance situation than others. More diverse, complex and dynamic fisheries are always more difficult to handle, but this does not mean that their governability level is necessarily low. After all, governability is about the governing system's capacities to handle the problems that plague the fisheries system. In many instances, as subsequent chapters will argue, these problems are 'wicked' and not at all easy to address. But some governors and governing systems are able to deal with thorny issues and arrive at acceptable solutions, while others – whose situation is ostensibly straightforward – have a more than difficult time. It is these differences, and the processes that help to assess and define acceptable ways forward, that form the topic matter of the book.

References

Armitage, D., & Johnson, D. (2006). Can resilience be reconciled with globalization in the increasingly complex conditions of resource degradation in Asian coastal regions. Ecology and Society, 11, 1. Retrieved from http://www.ecologyandsociety.org/vol11/iss1/art2

Arnstein, S.R. (1969). A ladder of citizen participation. Journal of the American Institute of Planners, 35, 216–224.

Bavinck, M. (2001). Marine resource management: Conflict and regulation in the fisheries of the Coromandel Coast. New Delhi: Sage.

Bavinck, M. (2005). Understanding fisheries conflicts in the South – A legal pluralist perspective. Society and Natural Resources, 18(9), 805–820.

Bavinck, M., & Chuenpagdee, R. (2005). Current principles. In J. Kooima, M. Bavinck, S. Jentoft, R. Pullin (Eds.), Fish for life: Interactive governance for fisheries (pp. 245–264). Amsterdam: Amsterdam University Press.

Bavinck, M., Chuenpagdee, R., Diallo, M., van der Heijden, P., Kooiman, J., Mahon, R., Williams, S. (2005). Interactive fisheries governance: A guide to better practice. Delft: Eburon.

Bavinck, M., & Salagrama, V. (2008). Assessing the governability of capture fisheries in the Bay of Bengal - a conceptual enquiry. The Journal of Transdisciplinary Environmental Studies, 7(1), 13 pp.

Berger, P.L., & Luckmann, T. (1966). The social construction of reality: A treatise in the sociology of knowledge. Garden City: Anchor Books.

Berkes, F., Colding, J., Folke, C. (Eds.). (2003). Navigating social–ecological systems: Building resilience for complexity and change. Cambridge: Cambridge University Press.

Berkes, F., Hughes, T.P., Steneck, R.S., Wilson, J.A., Bellwood, D.R., Crona, B., Folke, C., Gunderson, L.H., Leslie, H.M., Norberg, J.H., Nyström, M., Olsson, P., Österblom, H., Scheffer, M., Worm, B. (2006). Globalization, roving bandits, and marine resources. Science, 311, 1557–1558.

Beyers, J., Eising, R., Maloney, W. (2008). Researching interest groups politics in Europe and elsewhere (Introduction Special Issue). West European Politics, 31(6), 1103–1128.

Boulding, K.E. (1956). The image. Ann Arbor: University of Michigan Press.

Bourdieu, P. (1986). Three forms of capital. In J.G. Richardson (Ed.). Handbook of theory and research in the sociology of education (pp. 241–258). New York: Greenwood Press.

Bryson, J.M. (2004). What to do when stakeholders matter: Stakeholder identification and analysis technique. Public Management Review, 6, 21–54.

Buanes, A., Jentoft, S., Karlsen, G.R., Søreng, A. (2004). In whose interest? An exploratory analysis of stakeholders in Norwegian coastal zone planning. Ocean & Coastal Management, 47, 207–223.

Bush, S., & Oosterveer, P. (2007). The missing link: Governance and trade in the space of place and the space of flows. Sociologia Ruralis, 47(4), 384–399.

Castells, M. (1996). The network society. Oxford: Blackwell.

Charles, A.T. (2001). Sustainable fishery systems. Oxford: Blackwell Science.

Chuenpagdee, R., Degnbol, P., Bavinck, M., Jentoft, S., Johnson, D., Pullin, R., Williams, S. (2005). Challenges and concerns in capture fisheries and aquaculture. In J. Kooiman, M. Bavinck, S. Jentoft, R. Pullin (Eds.), Fish for life: Interactive governance for fisheries (pp. 25–37). Amsterdam: Amsterdam University Press.

Chuenpagdee, R., Kooiman, J., Pullin, R.S.V. (2008). Assessing governability of capture fisheries, aquaculture, and coastal zones. Transdisciplinary Environmental Studies, 7(1), 20 pp.

Cilliers, P. (2002). Why we cannot know complex things completely. Emergence, 4(1), 77–84.

Coen, D. (2007). Empirical and theoretical studies in EU lobbying. Journal of European Public Policy, 14(3), 333–345.

Coleman, J. (1990). Foundations of social theory. Cambridge: Belknap Press.

Costanza, R., Andrade, R., Antunes, P., van den Belt, M., Boesch, D., Boersma, D., Catarino, F., Hanna, S., Limburg, K., Low, B., Molitor, M., Pereira, J.G., Rayner, S., Santos, R., Wilson, J., Young, M. (1998). Principles for sustainable governance of the oceans. Science, 281, 198–199.

Degnbol, P., Gislason, H., Hanna, S., Jentoft, S., Nielsen, J.R., Sverdrup-Jensen, S, Wilson, D.C. (2006). Painting the floor with a hammer: Technical fixes in fisheries management. Marine Policy, 30, 534–543.

DeLeon, P., & Resnick-Terry, P. (1999). Comparative policy analysis. Journal of Comparative Policy Analysis, 1, 9–22.

Ezzamel, M., & Reed, M. (2008). Governance: A code of multiple colours. Human Relations, 61(5), 597–615.

Ferlie, E., Lynn, L.E. Jr., Pollitt, C. (Eds.). (2005). The Oxford handbook of public management. Oxford: Oxford University Press.

Fligstein, N. (2001). The architecture of markets. Princeton: Princeton University Press.

Foley, M., & Edwards, B. (1999). Is it time to disinvest in social capital? Journal of Public Policy, 19, 141–173.

Giddens, A. (1984). The constitution of society. Berkeley: University of California Press.

Grafton, R.Q. (2005). Social capital and fisheries governance. Ocean and Coastal Management, 48(9–10), 753–766.

Gray, T.S. (Ed.). (2005). Participation in fisheries governance. Dordrecht: Springer.
Gutiérrez, N.L., Hilborn, R., Defeo, G. (2011). Leadership, social capital and incentives promote successful fisheries. Nature, 470, 386–389.
Habermas, J. (1989). The structural transformation of the public sphere. Cambridge, MA: MIT Press.
Halpern, D. (2005). Social capital. Cambridge: Polity Press.
Hanich, Q., & Tsamenyi, M. (2009). Managing fisheries and corruption in the Pacific Islands region. Marine Policy, 33(2), 386–392.
Hauck, M. (2008). Rethinking small-scale fisheries compliance. Marine Policy, 32(4), 635–642.
Hill, M., & Hupe, P. (2009). Implementing public policy. London: Sage.
Hood, C.C. (1983). The tools of government. London: McMillan.
Huxham, C. (1996). Creating collaborative advantage. London: Sage.
Jentoft, S. (1989). Fisheries co-management: Delegating government responsibility to fishermen's organizations. Marine Policy, 2, 137–154.
Jentoft, S. (2007). Limits to governability? Institutional implications for ocean and coastal governance. Marine Policy, 4, 360–370.
Jentoft, S., & Chuenpagdee, R. (2009). Fisheries and coastal governance as a wicked problem. Marine Policy, 4, 530–560.
Jentoft, S., & McCay, B. (2003). The place of civil society in fisheries management – A research agenda for fisheries co-management. In D.C. Wilson, J.R. Nielsen, P. Degnbol (Eds.), The fisheries co-management experience: Accomplishments, challenges and prospects (pp. 293–306). Dordrecht: Springer.
Jentoft, S., van Son, T.C., Bjørkan, M. (2007). Marine protected areas: A governance system analysis. Human Ecology, 35, 611–622.
Jentoft, S., Chuenpagdee, R., Bundy, A., Mahon, R. (2010). Pyramids and roses: Alternative images for the governance of fisheries systems. Marine Policy, 34(6), 1315–1321.
Jentoft, S., Chuenpagdee, R., Pascual-Fernandez, J.J. (2011). What MPAs are for: On goal formation and displacement. Ocean and Coastal Management, 54(1), 75–83.
Jentoft, S., Pascual-Fernandez, J.J., De la Cruz Modino, R., Gonzalez-Ramallal, M., Chuenpagdee, R. (2012). What stakeholders think about marine protected areas: Case studies from Spain. Human Ecology, 1–13. doi:10.1007/s10745-012-9459-6.
Jervis, R. (1989). The logic of images in international relations. New York: Columbia University Press.
Jervis, R. (1997). System effects: Complexity in political and social life. Princeton: Princeton University Press.
Johnson, D., Thorpe, A., Bavinck, M., Kulbicki, M. (2005). Links in the fish chain. In J. Kooiman, M. Bavinck, S. Jentoft, R. Pullin (Eds.), Fish for life: Interactive governance for fisheries (pp. 133–146). Amsterdam: Amsterdam University Press.
Johnson, D. (2010). Institutional adaptation as a governability problem in fisheries: Patron-client relations in the Junagadh fishery, India. Fish and Fisheries, 11, 264–277.
Kjær, A.M. (2004). Governance. Cambridge: Polity Press.
Kooiman, J. (2003). Governing as governance. London: Sage.
Kooiman, J. (2010). Governance and governability. In S.P. Osborne (Ed.), The new public governance (pp 87–104). London: Routledge.
Kooiman, J., Bavinck, M., Chuenpagdee, R., Mahon, R., Pullin, R. (2008). Interactive governance and governability – An introduction. The Journal of Transdisciplinary Environmental Studies, 7(1).
Kooiman, J., & Bavinck, M. (2005). The governance perspective. In J. Kooiman, M. Bavinck, S. Jentoft, R. Pullin (Eds.), Fish for life: Interactive governance for fisheries (pp. 11–25). Amsterdam: Amsterdam University Press.
Kooiman, J., & Chuenpagdee, R. (2005). Governance and Governability. In J. Kooiman, M. Bavinck, S. Jentoft, R. Pullin (Eds.), Fish for life: Interactive governance for fisheries (pp. 325–350). Amsterdam: Amsterdam University Press.
Kooiman, J., & Jentoft, S. (2009). Meta governance, values, norms and principles, and the making of hard choices. Public Administration, 87(4), 818–836.
Kooiman, J., van Vliet, M., Jentoft, S. (Eds.). (1999). Creative governance: Opportunities for fisheries in Europe. Aldershot: Ashgate.

Kooiman, J., Bavinck, M., Jentoft, S., Pullin, R. (Eds.). (2005). Fish for life: Interactive governance for fisheries. Amsterdam: Amsterdam University Press.

Kooiman, J., Bavinck, M., Chuenpagdee, R., Mahon, R., Pullin, R. (2008). Interactive governance and governability – An introduction. The Journal of Transdisciplinary Environmental Studies, 7(1).

Lewis, D. (2002). Civil society in African contexts: Reflections on the usefulness of a concept. Development and Change, 33(4), 569–586.

Lewis, D., & Opoku-Mensah, A. (2006). Moving toward research agendas on international NGOs. Journal of international Development, 18, 665–675.

Lindblom, C.E. (2001). The market system: What it is, how it works, and what to make. New Haven: Yale.

Lockwood, M., Davidson, J., Curtis, A., Stratford, E., Griffith, R. (2010). Governance principles for natural resource management. Society & Natural Resources, 23(10), 986–1001.

Lovan, W.R., Murray, M., Shaffer, R. (Eds.). (2004). Participatory governance. Aldershot: Ashgate.

Mahon, R., Bavinck, M., Roy R.N. (2005). Governance in action. In J. Kooiman, M. Bavinck, S. Jentoft, R. Pullin (Eds.), Fish for life: Interactive governance for fisheries (pp. 351–374). Amsterdam: Amsterdam University Press.

Mahon, R., McConney, P., Roy, R.N. (2008). Governing fisheries as complex adaptive systems. Marine Policy, 32, 104–112.

Mahony, N., Newman, J., Barnett, C. (2010). Rethinking the public: Innovations in research, theory and politics. Bristol: Policy Press.

McGoodwin, J.R. (1990). Crisis in the world's fisheries. Stanford: Stanford University Press.

Meyer, D.S., Whitter, N., Robnett, B. (2002). Social movements. New York: Oxford University Press.

Mikalsen, K.H., Hernes, H.K., Jentoft, S. (2007). Leaning on user-groups: The role of civil society in fisheries governance. Marine Policy, 31, 201–209.

Mitchell, R., Bradley, R.L., Wood, D.J. (1997). Toward a theory of stakeholder identification and salience. Academy of Management Review, 22(4), 853–866.

Müller, K.B. (1996). The civil society-state relationship in contemporary discourse: A complementary account from Gidden's perspective. British Journal of Politics and International Relations, 8, 311–330.

Nash, J. (2005). Social movements: An anthropological reader. Oxford: Blackwell.

Olsen, J.P. (2005). Maybe it is time to rediscover bureaucracy. Journal of Public Administration Research and Theory, 161, 1–24.

Onyango, P., & Jentoft, S. (2010). Assessing poverty in small-scale fisheries in Lake Victoria, Tanzania. Fish and Fisheries, 11, 250–263.

Osborne, S.P. (Ed.). (2010). The new public governance? Emerging perspectives on the theory and practice of public governance. London: Routledge.

Osborne, D., & Gaebler, T. (1992). Reinventing government. Reading: Addison Wesley.

Papadopoulos, Y., & Warin, P. (2007). Are innovative, participatory and deliberative procedures in policy making democratic and effective? European Journal of Political Research, 46, 445–472.

Parkins, J.R., & Mitchell, R.E. (2005). Public participation as public debate: A deliberative turn in natural resource management. Society and Natural Resources, 18, 529–540.

Pascual-Fernandez, J.J., Frangoudes, K., Williams, S. (2005). Local institutions. In J. Kooiman, M. Bavinck, S. Jentoft, R. Pullin (Eds.), Fish for life: Interactive governance for fisheries (pp. 153–172). Amsterdam: Amsterdam University Press.

Peters, B.G. (2010). Meta-governance and public management. In S.P. Osborne (Ed.), The new public governance? Emerging perspectives on the theory and practice of public governance (pp. 36–51). London: Routledge.

Pierre, J. (Ed.). (2000). Debating governance. Oxford: Oxford University Press.

Raakjær, J. (2009). A fisheries management system in crisis – The EU Common Fisheries Policy. Aalborg: Aalborg University Press.

Rammel, C. (2007). Managing complex adaptive systems. Ecological Economics, 63, 9–21.

Rhodes, R.A.W. (1997). Understanding governance. Buckingham: Open University Press.
Rittel, H.W.J., & Webber, M.M. (1973). Dilemmas in a general theory of planning. Policy, 4, 155–169.
Roberts, N. (2004). Public deliberation in an age of direct citizen participation. American Review of Public Administration, 34(4), 315–353.
Ruddle, K., Hviding, E., Johannes, R.E. (1992). Marine resources management in the context of customary tenure. Marine Resource Economics, Marine Resources Foundation, 7, 4.
Sabatier, P.A. (Ed.). (1999). Theories of the public policy process. Boulder: Westview Press.
Shipman, A. (1999). The market revolution and its limits. London: Routledge.
Sørensen, E., & Torfing J. (Eds.). (2007). Theories of democratic network governance. Houndsmill: Palgrave.
Song, A.M, & Chuenpagdee, R. (2010). Operationalizing governability: A case study of Lake Malawi. Fish and Fisheries, 11, 235–249.
Suarez de Vivero, J.L., Rodriguez Mateos, J.C., Florido del Coral, D. (2005). International institutions. In J. Kooiman, M. Bavinck, S. Jentoft, R. Pullin (Eds.), Fish for life: Interactive governance for fisheries (pp. 197–216). Amsterdam: Amsterdam University Press.
Suarez de Vivero, J.L., Rodriguez Mateaos, J.C., Florido del Corral, D. (2008). The paradox of public participation in fisheries governance: The rising number of actors and the devolution process. Marine Policy, 32, 319–325.
Sumaila, U.R., & Dominguez-Torreiro, M. (2010). Discount factors and the performance of alternative fisheries governance systems. Fish and Fisheries, 11, 278–287.
Sundar, A. (2012). Alternatives to crisis: Social movements in global fisheries governance. Human Geography, 5, 2.
Swedberg, R. (2005). Principles of economic sociology. Princeton: Princeton University Press.
Thorpe, A., Johnson, D., Bavinck, M. (2005). System to be Governed: Introduction. In J. Kooiman, M. Bavinck, S. Jentoft, R. Pullin (Eds.), Fish for life: Interactive governance for fisheries (pp. 41–45). Amsterdam: Amsterdam University Press.
Torfing, J., Peters, B.G., Pierre, J., Sørensen, E. (Eds.). (2012). Interactive governance advancing the paradigm. London: Routledge.
Van der Schans, J.W. (2001). Governance of marine resources – Conceptual clarifications and two case studies. Delft: Eburon Press.
Van Tulder, R., Van der Zwart, A. (2006). International business-society management. London: Routledge.
Walker, B., Holling, C.S., Carpenter, S.R., Kinzig, A. (2004). Resilience, adaptability and transformability in social–ecological systems. Ecology and Society, 9(2), 5.
Williams, R. (Ed.). (2000). Explaining corruption. Cheltenham: Edward Elgar.
Wilson, J. A. (2006). Matching social and ecological systems in complex ocean fisheries. Ecology and Society, 11(1), 9. Retrieved from http://www.ecologyandsociety.org/vol11/iss1/art9
Wilson, D.C., Nielsen, J.R., Degnbol, P. (Eds.). (2003). The fisheries co-management Experience: Accomplishments, challenges and prospects. Dordrecht: Kluwer Academic.
Wolfe, A. (1989). *Whose leeper?* Social science and moral obligation. Berkeley: University of California Press.
World Bank. (1989). A framework for capacity building in policy analysis and economic management in Sub-Saharan Africa. Washington, DC: World Bank.
World Bank. (2004). Saving fish and fishers. Washington, DC: World Bank.

Part II
Meta Concerns

Chapter 3
Concerns and Problems in Fisheries and Aquaculture – Exploring Governability

Svein Jentoft and Ratana Chuenpagdee

Abstract Concerns, problems and issues in fisheries and aquaculture are, in many instances, undermined by the lack of a thorough examination of their nature and of the characteristics of the associated systems. Their persistence creates governability challenges, which have restricted effective policy interventions. In order to improve governability, we submit that there is a need to first recognize the complexity of the problems, and then conceptualize them in a way that points towards solutions. In so doing, it may reveal not only limits to governance but also the opportunities and possibilities that exist to enhance governability, i.e., the overall quality for governance.

Keywords Governability assessment • Interactive governance • Wicked problems • Fisheries • Aquaculture

Introduction

The second chapter of the book explains the key elements of the interactive governance theory, linking it to the overall issue of the governability of the system-to-be-governed, the governing system and the various governance orders and modes. The challenges of applying this theory and the governability concept are the focus of this third chapter.

S. Jentoft (✉)
Faculty of Biosciences, Fisheries and Economics,
University of Tromsø, Tromsø, Norway
e-mail: svein.jentoft@uit.no

R. Chuenpagdee
Department of Geography, Memorial University of Newfoundland, St. John's, NL, Canada
e-mail: ratanac@mun.ca

M. Bavinck et al. (eds.), *Governability of Fisheries and Aquaculture: Theory and Applications*, MARE Publication Series 7, DOI 10.1007/978-94-007-6107-0_3,

Here we explain a way to operationalize and apply it to examine concerns, problems and issues related to fisheries and aquaculture. First and foremost, the complex set of concerns related to fisheries and aquaculture governance needs to be recognized in order to enable an understanding of the nature of the problems that are to be solved. Only when this happens can we move into identifying the structural and functional aspects of fisheries and aquaculture governance determined by the characteristics of the systems-to-be-governed, the governing system, and their interactions. The chapter serves as an introduction to the following chapters in this section of the book, each of which elaborates on the relation of the basic concerns facing fisheries governance outlined in the *Fish for Life* volume, i.e., social justice, livelihoods, food security and ecosystem health (Kooiman et al. 2005), to governability. These concerns are value-laden, sometimes irreconcilably so, and therefore require a comprehensive and holistic governance approach. The conceptualization of the relationship between these concerns and the problems that must be solved, as well as the exploration of limitations and possibilities that may exist, are the first steps in improving governability.

The persisting and re-occurring problems in fisheries and aquaculture suggest that there may be some features related to these resource systems and the governing mechanisms that make governance particularly challenging. In their discussion of planning, Rittel and Webber (1973) conceptualize these problems as "wicked." By this, they mean that the problem is neither simple nor easy to define, but requires a deliberative process, an aspect that is also suggested in the interactive governance theory (Kooiman 2003). As is commonly perceived as ideal in fisheries management, the planning process starts by defining the problem, setting the goal, seeking and analysing information, outlining the available alternative solutions, calculating their relative merits, making and implementing a decision, and, finally, evaluating outcomes and, if needed, making modifications (cf. Gilmore and Camilius 1996). This systematic methodology is typical of engineering or scientific inquiry, and, according to Rittel and Webber (1973), may work for what they call "tame" (or benign) problems. It does not, however, apply to societal problems in real life situations. These are messy and uncertain, and therefore require a process that allows for interactive communication and learning.

In the following sections, we draw on Rittel and Webber (1973) to examine how fisheries and aquaculture concerns and problems can be conceptualized and how they can give rise to governance challenges. We then present the underlying hypotheses for assessing governability and provide an explanation of the ways in which key attributes may cause fisheries and aquaculture systems to be more or less governable.

Conceptualization of Concerns and Problems

Concerns are 'meta-order' governance issues that are deeply situated in social values, norms and principles about things we care for and consider important (Kooiman and Jentoft 2009). Thus, ecosystem health, social justice, livelihoods and food security are aspects that require governance attention, most likely all at once, and addressing one of them alone in any instance may result in the worsening of others. This is why fisheries and aquaculture governance must transcend academic disciplines and discourse, which tend to emphasize a particular concern while ignoring others. For instance, conservation biology may be more interested in the health of the marine ecosystem, but not of the people depending on it for their livelihoods. The latter belongs to social science discourse. Consequently, marine protected areas are promoted primarily by biologists as an ecosystem conservation tool, while their social impacts are often ignored (Christie et al. 2003).

Addressing these multifaceted concerns in fisheries and aquaculture is a major challenge and creates problems for governance. These problems are partly related to how they are defined and what solutions are possible. The conceptualization of these problems is socially constructed, meaning that social processes that "set" a problem must come before it can be solved (Schön 1983). Fisheries stakeholders are numerous and they have competing and, more often than not, conflicting interests, values and worldviews. Thus, in practice it is close to impossible to achieve full consensus about what the problems are and where the solutions lie. Even when they share similar interests, different stakeholders tend to frame the problem differently. Moreover, scientists' perspectives of the 'fisherman's problem' (McEvoy 1990) and what it takes to solve it varies among disciplines. So too, the solutions they prescribe may be different (Brady and Waldo 2009). Where fisheries biologists see fish, social scientists see people. Where governments see sectors, economists see markets, and sociologists see communities and social groups. This is also reflected in the concerns that each group would consider most important, thus differentiating what they want the governing system to concentrate on.

Many of the problems in fisheries and aquaculture tend to reappear and are not solved once and for all (Jentoft and Chuenpagdee 2009). Instead, they need to be attended to on a continuous basis. For example, protecting ecosystem health, sustaining adequate resource availability, and securing food and livelihood activities for communities and consumers require long and committed governance efforts (see Chap. 7 by Pascual-Fernandez and Chuenpagdee, Chap. 5 by Johnson, and Chap. 6 by Pullin, this volume). This implies that the governing system must perform its tasks and evaluate the outcomes on a regular basis. The difficulty observed worldwide with the implementation of fisheries and aquaculture regulations suggests that effective governance is an on-going struggle that requires adaptation and innovation. In particular, it calls for creative interactions along the lines suggested by the interactive governance perspective.

Yet, interactive governance is no panacea. It does not guarantee success. Rather, it requires rigorous assessment of what makes a system more or less governable in order to help improve governance in certain contexts and to provide general insights and lessons that may be broadly applicable. Since fisheries and aquaculture governance is largely about hard choices (Kooiman et al. 2005), one challenge and concern may be addressed in ways that may help or hinder the resolution of other problems. For instance, as governors deal with resource conservation issues, they may complicate another problem, like that of alleviating poverty (Chap. 9 by Onyango and Jentoft, this volume). As shown around the world, it is difficult to both conserve and utilize resources at the same time and in a socially just manner (see Chap. 4 by Jentoft, this volume). Even when a decision could be made either to conserve or to utilize, the question of who benefits and who loses from such decision remains. It goes without saying that winners and losers would tend to frame the problem differently and have very different ideas of whether the problem is actually solved. Thus, equity and justice, as far as benefit and burden sharing is concerned, are among the criteria on which to make judgements with regard to determining whether a problem has been addressed. However, what constitutes equity and justice is equally difficult to determine and agree on. The sharing of burdens and benefits in ways that stakeholders would accept as fair and just is not only an ethical and moral issue, it also has practical implications. This is largely because it will determine what stakeholders are willing to abide by and accept. At the end of the day, they determine the degree to which a governance system will work or not (Jentoft and McCay 1995).

One of the main features of fisheries and aquaculture is embeddedness (Hanna and Jentoft 1996). Their problems are situated in particular ecological, social and political contexts and are often symptoms of larger issues or problems within other problems. In other words, they may operate at various scales. Thus the level at which the problem should be addressed or dealt with first – at the producing or receiving end – is a key question. For example, from the ecosystem perspective, the 'fishing down the food web' phenomenon (Pauly et al. 1998) indicates the accumulating effects of continued fishing pressure and the shift of target species from large pelagic fish on top of the food chain to invertebrates. Rebuilding fisheries would require measures that not only address problems at the lower food chain, but also at the top. However, fishing also links to the alteration of natural ecosystems, and, in cases where destructive bottom-tending gears such as trawls and dredges are employed, the degradation of seafloor habitats (Chuenpagdee et al. 2003). This implies that the restoration and protection of habitats would need to be considered as part of the ecosystem management. From the social and policy perspective, modernization in fisheries has resulted in distributional issues between small- and large-scale fisheries (Butcher 2004), which have consequences on broader concerns like poverty, food security, social justice, gender bias and human rights. The effects of the decline in fisheries catches are more palpable on small-scale fishing sector and their communities than on the industrialized sector. Small-scale fishers are highly dependent on the resources and have limited financial assets and capacity that would enable them to explore other livelihood options. The remoteness of their location, the lack of capital, and the weak bargaining power in the market and in the policy realm ensure that small-scale fishers are highly vulnerable to changes that occur in the fish production chain (Chuenpagdee 2011). Addressing the problems within small-scale fisheries sector would thus require measures and interventions implemented at all scales.

Fisheries and aquaculture problems may have features that are known, but they occur in different situations and contexts. Hence, potential solutions cannot be the same across the board. For instance, a fisheries crisis, such as a resource in peril, observed in one country may well resemble a similar crisis in another country. However, when looking at the details, there are always some features of the crisis in each country that make them ecologically, socially, culturally and politically unique. This is why governability assessments must pay attention to contextual variables. This is also one of the arguments for the 'subsidiarity' principle in fisheries governance, which states that problems should be addressed (and responsibilities to solve them vested) at the lowest possible organizational level (Jentoft and Eide 2011). However, it does not suggest that the decentralization of authority from high to low levels is always necessary or wise. In fact, there are issues which the local community is not the appropriate entity to deal with. The construction of infrastructure for post-harvest activities and overall legislative frameworks, which requires active involvement and financial backing of governments, serves as a case in point.

On the whole, fisheries and aquaculture problems have a number of properties that governors would recognize as being wicked (Jentoft and Chuenpagdee 2009). One of them is that "[t]he formulation of a wicked problem *is* the problem!" (Rittel and Webber 1973, 61). In other word, a problem may be felt but people may disagree in how to interpret it. For instance, explanations for low fishery yield are not always clear and consensus about the nature of the problem is hard to find. There is often a "blame game" of who causes it and who is responsible for solving it. In most cases, there is no single natural or social variable that explains outcomes. Even with the best of science, it is still difficult to provide clear answers with regard to problem definitions and solutions. The objective of sustaining a particular resource base, for instance, leaves us with a number of questions. At which level should the resource be sustained? How large should this stock biomass be? If a recovery plan is needed, how fast should it proceed? How big should a protected area be? Should the state of the ecosystem be the only concern, or are there other concerns that are also important, and if yes, how should they be prioritized? Which concerns should be given the greatest attention and resources? These are questions where science falls short. Science may help raise critical questions, but on its own it may not provide answers that lead to decisions. Other concerns may play a role, different rules may apply, and conflict may be a major driver. On political issues, stakeholders tend to have different views and priorities, many of which are not based on knowledge as much as they are on values, beliefs and interests. Here, the problem definitions and the solutions that one may support may not only reflect a political stance, but such a stance will also determine which questions are asked and which concerns are given priority (Turnbull 2006).

Operationalizing Governability Assessment

The governability concept recognizes that there are limits to how systematic and effective a governing system can be in addressing challenges and concerns that occur within the system-to-be-governed. With limits of governability, one must assume that governance does not always go according to plan, and that governors often have to accept solutions that are less than ideal. However, the limits of governability are not necessarily structural. They can sometimes be related to how the systems function. As the case may be, there is a potential for either marginal or radical governability enhancement, through governance interventions aiming at changing the way the systems are structured and working. What these limits are and how they can be influenced is a governance research issue.

When assessing the limitations and opportunities for enhancing governability, one might expect to encounter the following five hypotheses: (1) the systems-to-be-governed may be too diverse to be handled from top-down mode of governance. (2) Diversity suggests that no 'one size fits all' situation exists. Problems are essentially unique and governance solutions can therefore not be generalized. Instead, solutions must be built on local knowledge and decision making involving "those who know." As mentioned, local level management and devolution of authority and power may not work in all instances, such as in communities with weak leadership, low capacity and high heterogeneity and fragmentation (Archeson 2006). (3) The more complex the system is, the more governance would need to emphasize the structures of relationships and the ways in which they may hinder or facilitate constructive interaction and collective problem solving. Similarly, (4) the dynamics of the system requires knowledge of the internal interactions within the natural and social systems-to-be-governed, and the governing system; understanding the ways in which they affect the ability to govern and the overall quality of governance. Finally, (5) scale issues draw attention to boundary setting in addition to the spatial and temporal distribution of the system components, especially at the border or in overlapping areas where the majority of interactions take place.

Diversity, complexity, dynamics and scales are traits that may complicate the governability of fisheries and aquaculture, sometimes to the extent that the goals of governance are hard to achieve. They make systems less controllable and predictable. Their developments are often non-linear and information is never complete, making governance outcomes inherently uncertain (Degnbol and McCay 2007). As a consequence, they require a governance approach that is flexible and adaptive rather than one that emphasizes control and stability (Mahon et al. 2008). Conflicts within and between systems tend to persist over time and are therefore energy consuming. The same is true when goals are [re]negotiated among participants in the governing process. Governability assessments must therefore recognize that goals are not given *ex ante*, but are themselves governance outcomes. Under such circumstances, governance is not so much about exercising authority from the top-down as it is about political brokerage, where operating goals are at best imperfect compromises (cf. Jentoft et al. 2011).

At a general level, governability is an outcome of the structure that defines and the processes that occur within and between the governing system and the system-to-be-governed. Interactive governance theory broadens the perspective of governability to something that is not solely an issue and a responsibility of the governing system, such as a government agency. Rather, governability depends on the ability of these systems to deliver – individually and in concert – on the challenges and demands resulting from their diversity, complexity, dynamics and scale. Hence, in order to enhance governability, a wide range of instruments and mechanisms would be drawn from and applied within all systems. Since both structural and functional governability can be influenced by acts of governance, governance is basically about promoting governability.

Structural Governability

From the interactive governance perspective, governability is explored as a measure of how governable fisheries and coastal systems are given the particular features of the natural and social systems-to-be-governed, the governing system(s), and the interaction between them. For the system-to-be-governed, we distinguish between two sub-systems, the natural ecosystem and the socio-economic system. From a governability perspective, we focus on their separate qualities as well as the relationship and interaction between them. While the marine environment has intrinsic value on its own, it also provides a wide range of ecological goods and services to organisms within the system, not least humans. The social system affects change in this natural system, but it is also dependent and therefore impacted by these changes, which set limits to resource users' potential. The governing system aims to influence the interaction between the natural system and the socio-economic systems that it governs.

The natural system-to-be-governed in the fisheries context refers to the marine environment (including the geological, physical, and chemical properties), the flora, fauna and other organisms contained within the ecosystem, the habitats within which they live (including the water column), and oceanographic currents such as upwelling and storm surges. Natural ecosystems are related to the biophysical and geo-morphological environment, living and non-living organisms, vegetated and non-vegetated habitats. The external influences that may alter the natural system come from open oceans, coastal zones, as well as from upland areas.

The socio-economic system-to-be-governed comprises of direct and indirect resource users (e.g., commercial and recreational fishers fish farmers, shellfish gatherers, divers, tourists, developers, oil and gas explorers, etc.), the social relations that they form, and interactions that occur among them. These various users may share common values, but more often than not their interests vary to the point that their activities cause competition, conflicts and resource degradation. In the simplest form, some of their use areas may overlap, and tension arising from this may be lessened through proper zoning. Some activities may, however, completely displace others. For example, areas used for mariculture are no longer accessible for small-scale fishing. Likewise, the development of an exclusive beach resort often prohibits others from enjoying the area. Additionally, there are issues like externalities such as those caused by the use of destructive fishing gears by a group of fishers, which degrade ecosystems and create loss of income and livelihoods for others. The social relations and interactions among various groups may be more or less structured, institutionalized and exposed to interference by governing bodies through, for instance, numerous regulations and rules. The social system-to-be-governed can also form governing systems themselves, as is the case with user organizations or communities.

The governability of the social system-to-be-governed may be looked at as a three stage process. In the context of halting ecological degradation, for instance, the governing system must work with and through the socio-economic system by influencing user-behaviour such as fishing patterns. However, it is also a process with a feedback loop, meaning the governing system is held accountable by the social system-to-be governed. In a democratic system, representatives of the social system-to-be governed are the ones with the power to interfere with the ways in which the governing system works. When resource crises hit, the governing system is questioned and governance reform may follow.

Governance theory argues that the system-to-be-governed and the governing system must be compatible in order to be mutually responsive. Thus, for instance, diversity within the system-to-be-governed would require that the governing system be contextually sensitive and therefore it also becomes structurally diverse. Similarly, governing mechanisms need to be tailored to the particularities of the targeted system-to-be-governed. The greater the variations between different system-to-be-governed, the more diverse the governing systems may need to be.

In the same vein, the system's complexities, dynamics and scales demand an adequate response requiring the governing system to mirror those traits (Chap. 10, Scholtens and Bavinck, this volume). Hence, governability will hinge upon the extent to which the governing system achieves this (cf. Jentoft 2007). This is not a matter of natural adaptation, but of deliberate planning and structural design by societal actors such as legislative bodies, planning agencies and civic organizations. This can happen alone or, preferably, in concert. We say preferably because governing capacity and interaction is enhanced through collaboration. Governing interactions are about the two-way mediation between the governing system and the systems-to-be-governed. These interactions identify the various ways in which the governing system is sensitive to the diversity, complexity, dynamics and scale within the systems-to-be-governed, and the ways in which the governing system responds accordingly. These connections structure the ways in which information is collected and communicated, representation is organized, and stakeholders participate, as well as the manner in which learning occurs among the actors involved in both systems. Furthermore, governing interactions can enable adaptation and capacity building and reveal the negotiation process of setting goals and objectives, resolving conflicts, and achieving compromise.

Functional Governability

Governability lies within the inherent and constructed traits of fisheries and aquaculture systems. The diversity, complexity, dynamics and scales of these systems present real challenges to governance, thus affecting their governability. Diversity, whether it regards resource units or relevant stakeholders, is about the heterogeneity of system's elements and its variability. A marine ecosystem in the tropics, especially those containing coral reefs, is much more diverse in terms of ecological biodiversity than one situated in a temperate zone. These areas tend also to be more heterogeneous. For example, they contain subsistence, commercial, small-scale and large-scale fishing sectors using different types of boats and gears and targeting multiple species. This kind of diversity calls for fine-tuned, particularistic governing approaches. For instance, in order to fully comprehend, represent and control the issues at hand, governance must occur where the problems are felt or where the opportunities appears, and be conducted in cooperation with concerned stakeholders.

Complexity raises the issue of handling interrelationships and interdependencies. Any intervention into the system will have an effect on these linkages and may thus change the way the whole system works. Stakeholders sometimes interact predictably and with patterns that are easy to understand. For example, once landed, catches from large-scale fisheries are typically sold to the main wholesalers in accordance with prior arrangements. In this case, vessel owners have long and established relationships with certain buyers and negotiations between them are minimal. Sometimes these arrangements are institutionalized through vertical integration where transactions are internalized within one and the same enterprise. In small fishing villages where the majority of fishing is conducted by individual fishers, the marketing system can be very complicated. Fishers or their wives and children may sell their own catches to independent buyers or they may sell to certain buyers, often as part of the loan agreement. One governability issue is to anticipate and contain systemic, secondary effects. Another is that systems also have latent functions, which are often difficult to discern and account for. There is a risk of unforeseen consequences and feedbacks that interfere with the way the system works. For instance, when a common property is replaced by private property, it may leave community members to fend for themselves on an individual basis.

Dynamics refers to the fact that systems may be unstable and that they evolve over time. A major source of these dynamics is the interaction that occurs within and between systems, as actors respond to each other and to shifting circumstances. Governability here would then be expressed as the capacity and capability of systems to cope with internal demands and to mediate and accommodate external drivers. Internally we can think of conflicts between stakeholders giving rise to constructive or destructive interactions, either of which generates dynamics in the system. Another example is how a governing system may lag behind the development in the system-to-be-governed, causing low governability. A system-to-be-governed that is able to resist or evade governing actions may similarly be a sign of low governability. A governability assessment must therefore focus on how these systems function *in vivo*, as a living organism with uncertain and unpredictable outcomes.

Further, fisheries and aquaculture as natural systems, and the social and governance systems related to them, come in varying spatial and temporal scales. The matching of scale in the operation of these systems is therefore another important aspect that determines governability. Notably, a certain degree of overlap among natural, social, economic, and political boundaries is expected for the system to function. The extent to which this overlap occurs, particularly in the case of multiple jurisdictions, is a governability issue. The extent of the overlap is likely a determining factor to the governability problem. How it plays out at the end of the day, and how it may be addressed, is an empirical question.

The fisheries and aquaculture governing system is a decision-making machine that must relate to the diversity, complexity, dynamics and scale of the system-be-governed.

The diversity implies on-site presence. The complexity requires an eye for interaction, by-effects and latent functions. The dynamics do not afford the governing system a rest and are such that change does not necessarily converge towards equilibrium. Nor are they such that change is always cyclical. Instead, things often change in lasting ways. As a result, the idea of adaptive governance and corrective feedbacks with built-in learning does not always work well. Adaptive governance exists within limits, whereas transformative governance is about surpassing or stretching them. Adaptation also suggests marginal rather than fundamental, radical change and that a new solution is sought within proximity of old ones. Change does not occur automatically, as in the case of a natural evolutionally process, nor does it occur easily, as it would in the event of a re-adjustment of the governing mechanism. Instead, change often comes as a consequence of a decision-making process ridden with conflicts and political log-rolling, where winning outcomes are determined by power as much as an inherent collective rationality.

Conclusion

Governors, be they individuals or agencies, operate within systems that are inherently diverse, complex and dynamic, restricting what they can possibly achieve. Their ability to govern is limited, and falling short of expectations is part of the norm. Yet, these limitations are also opportunities; they are not necessarily given once and for all, but may be subject to governance initiatives and reforms such as institutional redesign, organizational innovation and the development of new management tools. Sometimes these marginal adjustments of routines, which Kooiman (2003) labels "first" order governance, will suffice. Other times they require a more thorough overhaul ("second-order" governance), such as institutional reform. In some instances, however, governors must start over again by rethinking the basic principles, images and values that underpin governance in the first place and the ways in which they provide guidance for institutional design and operation ("meta-order" governance). Such is the case when governance failure becomes critical. A governability assessment must include a systematic search for possible malfunctions among the diversity of the system components and their complex connections. Governors must be prepared for the fact that things will take time; governance is often a slow process, because it involves multiple stakeholders who need to be convinced that there is a problem and that there is a need for change. If they are not convinced they will tend to resist the call for change.

The beginning of the governability assessment described in this chapter enables the governor(s), including researchers, to identify what the problem is, how the system works, and what might possibly explain why it does not work. The goals of governance are assumed to be negotiated internally as part of interactions; they are not predetermined as something that the system necessarily tries to achieve. As we anticipate that goals are themselves the outcomes of the interactions structured by the particular systems under scrutiny, they should be assessed empirically. In these interactions, power is expected to play an important role. In the first place, power influences what goals are to be established and who sets them. This implies that the governing system is designed to correspond with the properties of the systems-to-be-governed, the structural features of the governing system, both issues of governance in and of themselves, and the ways these systems interact. It is precisely for this reason that addressing governability must recognize the limits and opportunities of particular systems rather than expecting them to perform beyond what they can actually and potentially do.

References

Archeson, J.M. (2006). Institutional failure in resource management. Annual Review of Anthropology, 35, 117–34.

Brady, M., & Waldo, S. (2009). Fixing problems in fisheries – Integrating ITQs, CBM and MPAs in management. Marine Policy, 33, 258–263.

Butcher, J.G. (2004). The closing of the frontier: A history of the marine fisheries of Southeast Asia ca 1850–2000. Singapore: Institute of Southeast Asian Studies.

Christie, P., McCay, B.J., Miller, M.L., Lowe, C., White, A.T., Stoffle, R., Fluharty, D.L., McManus, L.T., Chuenpagdee, R., Pomeroy, C., Suman, D.O., Blount, B.G., Huppert, D., Eisma, R.L.V., Oracion, E., Lowry, K., Pollnac, R.B. (2003). Toward developing a complete understanding: A social science research agenda for marine protected areas. Fisheries, 28(12), 22–26.

Chuenpagdee, R. (2011). A matter of scale: prospects in small-scale fisheries. In R. Chuenpagdee (Ed.), Contemporary visions for world small-scale fisheries (pp. 21–36). Delft: Eburon.

Chuenpagdee, R., Morgan, L.E., Maxwell, S.M., Norse, E.A., Pauly, D. (2003). Shifting gears: Assessing collateral impacts of fishing methods in the U.S. waters. Frontiers in Ecology and the Environment, 1(10), 517–524.

Degnbol, P., & McCay, B.J. (2007). Unintended consequences of ignoring linkages in fisheries systems. ICES Journal of Marine Science, 64(4), 793–797.

Gilmore, W.S., & Camilius, J.C. (1996). Do your planning processes meet the reality check? Long Range Planning, 29(6), 869–879.

Hanna, S., & Jentoft, S. (1996). The human use of the environment: An overview of social and economic dimensions. In S. Hanna, K. Folke, K-G. Mähler, K. Arrow (Eds.), Rights to nature (pp. 35–55). Washington, DC: Island Press.

Jentoft, S. (2007). Limits of governability: Institutional implications for fisheries and coastal governance. Marine Policy, 31, 360–370.

Jentoft, S., & Chuenpagdee, R. (2009). Fisheries and coastal governance as a wicked problem. Marine Policy, 33, 553–560.

Jentoft, S., & Eide, A. (Eds.). (2011). Poverty mosaics: Realities and prospects in small-scale fisheries. Dordrecht: Springer.

Jentoft, S., & McCay, B.J. (1995). User participation in fisheries management: Lessons drawn from international experiences. Marine Policy, 19(3), 227–246.

Jentoft, S., Bavinck, M., Johnson, D., Kaleekal, T. (2009). Co-management and legal pluralism: How an analytical problem becomes an institutional one. Human Organization, 68(1), 27–38.

Jentoft, S., Chuenpagdee R., & Pascual-Fernandez, J. (2011) What MPAs are for. On Goal Formation and Displacement. Ocean and Coastal Management, 54(1), 75–83.

Kooiman, J. (2003). Governing as governance. London: Sage.

Kooiman, J., Bavinck, M., Jentoft, S., Pullin, R. (Eds.). (2005). Fish for life: Interactive governance for fisheries. Amsterdam: Amsterdam University Press.

Kooiman, J., & Jentoft, S. (2009). Meta-governance for natural resources: Choices, values and principles. Public Administration, 87(4), 818–836.
Mahon, R., McConney, P., Roy, R.N. (2008). Governing fisheries as complex systems. Marine Policy, 32, 104–112.
McEvoy, A.F. (1990). The fisherman's problem: Ecology and law in the California fisheries, 1850–1980. Cambridge: Cambridge University Press.
Pauly, D., Christensen, V., Dalsgaard, J., Froese, R., Torres, F. Jr. (1998). Fishing down marine food webs. Science, 279, 860–863.
Rittel, H.W.J., & Webber, M.M. (1973). Dilemmas in a general theory of planning. Policy Sciences, 4, 155–169.
Schön, D.A. (1983). The reflective practitioner: How professionals think in action. London: Temple Smith.
Turnbull, N. (2006). How should we theorize public policy? Problem solving and problematicity. Policy and Society, 25(2), 3–22.

Chapter 4
Social Justice in the Context of Fisheries – A Governability Challenge

Svein Jentoft

Abstract Social justice figures among the key concerns of fisheries governance. This is for various reasons; one being that it is an important value in and of itself. Fisheries governance as a distributional and potentially disempowering or empowering mechanism should be evaluated on the basis of whether it works in accordance with general justice principles, such as those pertaining to human rights. However, as is argued in this chapter, justice is also an important governability condition. Without justice, stakeholders are likely to revolt against governing efforts to sustain the resource or to promote development. This chapter draws on international experiences to illustrate what justice as a governance principle implies for fisheries.

Keywords Interactive governance • Social justice • Governability • Governance orders • Fisheries

Introduction

The concept of governability, defined by Kooiman (2008: 173) as "the overall capacity for governance of any societal entity or system," touches on issues of social justice in fundamental ways. A system-to-be-governed that is characterized by inherent inequities and justice disparities among its constituents is likely to experience tensions and conflicts that might lead to a lack of cooperation and resistance to governing interventions. A governing system is often held accountable for generating such problems or blamed for not correcting them.

S. Jentoft (✉)
Faculty of Biosciences, Fisheries and Economics,
University of Tromsø, Tromsø, Norway
e-mail: svein.jentoft@uit.no

M. Bavinck et al. (eds.), *Governability of Fisheries and Aquaculture: Theory and Applications*, MARE Publication Series 7, DOI 10.1007/978-94-007-6107-0_4,

This chapter demonstrates how justice issues tend to play out in fisheries governance by drawing on experiences from different parts of the world. It is argued that in order to enhance social justice, hence increasing governability, one needs to look at both the system-to-be-governed and the governing system. Emphasis must also be placed on the interaction between the two systems. More precisely, this chapter deals with two closely related questions: (a) How should we think of social justice from an interactive governance perspective; and, conversely, (b) how should we think of interactive governance from a justice perspective? The short answers to both questions are as follows: in the first instance, we would use the governance framework, as outlined and explained in Chap. 2 of this book and in Kooiman (2003), to analyze social justice issues that relate to fisheries. We apply this framework to get a grip on social justice as something that may inhibit or enhance governability, as it pertains to resource management. In the latter instance we draw on the academic literature on social justice to inform the governance framework and our search for potential and actual governability impediments and opportunities. For such an endeavor, there is a vast reserve of literature to learn from.

In arguing both for the need to address the ways in which the interactive governance theory may help to elucidate justice issues in fisheries governance, as well as the manner in which justice principles may be applied in fisheries governance, this chapter explores the linkages between governability and social justice. Governability may be enhanced by the implementation of sound justice principles; justice being something governance systems should aim for from both a normative and a functional perspective, i.e. for its own sake and for the workings of the system-to-be-governed and the governing system. In this chapter, I first briefly summarize social justice theory and introduce its basic conceptual characteristics. Next, I examine the relevance of this discourse for fisheries governance as it pertains to the three governance orders – first, second and "meta." Thereafter, I discuss governability as a condition for making governance systems work, and how a governability assessment can be applied in research focused on social justice in fisheries. I argue that in enhancing governability, one needs to pay as much attention to procedural justice, i.e. how justice claims and claimants are recognized and included in the governance process, as one does to distributional justice and the fairness of the institutional design of governance systems.

Justice Concepts

There is more than one way of thinking about justice. So too, the term is difficult to define and can mean several things. Is treating people in the same way always just? Whose justice are we talking about – and with respect to what? Could privileged treatment (positive discrimination) in some instances be just? If yes, when? The idea that several justice concepts exist and that conflicting justice principles may apply to the same situation is a challenging one from a governability perspective. It is also easily observed in fisheries where the issue of open access, for instance, as well as the opposite idea of the importance of tenure, may be defended by referencing two equally laudable justice principles: that of not being excluded from resource extraction, and that of securing a viable livelihood free from destructive intrusion for oneself or for one's group or community. Let me now discuss this issue in a more orderly fashion.

When there is scarcity of resources, issues of distributional justice typically arise (Bojer 2006). Does everyone get a fair share? Does everyone get what they deserve – or think they deserve? Claimants to fisheries resources may find themselves having to settle for less than what they want or need. In the short run, and within a given total allowable catch (TAC), fisheries management is a zero-sum game; more for one means less for someone else. Therefore, distributional issues, i.e. when quotas are allocated or fishing territories are divided among fisher groups, often lead to social conflict and political strife among resource users. Distributional justice is closely related to how markets work. Markets do not automatically distribute resources to those most in need or those who have other legitimate claims to them.

Distributional justice must also be discussed in relation to power differentials and power dynamics. A recurrent justice issue in social science fisheries social research is how the "fish chain" produces winners and losers (Kooiman et al. 2005; Pullin (Chap. 6), this volume). For instance, there is now substantial literature on the role of middlemen in transactions with fishers (see Jentoft et al. 2010). Asymmetric relationships tend to occur because fishers lack alternatives; they have no other place to go and must therefore accept whatever price the middleman offers. Efforts to increase distributional justice would therefore need to target power relations that cause unfair distributional outcomes.

Social justice is also related to resource use and overexploitation. Justice is then both a contributing factor and an outcome. As Chuenpagdee et al. (2005, 35) argue, "To make a living when no alternative sources of employment are available and one's bargaining power is weak, the only response to falling prices is for fishers to increase their fishing effort." Those who drive fishing pressure beyond the point of sustainability are not necessarily those that have to live with the consequences. A case in point is the "roving bandits" who have no attachment to place and community and thus move around the world in search of un- or underexploited resources. When they are done with them and leave, fish stocks are ruined (Berkes et al. 2008). This certainly has ecological and management implications that, from a governability perspective, have proven difficult to deal with for national governments. From the perspective of local populations who suffer the consequences for their livelihoods, however, the problem is also about distributional justice.

It is important to note that there is more to justice than distributional aspects. Young (1990) argues that the "distributive paradigm" focuses too much on material goods and does not take into account group differences and the fact that some groups are oppressed and marginalized. According to Miller (1999), justice should therefore be conceived as a tripartite concept, emphasizing need, desert[1] and equality.

---

[1] By desert, we mean what we believe a person deserves. For an elaboration of this concept, see Pojman (2001).

In addition, people may have important concerns and therefore have the right to be recognized in the political process as, for instance, a legitimate stakeholder who should be involved and heard (Haldemann 2008). The latter is about the procedural – or democratic – element of justice in decision-making, which is central to interactive governance from a normative perspective. The implication of "broader and more authentic participation" is a demand for empowerment through more participatory democracy (Schlosberg 2003). This is emphasized, for instance, in the new European fisheries policy and implemented through the regional management councils, the so-called RACs (Degnbol and Wilson 2008). Accordingly, justice has an institutional and a processual dimension, where the first provides opportunity for the latter, but the latter does not necessarily follow directly from the former. Sen (2009) therefore argues that we need to be as concerned with process as we are with institutional design.

Even though it is possible to 'split' the concept of justice, as Miller (1999) does, its various parts relate to each other. Schlosberg (2003, 96) contends, "One must have recognition in order to have real participation; one must have real participation in order to get real equity; further equity would make more participation possible, which would bring further recognition, and so on" (cf. also Honneth 2001). It is not uncommon for small-scale fishers, regardless of their great number relative to the industrial sector, to be ignored by governments more focused on production and export earnings than on the livelihood conditions of fishing people. Despite being more numerous, they have less political clout; their views are not heard and their knowledge is perceived as less relevant to decision-makers than that of the industrial sector (Jentoft and Eide 2011).

Recognizing the human rights of small-scale fishing people is a place to start. From that follows a number of obligations, such as respecting their livelihoods, tenure and access rights, their rights to be informed about government policies and have a say in matters that affects them, rights to organize, and all the other things that are inscribed in the Human Rights declaration.[2]

Social justice cannot, however, be easily attributed to some universal standard. Rather, as Walzer (1983) points out, justice is sometimes contextual – or "local" (Elster 1992); justice principles are often meant for particular societal spheres. The principles of distributional justice are not the same for every good or burden, for every social and political community, or for every situation or circumstance. The justice-principles are not always stable either, even within the same sphere. Moreover, there is no one single set of appropriate criteria "against which all distributions should be measured" (Dahl 1989; 303). There is also the fact that justice means different things to different people or groups. All these caveats make justice a challenging empirical issue and something that governors need to know about and consider for the sake of governability (cf. Perusse Daigle et al. 1996).

Conflicts among user-groups make fisheries systems difficult to govern largely because people will tend to look out for their own rather than what is in

[2] See Bangkok civil society declaration http://sites.google.com/site/smallscalefisheries/statement (Accessed November 1, 2010).

the common good. A "what's in it for me" attitude does not bode well for interactive governance processes, but sets the stage for cumbersome negotiations. Not all conflicts are conflicts of interests, however. They may well be about the values, norms and principles that should underpin governance, including how to define and address issues pertaining to social justice. This problem is associated with the term "legal pluralism," which points to the fact that several normative orders may be applied to the same situation (Bavinck 2001; Jentoft et al. 2009). State and community law may ascribe to different norms in relation to resource management and distribution. If so, governability is not necessarily enhanced through the redistribution of resources, but instead through the reconciliation of legal principles that may or may not be possible in particular situations. Again, the issue of due process is important.

In order to resolve such conflicts a deeper understanding of the foundational idea behind these principles is required. Based on such an understanding, governance should, for the sake of governability, facilitate a process where reconciling conflicting ideas about justice is possible. This is also what Rawls (1973) aspires to capture with his concepts, the "original position" and "justice as fairness". These are referring to what people would consider to be fair if they were to decide on a justice principle "behind a veil of ignorance", that is without cognizance of their own and other people's interests in the case. This further suggests that the way in which justice impacts on governability must be studied empirically in concrete situations as the principles that are applied in, for instance, a particular fishery or community might be different from those applied in other fisheries or communities. These justice principles may not be explicitly expressed when the governing system was originally designed, but they are implicit in the way it actually works. They may not always be officially declared either, and when they are, they are not necessarily implemented. Uncovering the reasons for this is an important governability research issue, where "meta-order" (see below) values, norms, principles and rules cannot be taken at face value, but must be scrutinized in depth in empirical settings.

Justice at Different Governance Orders

Interactive governance theory stresses the need to address social justice principles at each of the three governance "orders" (Kooiman 2003). Justice at the "meta-order" is about the constitutive governance principles, principles that govern governance. Human rights principles belong here. Justice at the "second order" is about the design, mandates and procedures of institutions. One example is the so called "subsidiarity principle," which states that governing functions should be vested in the lowest possible organization. The "first order" is about day-to-day problem solving and opportunity creation. Here one should emphasize justice principles for the process of governing, such as transparency and cohesiveness. A normative principle for interactive governance is that governing at the three levels must be consistent with each other, because, as Kooiman (2003, 182) argues, governance orders cannot function without each other. The same can be said of justice principles, even if their relatedness is not always made explicit and reflected upon in real governing situations. The degree to which they actually are consistent is a research issue relevant to governability.

Meta Order

The fate of small-scale fisheries is a global concern. Will they survive under the pressures of globalization, industrialization, urbanization, climate change, and resource degradation? Are local fishing communities doomed? Is poverty inevitable? These were the questions at the forefront of an FAO meeting on small-scale fishing in Bangkok, Thailand in October 2008. There was a strong sentiment among participants that fishing rights for small-scale fishers are also human rights. In a comprehensive statement, representatives of the assembled civil society organizations declared "that the human rights of fishing communities are indivisible and that the development of responsible and sustainable small-scale and indigenous fisheries is possible only if their political, civil, social, economic and cultural rights are addressed in an integrated manner" (See Samudra, No. 51, 2008b).

Fishing rights, inspired by Garrett Hardin's idea of the tragedy of the commons, are typically perceived as a necessary management tool aimed at limiting the freedom of resource users and curtailing their resource extraction. The human rights perspective of fishing is a very different one: It states that people have some inviolable rights to begin with, and that these rights are intact regardless of what governments do or are willing to accept. As stated in the more than 60 year old UN Declaration of Human Rights, such rights are fundamental and universal. The decision of the UN Human Rights Commission regarding Iceland's fishing quota system testifies, however, that there is still some way to go until it is generally recognized that fishing rights can also be human rights (see Samudra, No. 49, 2008a).

In 2007, the General Assembly of the UN adopted a Declaration on the Rights of Indigenous Peoples. Even if the language pertaining to marine resource and sea space rights was considerably watered down from what was stated in the draft, it still has paragraphs that concern fishing rights. In the drafted text, paragraph 26 read as follows: "Indigenous peoples have the right to own, develop, control and use the lands and territories, including the total environment of the lands, air, waters, coastal seas, sea-ice, flora, fauna and other resources which they have traditionally owned otherwise occupied or used." In the wording that was finally approved, the direct reference to the ocean was removed. The same paragraph 26 now reads: "Indigenous peoples have the right to own, use, develop, and control the lands, territories and resources that they possess by reason of traditional ownership or other traditional occupation or use, as well as those which they have otherwise acquired."

In stark contrast to land and terrestrial resources, the altered language regarding the ocean and marine resources as open access and the reference to sea space as no-one's property is less reassuri ng for indigenous small-scale fishers. Will indigenous peoples have the same rights to fishing grounds as they do to forests and agricultural land? The revised language may be seen as a necessary compromise in order to save the Declaration. It is also more in line with the ILO 169's position on the matter. The Declaration proved to be a hard bargain, with the US, Canada, Australia and New Zealand voting against it.[3] Nonetheless, the Declaration does contain important principles regarding indigenous peoples' rights to livelihoods, culture, natural resources and self-determination. Although fishing rights were deleted in the final text, they would be assumed under "territories and resources". In the negotiations leading up to the decision, the letter *s* in peoples was a difficult one because it determines whether rights in question are individual or collective. In the final text, however, the *s* was retained.

Since the UN Declaration on the Rights of Indigenous Peoples is drawn from human rights legislation and principles that are universal, these rights have broader relevance than the Declaration might suggest. Non-indigenous populations share many of the same concerns and problems that the Declaration addresses. The statement of the civil society group presented at the Bangkok meeting actually reads as a parallel declaration to that of indigenous peoples, and it may thus be a first step towards a general formal recognition of small-sale fishing peoples' human rights. Even if declarations belong to what is called "soft law" and are therefore not as binding as a UN convention, they do create political space for those concerned. They also put pressure on governments to act upon them. If such a declaration is what small-scale fishing people opt for globally, a lesson from the process on the Declaration on the Rights of Indigenous peoples is that one should be prepared for a long haul. The Declaration took decades of struggle by the indigenous movement before it was finalized.

Small-scale fishing people depend on initiatives that work, and they need them fast (Jentoft and Eide 2011). Their communities and culture are not as resilient as we tend to believe. Rather they are vulnerable and not easily restored once broken. A culture lost is gone forever, as with bio-diversity or indigenous languages and knowledge. As many at the FAO conference in Bangkok argued, a Declaration for small-scale fishing people's human rights would be an instrument of committing governments to secure the 'rights to life' of small-scale fishing people.

Human rights principles, such as those inscribed in the UN Declaration for the Rights of Indigenous Peoples and the Bangkok statement, belong within the broader and classic discussion pertaining to natural rights; the rights that are "natural" to man. The idea is that people have inalienable rights, individually and collectively, that cannot be changed by "positive law." Philosophers like Aristotle, Rousseau, Hobbes, Aquinas, Locke and others all debated these rights. Natural rights were also invoked by the founding fathers and authors of the American constitution. In the case of indigenous peoples, we also talk about collective rights; the right to exist as a distinct people with a culture of its own. This further involves rights to maintain a

[3] Their arguments for rejecting the Declaration can be found at http://www.en-wikipedia.org/wiki/Declaration_on_the-Rights_of_Indigenous_Peoples

traditional livelihood, such as small-scale fishing in the case of the indigenous Sami of Norway (Jentoft and Karlsen 1997).

Second Order

A comprehensive concept of justice should include institutions and the ways in which they restrict, enable and guide human behavior. In the interactive governance framework (Kooiman 2003), the design and maintenance of institutions are at the second order of governance. With meta-order principles and norms, and the actions taken by the governors (first order), they co-determine the way governance works. Limits to, and opportunities for, governability are often of an institutional nature. Efforts to improve governability therefore require institutional reform of some sort, either of institutions that are inherent to the system-to-be-governed, those that make up the governing system, or those that regulate or facilitate their interaction.

Schattschneider's (1960) idea that "organization is bias" provides a relevant perspective for discussing social justice at this order. He argues that some issues and interests inevitably fall beyond organizational boundaries. There is thus a risk that justice will be ill served if some issues, interests and stakeholders who should have been inside the organization are excluded, and conversely if some who are inside should yield to others who are currently excluded. Still, moving boundaries so that issues, interests or stakeholders previously left out are integrated does not necessarily solve justice problems. There will always be someone whose interest falls outside the domain of the organization and who could therefore claim that they are being discriminated against. In other words, the closer the boundaries of the organization move towards an outsider, the easier it is for him or her to raise a justice claim. For a governing system, determining where boundaries are, i.e. on the issue of how inclusive it shall be, involves what interactive governance associates with a "hard choice" (Kooiman et al. 2005).

The question we must therefore ask is, how can a particular organizational bias be justified? Since there would be limits to how inclusive co-governing institutions can be, determining exactly where the limit should be drawn in order for decision-making procedures to be just is an institutional design issue in need of a principle. In the case of Norway, for example, the fisheries governing system that includes various stakeholder groups in a co-management process has been criticized for biasing in favor of some interests, while leaving others out. Historically, the Norwegian Fishers' Association, which does not represent more than 60% of all fishers, has been granted a privileged position in negotiations with the state, of being the one representative of not only all fishers, but the entire fishing industry. It is traditionally also the main industry advisor to government, and has been allowed to determine quota allocations between fisher groups. It has, however, been argued (from a justice perspective) that this privilege is discriminatory to other stakeholders, particularly those who have no organization to voice their interests. The Sami fishers have traditionally not had such an organization, and have therefore missed the clout to really make an impact on the policy of the Norwegian Fishers' Association and through this, that of the government. The change towards a market-based governance mode (vessels are bought and sold with their quota rights attached) in recent years does not seem to have altered the special partnership between this association and the state. Instead, it has changed the role and identity of this association, making it less concerned with community development and livelihood challenges, particularly among its least powerful members like the Sami fishers, and making justice issues in fisheries all the more acute (Mikalsen et al. 2007).

In Norway, the Sami have been among those criticizing this market-based management regime for not taking into account their rights and interests as an indigenous people. Still, as Eythòrsson (2003) concludes, the interests of the Sami people have been largely invisible in fisheries management. "Matters pertaining to the coastal Sami have been considered not merely irrelevant, but highly inappropriate" (Eythòrsson 2003, 159). The rather limited appreciation of the Sami presence in the fishery was clearly demonstrated when a new quota system was introduced in 1990 and almost no Sami fisher was qualified. The protests raised by the Sami Parliament resulted in the appointment of a Sami representative to the Regulatory Council (Davis and Jentoft 2001).

In Norway, the idea of giving preferential treatment on the basis of territorial and ethnic characteristics, which is an issue of "restorative" justice, i.e. a way of making up for historical injustice, has received mixed reception. The principle that maintains support suggests that fish resources are national property and, hence, their management should rest with the state in cooperation with groups such as the Fishermen's Association. Another principle with justice implications is that resource rights in Norway should be vested with the individual and not municipalities or communities. This is a conception of justice that emphasizes desert rather than dependency or need. It has also had consequences for the kind of management institutions proposed for addressing Sami interests and concerns. Rather than forming co-governance arrangements where the Sami can enjoy rights or self-determination with respect to fisheries regulations, the government insists on a more limited advisory function with the state at the helm.

First Order

Governance at this order is about the decision-making process, the day-to-day problem solving and opportunity creation, as well as the tools and techniques employed for these purposes. Principles that apply at this order are frequently referred to in fisheries governance. Here we are primarily interested in their justice implications, and, in the next instance, how they are linked to governability.

Transparency is a particularly important governing process principle. Do decision-making processes allow people to see what powerful institutions are up to? Is all information required or used in deliberating policies and arriving at decisions generally accessible? Transparency is a condition for another governing principle; that of meaningful and effective participation in decision-making processes by stakeholders. As Dahl (1989) argues, not only the substance of decisions matters, but also the procedure through which one arrives at them.

This is also in line with Sen's criticism of Rawls's "Theory of Justice" (1971). Sen argues that Rawls (1971) is too preoccupied with the institutional dimensions of justice (i.e. second order in interactive governance language) and not with those that apply to process (first order). It is also important to consider what happens to, and matters for people in their day-to-day lives, regardless of the principles that are written into the institutions that affect them. Furthermore, it is worth remembering that, regardless of outcome, people must have ownership of decisions. It is for this reason that, "Democracy has to be judged not just by their institutions that formally exist but by the extent to which different voices from diverse sections of the people can actually be heard" (Sen 2009, xii). In a similar vein, Dahl therefore says, "the democratic process is in itself a form of justice." This would also be the reason why stakeholders in fisheries governance are supposed to be involved in the decision-making process, why they should not be reduced to passive receivers of rules and regulations by some higher authority, and why the subsidiarity principle mentioned earlier in this chapter is also sound from a justice perspective. Even though the content of the rule and the outcome it produces are exactly the same, people can be expected to respect decisions and abide by rules if they are of their own as opposed to being imposed on them.

Thus, fisheries governance must not only allow for distributive justice, but also for procedural justice. Procedural justice is linked to governability through the legitimacy it provides. As Perusse Daigle et al. (1996, 19) state with regard to procedural justice and participation in fisheries resource allocation: "Providing individuals with an opportunity to voice their opinions and concerns regarding allocations leads them to believe they have been treated fairly… and increases satisfaction with decision makers." The move towards participatory, co-governance modes has for this reason gained momentum in fisheries in recent years (Wilson et al. 2003). Ways of involving stakeholders in regulatory decision-making, from consultation to devolvement of management tasks, are now being introduced in many countries in both the north and south. It is for instance part of the EU common fisheries governance reform. Thus, interactive governance is more than a normative theory; it is an empirical phenomenon that should be investigated from a critical perspective. How does it really work according to generally accepted standards of what constitutes just governing?

Notably, the trend towards interactive governance in fisheries is not likely to be explained by a sudden appreciation of social justice as an important value. Alternatively, given the diversity, complexity and dynamics of fisheries systems-to-be governed, there may be the assessment that the governability challenge is beyond the limits of what governments can do on their own. Another, and perhaps the most likely driver, is the expectation that stakeholder participation will improve compliance, which is an urgent problem in fisheries worldwide and a major reason why so many fish stocks are in peril. It is assumed that participation increases the legitimacy of governing systems and hence the willingness of stakeholders to follow the rules and accept outcomes (Jentoft 2000). However, regardless of what mechanism is behind this development, the impacts on justice should be highlighted. The process of participation needs critical scrutiny: who participates, how effective are they, who benefits? As with democracy in general, interactive governance designs may also be corrupted. Power must always be reckoned with in interactive fisheries governance, both its restricting and enabling aspects (Jentoft 2007). As Sen (2009) also argues in his critique of Rawls, although institutions are important in advancing justice and safeguarding against misuse of power, institutions are no guarantee that justice will be served despite their best intentions to do so. Thus, given the connection between justice and governability, a facet that will be explored below, institutions are not an assurance that governability will be ensured. There is no doubt, however, that they are a necessary part of the equation.

Governability and Social Justice

Justice concerns are present in the system-to-be-governed and the governing system, as well as in the way the two interact and affect governability as "a capacity for governance" (Kooiman 2008, 173). A system-to-be governed characterized by injustice is likely to be ridden by conflict and strife. That obviously makes governing more difficult, because stakeholders are likely to be uncooperative and resistant to governing intervention. The governing system may be held accountable for generating justice differences within the system-to-be governed, or blamed for not correcting them, causing legitimacy problems that may reduce governability. Thus governability is basically a relationship between the governing system and the system-to-be governed, where the capacity to enhance justice is determined by the interaction between the two.

In the following section, the relationship and interaction between the two systems and how they affect governability is illustrated by first discussing how property rights in fisheries raise concerns of social justice. Next I discuss how the legitimacy of property rights affects compliance and, hence, governability. Finally, I illustrate what potential for enhancing governability exists in restructuring the relationship between the governing system and the system-to-be governed in the form of a social contract that affords fisher-groups a larger role in fisheries governance.

Property Rights

In fisheries, property rights are relevant at both the second and the first order. To really understand their justice implications and why they are so controversial, however, they must be examined at the meta-order. At the first order, property rights are tools employed as steering mechanisms and incentives. First order governance involves governors who are armed with tools and techniques that, at their surface, may be considered neutral from a justice perspective. The supposition here is that if the goals are good, so too are their means. In reality, however, tools and techniques are as "political" as the goals and outcomes. In fisheries resource governance, property rights are among such tools. In fact they are often regarded as both the problem and the solution to sustainable fisheries. The problem of overfishing is perceived as a consequence of poorly defined property rights, whereas the solution is seen as a matter of getting the rights right – most commonly as private property. Thus property rights are embraced by institutions such as the FAO.

However, despite their presumed advantages, property rights invoke issues of social justice that pertain to "need, desert and equality." In fisheries, property rights are frequently contested when they are instigated, but over time they become something that is taken for granted. For people living under an open-access regime, property rights are often perceived as an alien and inappropriate concept. How can somebody acquire privileged ownership of a resource that is free for all to share? A person who is dependent on access to the resource in order to feed his family or keep his business afloat would experience this as an infringement on his ability to fulfill this need. He may even perceive it as a violation of his human right. That is not to say that open access systems are necessarily just. The "Freedom in a commons" that Garrett Hardin (1968) talked about sounds like justice and fairness, but may well lead to competition among users whereby the stronger fisher thrives at the expense of the weaker. There is, however, empirical evidence that under such circumstances local communities have often developed arrangements that keep competition among members in check, while granting them equal access and a fair share of the resources. In many instances, justice concerns are a more important driver than those of conservation (Jentoft 2004).

Justice problems are not only related to the outcomes of property rights, but also to their nature. We therefore need to ask what property rights are in the first place. The essential thing about a property right is not the relationship it establishes between a person who is the owner and the item that is owned, but the relationship it forms between people: the haves and the have-nots. Property rights can basically be understood as social relationships, and any change in property rights intervenes in existing social relations by differentiating categories of people. When one person benefits from acquiring a property right, others necessarily lose, because the owner is in a rightful position to deny others from enjoying the stream of benefits derived from that which is owned. It could therefore be argued that property rights are inherently inequitable. Also, justice issues typically emerge long before property rights are implemented, as people can anticipate their social and economic impacts, or at least how they themselves will be affected.

It is not for nothing that social scientists have long been concerned with the empowering and disempowering effects of property rights. Proudhon famously captured the quintessence of this problem in his classic treatise, "What is Property" (1840), with the statement: "Property is theft!" Fishing rights are often opposed by similar language. That is perhaps going too far, since property rights can also serve good purposes. They are often introduced in a democratic manner that is also just.

Hersoug (2006) argues that we therefore need to ask if fishing rights are empowering the right people, i.e. those that have legitimate justice claims. Property rights can certainly lead to unfairness, but they can also be employed for correcting inequities, as with securing tenure rights. Such rights can obviously be employed as a mechanism of positive discrimination in order to protect those in need of it, i.e. the marginalized and impoverished among fishers. When Marx stated, "from each according to his abilities, to each according to his need" (http://libcom.org/library/critique-of-the-gotha-program-karl.marx), he was alluding to social justice along such lines. In his case, they were the basic principles for societal governance.

Given that property rights are inherently inequitable due to their exclusivity and the way they define the relationships that people have among themselves, how are we to determine what property rights regime to embrace? Here, Rawls provides a principle as part of his "justice as fairness" concept that can serve as a Litmus test: "Social and economic inequalities should be arranged so that they are to the greatest benefit of the least advantaged persons" (Rawls 1971, 330). Accordingly, there are legitimate reasons to resist a particular property rights regime, unless it can be demonstrated in both theory and practice that the regime will increase the welfare of those most in need.

Functions of Justice

As mentioned above, a key governability issue in fisheries, or indeed in any form of societal governance, is how to make sure that those who are affected by rules and regulations also comply with them. The extent to which resource users do so will decide the outcome of fisheries governance, whether or not stated goals will be accomplished and natural resources and ecosystem services will be sustained. It is commonly assumed that poor governance performance in this situation can only be improved by stricter enforcement, investing more in surveillance and control of fishing operations, be it an increase in the frequency of inspections at sea or when catches are landed, or in the form of satellite sensing.

However, in a study of the abalone and rock lobster fishery in South Africa, Hauck (2008) observes that the compliance problem is more complex than this and that solutions are not as obvious as commonly thought. She argues that we also need to focus on the reasons that fishers give for non-compliance and that the solution to the compliance problem must (also) be sought in alternatives to stricter enforcement. We also need to assess the rules and regulations themselves, the laws from where these rules and regulations are derived, and the social and political processes that produced them. Hence, Hauck states that violations of rules in South Africa are not just a criminal justice issue, but also one of social justice. If the law favors some groups at the expense of others, there is little reason to expect that compliance will be high. Instead, people in this situation will insist that breaking the rules is morally justifiable, a form of civil protest, especially when fishers are poor and have no alternative sources of food or income to fishing. Should government decide to increase the level of control and instill stricter penalties, fishers would interpret it as an even greater injustice to which they are likely to respond in kind, leading the fishery compliance problem into a vicious cycle.

Hauck argues that in order to understand the compliance problem and to find ways to address it, a broader perspective on fisheries systems and governing mechanisms is required. Enforcement may well be part of the solution, but it is not all that is needed. Drawing on sociological and criminological literature, she states that fisheries management violations must also be understood from a normative perspective. Breaking the law is a moral problem and we need to raise the sociologically more interesting question of why people adhere to norms and regulations in the first place (Song and Chuenpagdee 2011; Jagers et al. 2012). Hauck (2008) even seeks to move beyond this to address the basic question of social justice, the way fisheries laws come into being and whose interests they serve. Thus, she concludes that, "while the monitoring and enforcement of rules have been identified as important for enhancing compliance, they need to be implemented through legitimate, socially valued institutions that are accepted and supported" (Hauck 2008, 215). From this follows a range of policies and management reforms that should be instigated in South Africa – and in other situations where compliance appears to be a problem.

Social Contract

In Norway, "quota ladders" were first introduced in the early 1990s as a response to a distribution conflict that followed the unexpected Barents Sea cod stock collapse, which led to the lowest TAC ever (Jentoft 1993). Since then, the ladder has been refined and made more comprehensive; it currently includes other species in addition to cod. Quota ladders are pre-negotiated allocation rules between fleet groups that take into consideration the changes in the TAC from year to year. Thus, when the TAC is low, the coastal fleet will get a larger share than the industrial, trawler fleet. When the TAC is less than 100,000 t, the coastal fleet gets 80%, with a gradual reduction of that share with an increasing TAC. When it exceeds 300,000 t, the share is 65%. There are also similar allocation formulas for coastal vessel groups.

The overall goal for the formulation of long-term allocation rules has been to create stability and predictability for the fishing industry, especially for full-time fishers. The demand for such allocation formulas has been raised by and negotiated within the Norwegian Fishers' Association. The Fisheries Ministry has largely accepted the outcome of those negotiations. Despite the growing prominence of long-term quota allocations in Norwegian fisheries management, a principled debate on what constitutes just distribution and participation among those affected has largely been missing. The argument has favored '*historical rights*' or entitlements, which implies that the ladder would not cause radical changes to the existing pattern of fishing rights and quota shares. Using historical catch as a basis for initial allocations promotes the desired stability, and continuing established practices is the least controversial within the fishing industry. Thus, the agreed upon ladder is intended to last, with the possibility of variation in allocation rules from one fish stock to the other.

The question that remains is how the idea of historical rights and the pattern of resource distribution among vessel groups can be related to various concepts of justice. Walzer's (1983) argument is that we should distinguish between different spheres of justice since society has no single principle of justice. The Norwegian experience with quota ladders suggests that it is difficult to draw straight lines between spheres, as actors within the industry base their activity on one principle (desert) that is perceived as dubious, if not totally illegitimate, by other stakeholders who are kept outside the management 'loop'. The latter attitude is common among small-scale fishers who do not belong to the Association. Fisheries governance, therefore, seems deemed to work according to several principles of justice, which the governing system must somehow attempt to reconcile.

Hernes et al. (2005) argue that the calibration of different justice principles necessitates a management process that is inclusive; one that allows for a broader group of stakeholders to be involved in the decision-making process. In this sense, democracy may be considered a crucial precondition for social justice. In Norway, the co-governance approach certainly allows for participatory democracy, but as mentioned above, it may still be criticized for being less than optimal from a democratic perspective as it poses limitations on participation. Still, the existence of such an organization speaking on behalf of a majority of fishers has made the Norwegian fishery more governable than it would have been without it. This holds true as well in comparison to fisheries of many other countries. The government knows who to talk to and is able to rely on the internal governance structure of the Association. On the other hand, individual members are secured representation in the governance process to an extent that they would not have been otherwise.

The quota ladder would still qualify as a "social contract" between different groups of fishers, and also between the fishing sector and the government. First of all, it is based on the principle that distributional issues like quota allocations to fleet groups should be delegated to the Association, and that the government – within limits – accepts what the Association decides. This is also a governability issue, because the Association is better positioned to produce the required commitment among members. In governance terms, this increases the governability of the governing interactions. Secondly, since TAC allocation is a zero-sum game with clear conflicts of interest, compromise is always difficult to obtain. If user groups can agree on some allocation rule that reflects their sense of justice and fairness, and then, with the support of the Association's authority, commit to them over a period of time until they are re-negotiated, the in-between period would be a peaceful one with high governability. If conflict cannot be avoided, it is better to limit fishers fighting each other to when the allocation is renegotiated rather than having them fighting each other all the time. There is no doubt that the quota ladder system helps to reduce the interaction costs of fisheries governance in Norway, and hence improves the governability of both the system-to-be governed and the governing system.

Governability Assessment

Social justice is an issue requiring philosophical and ethical reasoning, as demonstrated by Rawls (1971), Sen (2009) and others referred to in this chapter. However, social justice can also be made subject to empirical research. How does real life fisheries governance live up to universal justice principles and to what extent does this help improve governability? The assumption is that any governance system reflects, deliberately or not, some perception of what constitutes justice. The perception of justice could be universal, like the human rights principles that pertain to meta-order governance, or applicable to a particular sphere, like fisheries governance, or even to more "local" situations at the level of a fishery or community. Since, as Sen (2009) argues, justice is not just about institutional design but also about social process, including governing decision-making, justice requires that governance is interactive and "communicative" in the sense of Habermas (1987). That is, it is transparent and free of dominance of any group. Whether this is also how fisheries governance works in practice is an empirical question. The factors that may explain the proximity or distance of reality from these ideals in particular cases are an important concern.

Table 4.1 presents a framework that may help phrase such questions as they relate to the properties of the governing system, the system-to-be governed and the governing interactions. In line with the interactive governance, emphasis on diversity, complexity, dynamics and scale as common properties of both systems, is the assumption that these properties will somehow be reflected in the justice consideration made by decision-makers and stakeholders. There will presumably exist a diverse array of justice principles that somehow interact with each other. Some might be in conflict, while others might be mutually supportive and deduced from principles at higher scales, such as for "first principles" or the so-called "golden rule" (i.e. one should treat others as one would like others to treat oneself).

As shown in the table, justice principles are thought to inform the operationalization of the governing system, the system-to-be governed and the governing interactions. They are socially and culturally constructed codes for what constitute "justice as fairness" in social relationships and for the distribution of benefits and burdens. Like other governance principles, they emerge from deeply held social and moral values translated into norms that are made obligatory for governance (Kooiman and Jentoft 2009). One cannot know in advance, however, what these principles are, how they relate to each other regarding consistency, and how they actually inform governance decision-making. Each of these should be treated here as issues for empirical investigation.

The governability assessment framework helps to phrase research questions for all systems that are assumed to share the same structural properties (diversity, complexity, dynamics and scale). First of all, governance theory hypothesizes that several justice principles may be active for the same sector or for the same problem, that there may or may not be consensus with respect to their valuation among governing actors/institutions and stakeholders, and that they may or may not be compatible or

Table 4.1 Justice assessment matrix

	Governing system	Social system-to-be governed	Governing interactions
Diversity	Which justice principles are embraced by which governing actors/institutions?	Which justice principles are presently embraced by stakeholders and how do they affect their behavior?	Which justice principles influence how governing actors/institutions interact?
Complexity	How do different justice principles embraced by governing actors/institutions relate to each other?	How do the different justice principles embraced by stakeholders relate to each other, and what choices follow from them?	How do different justice principles influence the relationship between the governing actors/institutions and the stakeholders?
Dynamics	Have the justice principles embraced by governing actors/institutions changed over time, how and why?	Have justice principles affecting stakeholders changed over time; how and why?	Have the justice principles influencing the relationships and interactions between governing actors/institutions and the stakeholders changed over time, how and why?
Scale	How do justice principles embraced by governing actors/institutions relate across scales for various governing orders?	How do justice principles embraced by stakeholders relate across scale for different governance orders?	How do justice principles influencing the relationships and interactions between governing actors/institutions and stakeholders relate across scales for various governance orders?

commensurable at the level of the individual. This sets the stage for a governing process that involves stakeholders in communication and negotiation concerning conflicting principles. Their ability to involve stakeholders impacts on governability. That is, it affects the capacity of governing system to be effective and rational. This capacity would largely hinge upon its legitimacy, which is again dependent on the degree to which the justice principles that underpin the governing system's performance concur with the principles that operate within the system-be-governed and between the three orders of governance. Governance theory suggests that this depends largely on the relationship between the governing system and the system-to-be governed and the principles of justice supporting their interaction. The table summarizes how these hypotheses can be converted into empirical research questions. Accordingly, for each system one should start by detecting who embraces which justice principles. Governance theory sees this as a matter of diversity. Complexity, on the other hand, is about how these principles relate to each other, and how they add up in a way that provides consistency in guiding individual or collective choice. The less they do so, the harder the choices and the greater the governability problem. This is because the choices that are made would be risky from a legitimacy perspective. With respect to dynamics, governance theory assumes that principles are unstable, and that they may change over time as a consequence of the shifting system components, relationships, interactions and boundaries (Chuenpagdee and Jentoft, Chap. 18, this volume). How and why they change, what kind of social interactions they derive from and lead to, is an interesting governability research issue. Finally, the issue of scale invites questions with regard to the orders of governance and how they cut across system boundaries. The governing system and system-to-be governed do not exist in isolation, but are part of the society at large. Thus, although one would expect to find justice principles that are specific to fisheries, one would not expect to find justice principles that conflict, at least not radically, with those that work for society as a whole. In other words, what constitute good governance principles in the fisheries sector would largely be in harmony with those that exist for society, as major discrepancies would not be sustainable. If not for anything else, national or international legislation would ensure it. This is what happens when the human rights principles agreed to by the UN are written into national legislation and made obligatory for the ways in which governments deal with the fishing rights of indigenous peoples. It should then be discerned if there is inconsistency between justice principles applied at different orders.

Conclusion

Social justice is among those qualities of governance systems that have value in and of themselves. It is an intrinsic condition taken for granted and pursued for its own sake. Still, it has governability implications, and therefore retains a functional purpose. Unavoidably, when missing, a governance system would suffer from a legitimacy deficit. Stakeholders will question its design and operation. They will take offence and voice their opposition to those governors who they hold accountable for it. They will also be less willing to abide by the rules and regulations. In some instances, they prefer the latter response to the former. Thus, one may conclude that for governance systems to be governable, they must ascribe to values, norms and principles that stakeholders support. Social justice is no doubt one of these values, perhaps even the most basic one, whose derived principles, applications and outcomes will always come under scrutiny. Stakeholders are likely to oppose governance mechanisms that they consider unfair to them, regardless of the positive merits these mechanisms may otherwise have. Conversely, they may be willing to make major sacrifices, like accepting strict regulatory measures that may hurt them economically, if the burden the restrictions impose is fairly shared.

Sometimes several and conflicting justice principles apply to the same situation. This situation is explored as a case of "legal pluralism" (Bavinck 2001; Jentoft et al. 2009). Legal pluralism confronts individuals with the difficult choice of which principles and rules to follow. The complexity of the governance issue is enhanced when user groups disagree on which principles and rules they should support. In such a situation governability is assumed to be low. Even if, for instance, fishers should agree that rights, desert, equality and need are all relevant as allocation principles, they often value them differently. Interactive governance as a normative theory recommends a process through which stakeholders can possibly reach some agreement on which values are paramount, how they should be weighted, and which solutions follow from them. According to this theory, these are issues that should ideally be raised prior to the actual implementation of policies; they are step-zero issues in the governance process (Chuenpagdee and Jentoft 2007). How principles are communicated and deliberated, and how this process may affect governability at some later stage, is an obvious topic for research.

Justice as a quality of the social system-to-be-governed is about the relations that determine the distribution of costs and benefits among its constituent members. Property rights are examples of such a relationship, as they determine who is entitled to what, and who has the right to exclude others from the gains arising from a particular property. Following Schattschneider's idea that "organization is bias," governance modes determine who is eligible to sit at the decision-making table and who is not. When interacting with the system-to-be-governed, the governing system engages with those relations. While it may help to correct for relations that produce injustice, it also risks entrenching or exacerbating such injustice. The latter is the basis for the criticism raised against property rights as a fisheries governance instrument from among small-scale fishing groups and the request that property rights be calibrated to broader human rights principles, as described above. Similar criticisms can also be leveled at co-governance.

Governability is about the capacity of the governance system to impose principles and instruments that correct for differences in justice within the system-to-be-governed. It is fundamentally about the restricting and enabling role of power. Does the governing system have the muscle it needs to control the power that rules the system-to-be-governed? In some instances the answer is no, and governability will consequently be low. In other instances, the governing system has the power, but refrains from using it. In situations like the case of the Raw Fish Act that was introduced in Norway in 1938, it can radically empower small-scale fishers vis-à-vis fish-buyers (Jentoft 2011).

Regardless of power differentials, interactive governance theory advocates a participatory process characterized by transparency, accountability, cohesiveness and inclusiveness. These governance principles are also about social justice, i.e. the right to be recognized and to have a say in the decision-making on issues within which people have things at stake. Thus, for enhancing governability, procedural justice is as much about governability as it is about distributional justice, and procedural justice is as important as the institutional design of governance systems.

References

Bavinck, M. (2001). Marine resource management: Conflict and regulation in the fisheries of the Coromandel Coast. London: Sage.
Berkes, F., Hughes, T.P., Steneck, R.S., Wilson, H.A., Bellwood, D.R., Crona, B., Folke, C., Gunderson, L.H., Leslie, H.M., Norberg, J., Nyström, M., Olsson, P., Osterblom, H., Scheffer, H., Worm, B. (2008). Globalization, roving bandits and marine resources. Science, 311(5767), 1557–1558.
Bojer, H. (2006). Distributional justice: Theory and measurement. London: Routledge.
Chuenpagdee, R., & Jentoft, S. (2007). Step zero for fisheries co-management: What precedes implementation. Marine Policy, 31, 657–668.
Chuenpagdee, R., Degnbol, P., Bavinck, M., Jentoft, S., Johnson, D., Pullin, R., Williams, S. (2005). Challenges and concerns in capture fisheries and aquaculture. In J. Kooiman, M. Bavinck, S. Jentoft, R. Pullin (Eds.). Fish for life: Interactive fisheries for fisheries. Amsterdam: Amsterdam University Press.
Dahl, R.A. (1989). Democracy and its critics. New Haven: Yale University Press.
Davis, A., & Jentoft, S. (2001). The challenge and promise of indigenous peoples' fishing rights – From dependency to agency. Marine Policy, 25(3), 223–237.
Degnbol, D., & Wilson, D.C. (2008). Spatial planning on the North Sea: A case of cross-scale linkages. Marine Policy, 32, 189–200.
Elster, J. (1992). Local Justice. How institutions allocate scarce goods and necessary burdens. New York: Russel Sage Foundation.
Eythòrsson, E. (2003). The coastal Sami: A 'pariah caste' of the Norwegian fisheries? A reflection on ethnicity and power in Norwegian resource management. In S. Jentoft, H. Minde, R. Nilsen (Eds.), Indigenous people: Resource management and global rights (pp. 149–162). Delft: Eburon.
Habermas, J, (1987). The theory of communicative action. Cambridge: Polity Press.
Haldemann, F. (2008). Another kind of justice: Transitional justice as recognition. Cornell International Law Journal, 41(2), 675–737.
Hardin, G. (1968). The tragedy of the commons. Science, 162, 1243–1248.
Hauck, M. (2008). Rethinking small-scale fisheries compliance: From criminal justice to social justice. PhD thesis, Department of Environmental and Geographical Science, University of Cape Town.
Hernes, H. K., Jentoft, S., Mikalsen, K.H. (2005). Fisheries governance, social justice and participatory decision-making (pp. 103–118). In T. Gray (Ed.), Participation in fisheries governance. Dordrecht: Kluwer.
Hersoug, B. (2006, November 3–7). Opening the tragedy. SAMUDRA Report No. 45.
Honneth, A. (2001). Recognition or redistribution: Changing perspective on the moral order of society. Theory, Culture & Society, 18(2–3), 43–55.
Jagers, S., Berlin, B., Jentoft, S. (2012). Why comply? Attitudes towards harvest regulations among Swedish fishers. Marine Policy, 36(5), 969–976.
Jentoft, S. (1993). Dangling lines: The fisheries crisis and the future of coastal communities. The Norwegian experience. St. John's: ISER Books.
Jentoft, S. (2000). Legitimacy and disappointment in fisheries management. Marine Policy, 24(2), 141–148.
Jentoft, S. (2004). The community in fisheries management: Challenges, opportunities and risks. In B. Hersoug, S. Jentoft, P. Degnbol (Eds.), Fisheries development: The institutional challenge (pp. 93–129). Delft: Eburon.
Jentoft, S. (2007). In the power of power: The understated aspect of fisheries management. Human Organization, 66(4), 426–437.
Jentoft, S. (2011). Fisheries development as freedom. In R. Chuenpagdee (Ed.), World small-scale fisheries: Contemporary visions (pp. 367–379). Delft: Eburon Academic Publishers.
Jentoft, S., & Eide, A. (Eds.). (2011). Poverty mosaics: Realities and prospects in small-scale fisheries. Dordrecht: Springer.

Jentoft, S., & Karlsen, G.R. (1997). Saami fisheries, quota management, and the rights issue. In G. Pàlsson, & G. Petursdottir (Eds.), Social implications of quota systems in fisheries (pp. 147–164). Copenhagen: Nordic Council of Ministers, Tema Nord Fisheries.

Jentoft, S., Bavinck, M., Johnson, D., Kaleekal, T. (2009). Co-management and legal pluralism: How an analytical problem becomes an institutional one. Human Organization, 68(1), 27–38.

Jentoft, S., Onyango, P., Islam, M.M. (2010). Freedom and poverty in the fishery commons. International Journal of the Commons, 4(1), 345–366.

Kooiman, J. (2003). Governing as governance. London: Sage Publications.

Kooiman, J. (2008). Exploring the concept of governability. Journal of Comparative Policy Analysis, 10(2), 171–190.

Kooiman, J., & Jentoft, S. (2009). Meta-governance: Values, norms and principles, and the making of hard choices. Public Administration, 87(4), 818–836.

Kooiman, J., Bavinck, M., Jentoft, S., Pullin, R. (Eds.). (2005). Fish for life: Interactive governance for fisheries. Amsterdam: Amsterdam University Press.

Mikalsen, K.H., Hernes, H.K., Jentoft, S. (2007). Leaning on user-groups: The role of civil society in fisheries governance. Marine Policy, 31, 201–209.

Miller, D. (1999). Principles of social justice. Cambridge: Harvard University Press.

Perusse Daigle, C., Loomis, D.K., Ditton, R.B. (1996). Procedural justice in fishery resource allocations. Fisheries, 21(11), 18–23.

Pojman, L. (2001). Justice as desert. Queensland University of Technology Law and Justice Journal, 1, 88.

Proudhon, P.-J. (1840[1890]). What is property?. New York: Humboldt Publishing Company.

Rawls, J. (1971/1973). A theory of justice. Oxford: Oxford University Press.

Samudra. (2008a). No. 49, Tripartite Report of ICSF. http://www.springerlink.com/content/100370

Samudra. (2008b). No 51, Tripartite Report of ICSF. http://www.icsf.net/icsf2006/ControllerServlet?handler=SAMUDRA&code=viewFullIssue&issueno=51&language=EN&subsiteId=2o

Schattschneider, E.E. (1960). The semi-sovereign people. New York: Holt, Rinehard and Winston.

Schlosberg, D. (2003). The justice of environmental justice: Reconciling equity, recognition, and participation in a political movement. In A. Light, & A. de-Shalit (Eds.), Moral and political reasoning in environmental practice. Cambridge, MA: The MIT Press.

Sen, A. (2009). The idea of justice. London: Penguin.

Song, A.M., & Chuenpagdee, R. (2011). Conservation principle: A normative imperative in addressing illegal fishing in Lake Malawi. MAST, 10(1), 5–30.

Walzer, M. (1983). Spheres of justice: A defence of pluralism and equality. Oxford: Basil Blackwell.

Wilson, D.C., Nielsen, J.R., Degnbol, P. (Eds.). (2003). The Fisheries co-management experience: Accomplishments, challenges and prospects. Dordrecht: Kluwer Academic.

Young, I.M. (1990). Justice and the politics of difference. Princeton: Princeton University Press.

Chapter 5
Livelihoods in the Context of Fisheries – A Governability Challenge

Derek Johnson

Abstract As a study of how fishers make a living, livelihoods analysis is a core interest for fisheries governability. An understanding of the employment that fisheries provide, the livelihoods strategies that fishers pursue, and the sense of well-being that fishers derive from and associate with their ways of making a living is essential for policy that works to support fisheries. Given the diversity, complexity, dynamics and scale dimensions of livelihoods in fisheries, however, the translation of that understanding into policy is a wicked problem. This chapter argues for five reference principles that respond to the wickedness of livelihood governability: inclusion, reflexivity, attention to particularity, fostering adaptive capacity, precaution, and social justice.

Keywords Livelihoods • Employment • Wellbeing • Wicked problems • Principles

Introduction

Popular images of fishing commonly evoke romantic notions of fishers' livelihoods: fishermen in sou'westers hauling in gear on slippery decks in horrible weather; lithe, muscular men pushing off their catamarans through heavy surf; and, sometimes, foul mouthed fish mammies bargaining hard over the price of their fish. These and innumerable other more mundane tasks are central to the material world of fishers' livelihoods. They also hint, however, at how the realities of making a living

D. Johnson(✉)
Department of Anthropology, University of Manitoba,
Winnipeg, MB, Canada
e-mail: derek_johnson@umanitoba.ca

M. Bavinck et al. (eds.), *Governability of Fisheries and Aquaculture: Theory and Applications*, MARE Publication Series 7, DOI 10.1007/978-94-007-6107-0_5,

in fishing are intertwined with conceptions of the practice of fishing. These conceptions are, in turn, deeply important in shaping the contours of livelihoods in different fisheries, and for different fishers and their households. As a governability problem, sustaining livelihoods in fisheries involves both material and interpretive considerations. Fisheries governance that seeks to support fishers' abilities to make a living must do so in ways that recognize local perceptions of what is right, satisfying, worthwhile, and appropriate.

An understanding of employment and livelihoods in fisheries that takes this position is in keeping with research on livelihoods that is now 25 years old (Chambers 1987). This chapter links livelihoods and wellbeing approaches in fisheries to interactive governance theory and, specifically, the notion of governability. My intent is thus to address the key question of how livelihoods in fisheries can be seen as a governability concern. What is the place of livelihoods in the larger problem of fisheries governability? In answering, I argue that a livelihoods perspective for governance not only provides insights into how people make a living, but it also reveals how they feel about their work. These considerations are important for governability, because the material, relational, and subjective dimensions of work in fisheries (Gough et al. 2007; Coulthard et al. 2011) have a significant influence on possibilities for change. More than this, however, the chapter highlights the connections between livelihoods, wellbeing, and interactive governance when the three perspectives are brought together. My principal argument here is that there is a logical fit among the perspectives due to their similar conceptual foundations.

Governability in fisheries is a wicked problem, and sustaining fisheries livelihoods is no exception to this. Following Jentoft and Chuenpagdee's (2009) logic, sustaining livelihoods in fisheries is a wicked problem for the following reasons: (1) the perception of a desirable and meaningful livelihood varies among the many actors and institutions related to it; (2) livelihoods in fisheries are constantly subject to modification and are never definitively ideal or sustainable; (3) problems with livelihoods are complicated by being embedded in larger social, cultural, political, economic, and ecological systems; (4) the diversity, complexity, and dynamics of fisheries livelihoods makes each case distinct; and, (5) fisheries governance, even that which is livelihoods sensitive, generally creates irreversible change for the ways in which fishers and aquaculturalists make their living.

In this chapter, I propose two ways to conceptually extend the understanding of livelihoods in fisheries as a wicked problem. First, linking livelihoods to the notion of wellbeing strengthens the livelihoods perspective, while also making useful connections to interactive governance theory's approach to governability. Second, in keeping with the meta-governance concern of interactive governance theory, I show that livelihoods approaches, wellbeing, and interactive governance are linked foundationally by their basis in a shared set of principles. These are inclusion, reflexivity, attention to particularity, fostering adaptation, precaution, and the overarching principle of social justice.

Employment and Livelihoods

Employment

Although my primary concern in this paper is with livelihoods in fisheries, I also reference employment as one way of emphasizing the material importance of fisheries and aquaculture globally. The distinction between employment and livelihoods is one of definitional fullness and complexity; employment being the narrower term and livelihoods the richer one. Employment is an economic measure of the jobs in different sectors of the economy, while livelihoods includes employment but contextualizes it within a range of other considerations.

Employment in fisheries does however require some further elaboration. Even as a seemingly straightforward concept, the term has important qualifications. Employment varies a great deal, for example straying far from the nine to five norm of work in industrial economies. Work at sea follows the rhythms of the seasons and the ecosystem, which diverge from the regularities of calendar and clock time. Early morning departures are the norm and work on board fishing vessels often involves long, gruelling hours of setting and hauling in gear and cleaning the catch and gear afterwards. Shore work also follows a schedule that varies with the time of departure and arrival of the boats, size of the catch, and the duties of provisioning, maintenance, and marketing. The onshore-at sea division further reflects the gendered character of employment in fishing; men's employment and women's employment are usually of a different nature with divergent forms of remuneration. The shore-sea distinction also points to the linkages of employment in the fisheries sector. The work of catching fish is only the first stage of employment in the fish chain and is generally only the smaller part of employment generated by the sector. Employment multiples of five to one are normal in calculating the employment impact of fishing (Béné 2006).

As Béné (2006) points out, it is standard in treatments of small-scale fisheries to talk about the importance of fisheries from an economic point of view. Fisheries are said to generate high levels of employment and serve as an important economic driver, particularly in developing countries. Béné (2006) goes on to note that when these statements are subject to closer scrutiny, fisheries only make relatively minor contributions to national economies, with the exception of a few countries such as Senegal. Nonetheless, fisheries and aquaculture, particularly in their small-scale forms, do offer direct and indirect employment for large numbers of people globally. Estimates of this number range from 29 to 100 million (Béné 2006, 22–23). According to global data compiled by the FAO (2009), forty-three and a half million people were directly employed in fisheries and aquaculture in 2006. Of these, the vast majority (86%) were in Asia and most (60%) only worked part-time (FAO 2006, 2009, 23). While aquaculture provided only 8.6 million jobs in 2006, it is growing faster than capture fisheries employment, which has slowed considerably and even gone into decline in countries like Japan and Norway (FAO 2009, 23–25). When dependents are included, the significance of employment in fisheries and aquaculture is even more evident; the FAO (2009, 23) estimated that the total number of people dependent on these sectors was approximately 520 million in 2006, or 7.9% of the world's population (FAO 2009, 25). These figures, along with other reasons such as food security (see Chap. 6), clearly show that it is in society's material interest to be concerned with the state of fisheries and their governability.

Livelihoods

The notion of livelihoods was elaborated within the literature on international development as part of attempts to refine the understanding of and approaches to poverty. Rather than focus on poverty in fisheries, however, our intent in this chapter is to trace some of the implications of livelihoods thinking for fisheries governability more broadly (on poverty in small-scale fisheries see Béné 2003; Thorpe et al. 2007; Jentoft and Eide 2011). In this section I provide an overview of the livelihoods concept, critical comments on it, and indicate how coupling livelihoods with wellbeing may address some of its shortcomings. I leave the full application of livelihoods to fisheries governability for the next section of the paper and limit myself here to indicating how livelihoods and wellbeing concerns are as pertinent to fisheries as they are to other sectors.

Attention to livelihoods in the development literature was an explicit attempt to extend the work of Amartya Sen and Robert Chambers on the multidimensional nature of poverty (Béné 2003, 957–958). The seminal work of these two scholars challenged earlier approaches to poverty that are narrowly focused on income-based measures and complement a social science understanding of the social complexity of small-scale fisheries (Johnson 2006). With reference to famine initially, Sen (1981) showed how hunger and poverty cannot be explained by limited resource endowments, but rather by problems of distribution that result from institutionally entrenched differences in power. He introduced the idea that power differences constrain people's entitlements, or their abilities to secure the various resources necessary to ensure their health, wellbeing, and even survival. In later work, he extended his analysis into the capabilities approach, which seeks to identify the social structural factors that constrain entitlements and the development interventions that can counter them (Sen 1992, 1999). While Sen showed how poverty acts to constrain possibilities for individual action, Chambers emphasized the importance of understanding the particularities of the experience of poverty. He argued that approaches to poverty within development rely upon expert driven assumptions and templates that generate large-scale poverty reduction programs that often miss the targeted poor by wide margins. One of Chambers' key contributions was to argue for rapid rural appraisal as a methodological approach for tapping into diverse experiences of and explanations for poverty (Chambers 1983).

Livelihoods approaches have sought to draw together and formalize these insights into frameworks for the assessment of poverty. Although the Institute for Development Studies approach spearheaded by Chambers may be considered the primary point of reference, there are numerous variations of the livelihoods approach with different categorizations and weightings of incorporated variables. Following Sen and Chambers, livelihoods approaches share the basic assumption that, like poverty, livelihoods are diverse, multidimensional and comprised of varying sets of assets or capitals that people employ to cope with threats to their wellbeing. In keeping with Chambers' actor-centered approach, livelihoods approaches treat the poor as creative agents who seek to use their resources, regardless of how limited they might be, to meet their needs as effectively as possible. Livelihoods approaches may be applied at multiple scales (Scoones 1999), but are most commonly applied at the household level (White and Ellison 2007, 160). This is sensible, because most people organize production, distribution and reproduction at that level. Scoones (1999) notes that while livelihoods approaches are intended for use in developing country contexts, they are universally applicable. Livelihoods approaches also have origins in the sustainability discourse, with sustainable livelihoods a common pairing of concepts. This connection points to the idea that resilient livelihood strategies are those that can cope with threats and resist shocks. Resilient livelihoods are evidenced in fisheries by features such as the multiple gears and diverse knowledge that fishers possess, and the common integration of non-fisheries activities into their livelihood strategies (Allison and Ellis 2001; Allison and Horemans 2006). The connection also indicates awareness of the fact that many rural livelihoods have significant and direct reliance on ecosystem services, and that there is a relationship between sustainable livelihoods and the sustainability of resource use. These are, of course, obvious features of fisheries.

Livelihoods approaches are typically organized around frameworks that establish a logical relationship among the key elements deemed to constitute a livelihood (Bebbington 1999, 2029–2030; Scoones 1999, 4; Allison and Ellis 2001, 379; Allison and Horemans 2006, 759; Bebbington et al. 2007, 181). These are schemas that attempt to capture the dynamic relationship between the creative attempts of actors to construct their livelihoods in the context of a variety of structuring forces. The exercise of agency is interpreted through the use people make of their livelihood assets. The range of these assets and how they are identified, alternatively as capitals or resources, varies from framework to framework. Among the examples considered here, the commonly shared assets are natural, human, and social capital. In addition to these, some schemas lump together physical and financial capital, while others separate them. Political and cultural capital are additional variables that are occasionally used. The configuration of assets that different authors use represents different ways of splitting up the important variables that influence livelihood sustainability. Each type of asset refers to the resources that individuals and households may draw upon. They include, among others, access to social networks, natural resources, and money and material goods; personal health, skills and knowledge; and the degree to which one can lay claim to or mobilize socially significant meanings and symbols. Assets are not mutually exclusive and, instead, are deeply interconnected (cf. White and Ellison 2007). In small-scale fisheries, for example, the social capital of kinship networks may facilitate access to financial capital in the form of loans or to natural capital in the form of access to prime fishing sites.

A major impetus for the multiple asset basis of livelihoods was the need to challenge the idea that rural individuals or households depend solely on a single natural resource dependent sector like agriculture or fisheries (Ellis 2000). Rather, and increasingly so as the pace of development increases, livelihoods transcend particular sectors. Even within specific sectors, livelihoods build on multiple opportunities from extraction to processing to trading to credit. Livelihoods approaches focus on how and why individuals and households choose to invest in and balance their different assets in order to make a living. The combinations they use are identified as the set of different possible livelihood strategies.

Livelihoods are constrained and given opportunity by the broader structural context within which they take place. These structures are institutional in the broad sense of being the economic, social, cultural, and political conditions that operate at multiple scales from the hamlet to the region and beyond. Thus, as Ram (1992) has eloquently demonstrated for South Indian fisheries, livelihood opportunities for women are generally significantly different from those for men. At a broader scale, Greenberg (2006) has shown the power of economic and political factors to structure livelihood opportunities in Mexican fisheries. Structuring influences also include changes over time of a non-institutional nature, although these changes may have institutional causes and may be filtered by institutional structures. In the vulnerability and risk literature, these are divided into shocks, trends, and seasonality (Allison and Horemans 2006), each referring to types of pressures, sudden, intensifying or recurring, that test livelihood resilience.

This last point provides a connection between the core structure-agency logic of the livelihoods approach and the natural environment. As mentioned above, sustainable livelihoods is a very common conceptual pairing that reflects the frequent application of livelihoods approaches to contexts where households are dependent on ecological goods and services. One ‘bracketed’ livelihoods approach (see below) is to see households as adapting to the vagaries of an external natural environment. Increasing the sustainability of livelihoods could then involve a number of strategies, whether through building human and social capitals that permit the construction of alternative income generating activities or through strengthening social and political capital so that those at risk can pressure or work with the state to strengthen institutional and infrastructural defenses against threats. Vulnerability reduction in this sense connects livelihoods analysis to complex socio-ecological systems literature and the notion of resilience. As indicated below, however, there is still more work to do to make livelihoods approaches more effectively cross-scale in their reach.

The primary challenge to, and the root of critiques of, livelihoods approaches is with the interplay of assets representing agency and the institutions and other factors representing structure. Bebbington et al. (2007, 183) note that most livelihoods analyses tend to ‘bracket’ the structural side in order to focus on the analytically original part of the livelihoods approach; the interplay of the capitals. This is problematic, as assets and institutions are in ongoing, mutually constituting relationships. Thus, the perception of the opportunities that an asset may present depends on the specific context. In different fisheries or for different individuals within a specific fishery, for example, debt relationships with moneylenders that involve social and financial capital may be either liberating, oppressive or a combination of both (Johnson 2010). An important shortcoming of livelihoods approaches is the lack of acknowledgement that the decision to disentangle structure and agency in the capitals-institutions framework is often accomplished by freezing the interplay of structure and agency in constituting the possibilities for action.

Two recent book chapters address this limitation of livelihoods approaches (Bebbington et al. 2007; White and Ellison 2007). Bebbington et al. (2007) argue that livelihoods approaches have engaged in two disconnecting practices. The first of these is the 'bracketing' mentioned in the previous paragraph. The learning that needs to take place in this regard is for those development agents who employ livelihoods approaches to be much more attuned to the ways in which local livelihoods strategies are informed, in a dynamic manner, by broader political economic relationships. The second disconnection is between the perspectives or worldviews of development agents like non-government organizations and the perceptions of, or perceptions of the possibility of, alternative paths of development. Certain options seen as desirable by non-government organizations may be seen either as undesirable or unfeasible by target populations.

While Bebbington et al. (2007) see the livelihoods approach as needing some reorientation, White and Ellison (2007) argue for a more thoroughgoing rethinking grounded in a social constructionist position. Like Bebbington et al. (2007), they flag the disconnection between development agents and the targets of development. To White and Ellison (2007) this results from the failure of outsiders to comprehend that their positions, like those of the development subjects whose lives they seek to improve, are constituted by social, cultural, and political relationships. Without this self-awareness, livelihoods approaches simply become inflections of dominant economic perspectives in development. The risk, as it relates to livelihoods, is with the terms and categories that livelihoods approaches use. They must not be seen as employing standard templates with universally applicable categories of capitals. Instead, contextually sensitive livelihoods approaches would acknowledge that the resources that people employ are not fixed in meaning and are thus assets in struggles for power.

There are three implications of White and Ellison's (2007) critique for livelihoods approaches. First, development practitioners – or fisheries managers – must make more effort to reflexively understand their own positions. Second, there must be acknowledgment that the boundary within livelihoods approaches between capitals and context is artificial and that, in reality, all capitals are actually resources that are given their meaning and power by locally relevant conditions. Third, each livelihood resource itself is not a discrete, universal category. Instead, resources may be given meaning by multiple categories. White and Ellison (2007) give the example of land, which can be given meaning as natural, social, cultural, and political capital (White and Ellison 2007, 167). With reference to work such as that by Bavinck (2001), it is easy to see how maritime space similarly can be a powerful social construction where different groups of fishers and government agents have divergent perceptions of marine space. Despite these layers of complexity in understanding livelihoods construction, however, it remains important to make the effort to apply and further refine livelihoods approaches. Careful scrutiny of how development practitioners or fisheries governors operationalize livelihoods must be matched with similar care for how local perceptions are significant in the ongoing building of livelihoods.

The chapters by Bebbington et al. (2007) and White and Ellison (2007) are in an edited volume on the topic of wellbeing in development (Gough and McGregor 2007), and are concerned with how wellbeing may complement livelihoods perspectives. Wellbeing may be thought of as the meta-goal of development; it is the objective that development understands as 'good change' (Chambers 1997) and seeks to pursue. Livelihoods as a development strategy therefore seek to enhance wellbeing. More than just specifying the goal, however, wellbeing also provides important complementary insights that enhance the livelihoods approach. Like livelihoods, wellbeing is both the end and a means. As individuals, households, and communities increase their wellbeing, they also increase their capacities to aspire to improve their wellbeing further. Greater health, improved knowledge, deepened social networks and other factors all contribute to wellbeing, while also strengthening the basis for the further pursuit of wellbeing. Wellbeing is also a development objective and a way of thinking about development. As with livelihoods approaches, wellbeing as a perspective emphasizes the multiple dimensions of what constitutes wellbeing, their changing intersections, and the bases they provide for agency and motivation.

More than just reinforcing livelihoods approaches, however, wellbeing also challenges them to be more attentive to the social and cultural construction of what good change represents. It pushes the livelihoods perspective to more consistently examine and seek to bridge the disconnections that separate development agents from development subjects and livelihood resources from context. The wellbeing approach argues that notions of wellbeing are given meaning in the context of particular social relationships and the cultural ideas that inform them. While there is a degree of objectivity to wellbeing in such forms as general human needs for nutrition, shelter and health, even these conditions may vary considerably. As humans, we are capable of adapting to conditions of great deprivation or relative opulence and consequently shift our standards of the normal preconditions of wellbeing (Biswas-Diener and Diener 2001). Equally, the weighting of resources that contributes to wellbeing varies by context. Coulthard (2008) has, for example, shown how the Pattinavar fishers of Pulicat Lake emphasize the social and cultural importance of attachment to traditional fishing methods as a determinant of their wellbeing. This disadvantages them in relation to newcomer groups in the fishery who are free of such constraints and use other fishing methods that are more productive in the current ecologically degraded lagoon context (Coulthard 2008). The case illustrates how particular conceptions of wellbeing link to livelihood strategies and gives insight into how rationality is socially and culturally specific.

Bearing in mind the overarching idea that social relations and meanings shape the possibilities for action using resources, wellbeing can be operationalized in three dimensions (Gough et al. 2007; Coulthard et al. 2011). First, wellbeing seeks to identify the degree to which objective and subjective needs at the individual or household level are met. Needs are understood to include the requirements necessary for wellbeing in a particular context. Second, wellbeing addresses the degree to which individuals are able to act in socially meaningful ways to achieve their wellbeing. It is conceivable that an individual's needs may appear to be met but that social restrictions on their action prevent them from working to achieve that to which they aspire. Third, wellbeing is concerned with whether individuals are satisfied with the outcomes that they are achieving. With reference to this aspect of wellbeing, we can understand situations where individuals are superficially well off, yet continue to be dissatisfied. This outcome is common where material wants are taken care of but insufficient attention has been paid to addressing spiritual, relational or intellectual satisfactions. These three dimensions of wellbeing can be paraphrased, as in the introduction to the chapter, as the material, relational and the subjective.

For the remainder of the chapter, I discuss the implications of livelihoods for governability with a stretched definition of livelihoods that assumes the incorporation of a wellbeing dimension. At times I will explicitly draw out the analytical emphasis of wellbeing; otherwise I will leave it implied. I recognize that this is provisional hybridization and that more theoretical work needs to be done on how the two perspectives are complementary. As I note below, however, there are solid epistemological grounds for linking the two.

Livelihoods and the Governability of the System-to-Be-Governed

Governability is at once about describing the contours of a System-to-be-Governed – assessing, in effect, the systemic characteristics that either obstruct or facilitate solutions to perceived problems – and a normative estimation of the key elements that can contribute to improved governability. These two dimensions of governability are interwoven: what governability analysis looks for as constituting a governable system is influenced by the normative variables that it prioritizes. Nonetheless, in considering livelihoods as a governability concern, I separate the two aspects of governability in this and the following section in order to facilitate my presentation of the task. As I argue at the beginning of this chapter, the challenges that livelihoods pose for governability are illustrated well by the wicked problems idea. Additionally, governability, livelihoods and wellbeing thinking share similar normative concerns with respect to inclusion, reflexivity, particularity, precaution and social justice. This foundational complementarity makes them natural partners in efforts to broaden the terms of fisheries governance. This section addresses the first argument with reference to the System-to-be-Governed, while the following section deals with the second argument with reference to the Governing System and Governing Interactions.

The development of the idea of governability, as it relates to fisheries, links to a broader perception that there is a need to develop tools that that can help us better understand fisheries systems, and especially small-scale fisheries systems, in the wake of the failures of earlier simplifying models of fisheries and fisher rationality (Berkes et al. 2001; Charles 2001; Andrew et al. 2007). The foundations of the governability tool for analysing fisheries systems is the threefold division into the System-to-be-Governed, the Governing System, and Governing Interactions. Each of these are characterized by diversity, complexity and dynamics and subject to cross-scale interactions. Livelihoods are part of the System-to-be-Governed, but their characteristics influence the Governing System and Governing Interactions, just as these latter systems shape the former. As with the System-to-be-Governed, more generally, livelihoods in fisheries and aquaculture are diverse, complex, dynamic, and influenced by factors at different scales. These governability characteristics have consequences that make livelihoods governance a wicked problem.

Diversity refers to the qualitative differences among elements in a given system (Kooiman 2002, 232 and 194–196). In fisheries and aquaculture, livelihoods diversity is evident both between and within different systems. Small-scale fisheries are sufficiently diverse that they challenge widely held assumptions about fishing. Béné (2003) and Thorpe et al. (2007) have shown, for example, that the images of fisheries as poverty-ridden refuges of the last resort, although sometimes true, are often inaccurate. Coastal small-scale fishers in the central region of Gujarat's fishing industry, for example, are generally better off than their average agriculturalist counterparts (Johnson and Sathyaplan 2006; Sathyapalan and Johnson 2008). Similarly, fishers on the large lakes of south central Africa, while not rich, are generally better off than their non-fishing agricultural neighbors (Allison 2005). In some cases, the poorest rural dwellers do not have access to many forms of fishing, which can be lucrative (Béné 2003). Fisheries are also frequently assumed to be a sector in which fishers are largely autonomous. Yet, as the livelihoods in fisheries literature shows, the norm is rather for fisheries and aquaculture to be integrated with other sectors (Allison and Ellis 2001; Allison and Horemans 2006). In many parts of the world, such as Goa for example, farmers may become aquaculturalists on their land in the rainy season (Rubinoff 1999). Historically in Atlantic Canada (Johnson 1999), and presently in many parts of Africa and Asia, fishers and members of their households are also farmers, foresters, gatherers and workers in industrial or service sectors. Seasonal dependence on state transfer payments is also a standard element in Canadian fisher's livelihood strategies. Fishers' livelihoods may be buffered by highly developed forms of community-based governance or these may be virtually absent (Jentoft et al. 2009). Diversities between and within fisheries that affect livelihoods range across a host of these and other variables, such as technology, ecology, mobility, gender relationships of work, knowledge, market relationships, degree of subsistence, institutional development and so on. All of these factors reinforce the fourth dimension of fisheries as a wicked problem that each particular problem is distinct, requiring governors to constantly refine their knowledge and adapt their responses.

Complexity measures the density and extent of relationships within a system. The greater the density of relationships, and the wider the bounds of a system, the greater is its complexity (Kooiman 2002, 177–199). In terms of governability, livelihoods in fisheries and aquaculture are complex for many reasons, of which I single out three here. First, the diversity of options that constitute livelihoods result in numerous combinations that make the outcomes of governance interventions unpredictable. Second, livelihoods are formed in the context of relationships and in reference to value systems that create additional layers of complexity in understanding choices and actions. Third, fisheries and aquaculture livelihoods are cross-scale in nature: ecological, market, migration and other linkages make defining the system itself complex, not to speak of the isolating pathways of causality in decision-making around livelihoods choices.

Complexity often creates unexpected or paradoxical effects that challenge fisheries governability. In relation to the first type of complexity that I raised in the previous paragraph, a paradox of governance in fisheries is that attempts to reduce capacity often fail. This is true of boat buyback schemes, where the total number of boats may be reduced while fishing capacity is maintained, because the remaining fishers in the fleet develop ingenious means to expand capacity. They accomplish this by, for example, building bigger boats or increasing the power of their engines or the effectiveness of their gears. In this case, fishers use their intimate understanding of the possibilities for intensification afforded by their livelihoods to increase their effort.

With reference to the second form of complexity, fishers may also fail to respond in predictable ways to attempts to encourage their livelihood diversification to reduce fishing pressure. Research by Pollnac et al. (2001) has shown that fishers often fail to behave predictably when offered relatively lucrative alternative livelihoods (Pollnac et al. 2001; Sievanen et al. 2005). In some instances they may reorient to other activities like seaweed farming; in other cases new income sources may just facilitate intensification of fishing effort. Cases like these illustrate that simplifying assumptions about fishers' rational economic behavior, while convenient as a basis for management, are a poor guide to fishers' actual behavior. Fishers often value their work and are loath to leave it (Pollnac and Poggie 2008).

The work by Pollnac et al. (2001) also links to the third form of complexity, namely the unpredictability of cross-scale and cross-sector linkages. As the cases above attest, fisheries are often highly diversified. Fishers, however, have limits to their flexibility. Tourism development, for example, may be seen as a development strategy that is relatively benign for fisheries, while potentially providing important possibilities for livelihoods diversification. In some instances such interventions create new complementary opportunities for fishers (Pascual 2004), while in others they threaten fishers' very access to the sea (Boissevain 2004). The promotion of new coastal development related to tourism or any other new economic activity that intends to benefit fisheries should first be based on a thorough governability assessment that considers particular social, demographic and cultural factors. Social justice concerns need to be considered, as elite groups within a fishery may be well placed to benefit from new developments that disadvantage less powerful groups. The cross-sectoral linkages in fisheries reinforce the livelihoods' approach insight that fisheries governance and development cannot be considered a sector-specific problem alone. This is a clear connection to the third dimension of wicked problems that fisheries problems are embedded within broader scales, including inter-sectoral economic relationships. That different groups may disagree on the advantages of cross-scale connections, as in the case of the tourism example, illustrates the first dimension of wicked problems that problems are complicated by varying perceptions.

On one level, dynamics in interactive governance theory is just the recognition that change is a significant consideration for governability. Without paying attention to regular changes over the daily or seasonal cycle and to the prevalence of unpredictable change in fisheries, fisheries governance is ill equipped to understand fisher behaviors. This observation is one of the key insights that livelihoods approaches have drawn from international development with their demonstration that adaptation to trends, shocks and associated uncertainty is an important feature of household livelihoods strategies in fisheries (Allison and Ellis 2001). As both livelihoods approaches and interactive governance theory recognize, dynamics in fisheries are discontinuous; change takes place in fits and starts (Kooiman 2002, 200–203). In this sense of dynamics, livelihoods and interactive governance are linked to complex adaptive systems thinking and resilience (Gunderson and Holling 2002). Fishing households, or fisheries as a whole, that are better able to bounce back from unexpected changes are more resilient.

The strategies that fishers develop to cope with dynamics in fisheries are part of the landscape of governability that must be considered when developing governance interventions. Likewise, the forces that drive dynamics must be also considered. On the one hand, research has shown that livelihood diversification at the household level is extremely important as the basis for evolving strategies to adapt to seasonal changes and unexpected shocks (Marschke and Berkes 2006). On the other hand, adaptive strategies for sustaining livelihoods can also contain paradoxes for governability. Class relationships in fisheries can be highly unequal and have historically often come under attack as unjust. Yet, the cyclical, recurrent debt relationships reinforced by the social norms on which they are often based are also an adaptive system that responds to the uncertainty and risk of fishing (Ommer 1989; Platteau 1995). Indeed, these kinds of debt relationships are a strategy to sustain fisheries livelihoods that is equivalent to livelihood diversification in that both provide insurance against risk. In the case of debt relationships, it is a knotty governability problem to ascertain when they are functional or perverse (Johnson 2010). Fisher strategies to sustain livelihoods over time can also break down and their inability to adapt can be a constraint on governability. The previously mentioned Pulicat Lake case (Coulthard 2008, 2011) illustrates the ossification of a once successful adaptive strategy. Coulthard (2008, 2011) shows how the historical system of property rights in stake net fisheries failed to adapt to changing cross-scale social, economic and ecological conditions with serious consequences for the livelihoods viability of the Pattinavar caste. Dynamics are thus of considerable importance for fisheries governability; they direct the attention of governors to the features of specific fisheries that build or limit resilience. The Pulicat Lake example also shows the link between dynamics and the fifth type of wicked problem; solutions to fisheries problems create their own legacies that can result in perverse path dependencies or adaptive failures. This is an extremely important caution for fisheries governors to bear in mind: answers to problems in fisheries and aquaculture at one moment in time may later hamper governability. More generally, though, dynamics is about the second dimension of wicked problems that solutions to livelihoods shortcomings are only ever provisional in an unending effort to respond, reactively or proactively, to an always changing social-ecological environment.

In order to convey the interactive governance approach to livelihoods governability, I have sequentially presented diversity, complexity and dynamics. As should have become apparent in the examples above, that separation is artificial. This is evident first in that I did not attempt to separate out the fourth aspect of governability, scale, which makes appearances in the complexity and dynamics sections. Upon closer inspection, it is also apparent that I could not avoid some crossover of the other concepts. Thus, for example, diversity appears as a potentially desirable governability characteristic to cope with dynamics, while the complexity of fishers' knowledge (and perhaps their social complexity in knowledge sharing) means that they dynamically adapt to attempts by state authorities to limit their fishing effort. Needless to say, the intersection of diversity, complexity, dynamics and scale makes it important for governability to be seen as an approximation that should be subject to ongoing revision. In this sense, governability assessment is compatible with adaptive management approaches to fisheries governance.

Livelihoods Governability: Other Dimensions

The wicked problem of the governability of livelihoods in fisheries and aquaculture extends beyond the System-to-be-Governed. If sustaining and strengthening the livelihoods and wellbeing of fishing populations are held to be important, then it is also important that the Governing System has the capacity to understand the constructions of livelihoods and wellbeing in the System-to-be-Governed while fostering the ability to learn through Governing Interactions. Interactive governance argues that governance is grounded in normative principles at the meta, or third order, level (Kooiman et al. 2005, 241–244). In what follows, I make an argument for a first list of principles to underpin a wellbeing sensitive approach to the governance of livelihoods in fisheries. The principles are inclusion, reflexivity, attention to particularity, fostering adaptive capacity, precaution and social justice. Of these, social justice can be considered overarching. Both interactive governance and livelihoods approaches are oriented towards preserving employment and livelihoods for the largest number in a sustainable manner. This leads both approaches away from policies such as subsidies for fleet modernization or rationalization through individual transferable quotas that often benefit a few at the expense of many.

The remaining principles can be grouped together as part of a larger epistemological challenge from fisheries social science to the equilibrium-based bio-economic models that have dominated fisheries policy (Allison and Ellis 2001). The latter have increasingly been conceptually, ethically and empirically challenged by the continued erosion of global capture fisheries and catastrophic events like the 1992 Newfoundland cod moratorium. In this recent 'post-equilibrium' based fisheries social science, there is an acknowledgement that the diversity, complexity, dynamics and multi-scale nature of governability – in short its wickedness – require responses that accept uncertainty and seek to apply adaptive principles of learning-by-doing (Armitage et al. 2007; Schwach et al. 2007).

In terms of compatibility between the Governing System and the System-to-be-Governed, then, the connection between livelihoods, wellbeing, and governability shows a set of shared principles that should guide governance. Post-equilibrium fisheries governance recognizes that formal governors are not omniscient. This insight necessitates the *inclusion* of a diversity of stakeholders in data collecting and decision-making processes that consider the importance of livelihoods. *Reflexivity* is a synonym for adaptive management at the institutional and co-governance levels in that it asks governors to monitor their interventions and to be willing to ask themselves foundational questions about principles and purposes. Inclusion and reflexivity acknowledge the first dimension of wicked problems that suggests that different participants have different perspectives, and that ongoing efforts need to be made to ensure those perspectives are heard. Consistent with the fourth dimension of wicked problems, livelihoods approaches and the governability perspective share the assumption that strengthening livelihoods necessitates consideration of the *particularity* of each case. In seeking to balance support for livelihoods with other societal priorities (wicked problem 3), the Governing System needs to bring fresh eyes to each challenge. At the same time, the Governing System needs to remain always aware that particular problems evolve and that it is therefore perpetually necessary to foster its own *adaptive capacity* and adaptive capacity in the System-to-be-Governed (wicked problem 2). With regard to livelihoods, *precaution* asks governors to bear in mind the history of interventions in fisheries that have had major impacts on employment and livelihoods and seek to minimize the risk that future interventions imperil the basic principle of social justice (wicked problem 5). This must of course be done while also seeking to address sustainability concerns, a factor that brings us back to the complexity of wicked problems and hard choices in fisheries governance.

The discussion of the principles to which the Governing System needs to make reference to accommodate livelihoods and wellbeing concerns is also directly relevant for Governing Interactions. The ability of the Governing System to be attentive to the livelihoods needs and aspirations of fishers and aquaculturalists requires the development and maintenance of effective channels of communication between the Governing System and the System-to-be-Governed. The maintenance of these channels is subject to principles of social justice, inclusion and reflexivity. All stakeholders should have access to them and institutional mechanisms should exist to compensate for the differences in capacities that exist among different groups to make their voices heard.

The portrait I have painted thus far in this section is highly idealized and far from the messy, contested, and complex political economy of fisheries and aquaculture throughout the world. Nonetheless, such a rosy ideal is a useful reference point when thinking about alternative paths to fisheries governance that give greater weight to social justice and livelihood concerns. It also helps us to think about taking further steps towards the implementation of interactive governance systems responsive to the wicked challenges of fisheries and aquaculture governability.

As Chuenpagdee and Jentoft (2009) point out, one entry point to the empirical operationalization of livelihoods governability in fisheries is through the governability assessment matrix (Table 5.1). The advantage here of the matrix is that it forces a more precise application of the foregoing theoretical discussion. Each of the cells in the table not only represent summary points of the discussion from this and the previous section, they also point to important ways in which that theoretical discussion can be linked to more specific, research-relevant questions related to governability for livelihoods and wellbeing.

In contrast to Jentoft's (Chap. 4, this volume) first stage operationalization of social justice, the specification of broad research questions through the governability assessment matrix here has a different balance. In Table 4.1, in keeping with the ethical focus of his chapter, Jentoft's analysis concentrates on the ideal dimension of reference principles in his governability assessment. The questions in Table 5.1, in contrast, put significant weight on material and social relational dimensions as well, and do so in a way that is consistent with an analytical wellbeing perspective. They are also guided broadly by the social justice and post-equilibrium principles articulated in the first part of this section.

In terms of diversity, the governability assessment of livelihoods asks about the range of material livelihood practices that are present in a given context, the governing bodies tasked with supporting livelihoods, and whether governance interactions represent the full range of fisher or aquaculturalist groups and related governing interests. For complexity, which is concerned with the relationships among those involved in fishing and governing fishing, the matrix tries to identify how social relationships and cultural norms constrain or facilitate opportunities to engage in livelihoods and the governance of those livelihoods. Dynamics looks at the pattern of change in a given fishery's livelihoods arrangement, the institutions responsible for governing livelihoods, and the ability of those institutions to sustain constructive Governing Interactions to cope with change. Finally, governability assessment asks us to consider how livelihoods options are linked across scale, and how those linkages may benefit some more than others. Similarly, scale considerations are an important concern when assessing the capacity of the Governing System to formulate policy and take actions relating to livelihoods. They are also of importance when assessing which groups may have greater power over the decision-making process.

The general intent of this chapter, and the specific intent of the foregoing table, to integrate livelihoods and wellbeing into the understanding of governability are paralleled elsewhere. A paper by authors from the WorldFish Center (Andrew et al. 2007) proposes a first sketch of a tool for the 'assessment and diagnosis' of small-scale fisheries. The tool offers a possible and complementary entry point for thinking about how to implement a livelihoods sensitive interactive governance approach. Andrew et al. (2007) argue that while small-scale fisheries are highly diverse – they

Table 5.1 Livelihoods governability assessment matrix

	Governing system	Social system to be governed	Governing interactions
Diversity	Who are the governing authorities (individuals and institutions) responsible for livelihoods and wellbeing and what are their reference principles?	What are the constituent livelihood options and their combinations in the fishery of focus?	To what degree do the institutions governing livelihoods and wellbeing represent the diversity of actors engaged in diverse livelihoods options and the diversity of relevant governing agents?
Complexity	What are the relationships among guiding principles related to livelihoods and wellbeing and among the advocates of those principles?	How are possible livelihoods options and their combinations variably available to members of different social and economic groups?	How do governing principles related to livelihoods and wellbeing intersect with the interests of different groups in the fishery and the governors who represent those groups?
Dynamics	How and why have the governing institutions responsible for livelihoods and wellbeing and their reference principles changed over time?	What are the trends, shocks, and other temporal uncertainties associated with livelihoods in the focus fishery?	To what degree do institutional arrangements facilitate interactive discussion of principles to guide decision-making about livelihoods responsive to changing conditions?
Scale	How are the institutions and principles related to livelihoods and wellbeing influenced by broader political-economic contexts?	How are livelihoods options constrained or broadened by cross-scale economic, social, political, and cultural interactions? How are different individuals and groups variously positioned to take advantage of these?	How are the possibilities for governance of livelihoods and wellbeing shaped by broader political-economic influences?

only imply that they are complex and dynamic too – their importance needs to be recognized by the development of a standard yet flexible framework to assess and diagnose them (Andrew et al. 2007). There is, nevertheless, still a long way to go in terms of winning the argument that fisheries governance is more than management;

that it should also be about development that sees beyond sectoral divisions and recognizes that the pursuit of wellbeing requires a profound understanding of local realities that social science is best equipped to provide. One plank in the argument returns us social justice: an understanding of governability that includes the sensitivity of livelihoods and wellbeing approaches is better equipped to meet principles such as inclusion that constitute social justice. It will also hold a more realistic perspective on social diversity, complexity and dynamics, as well as the cross-scale and inter-sectoral relationships that make the application of fisheries governance so complex in terms of outcomes that respect core principles. The second dimension of the argument is that a broader conception of fisheries governance may also be in society's best interest, particularly in developing countries. Given the large populations involved in fisheries and aquaculture and their economic contributions, which are often significant at the regional level, a full understanding of fisheries and aquaculture governance, such as I propose here, is most promising in terms of the likelihood of creating sustainable fisheries that continue to contribute socially and economically.

Conclusion

There are two ways to argue for livelihoods having a central place in fisheries and aquaculture governability. The first of these is the social justice argument that the maintenance of employment and livelihoods in fisheries is important as part of society's moral responsibility to strive to ensure decent and meaningful lives for all its members. The second argument is practical. Given the large populations dependent on fisheries globally, their regional economic significance, and their contribution to regional and global food security and high value trade in foodstuffs, it is in society's interest to ensure their livelihood sustainability.

The argument for the importance of livelihoods in fisheries and aquaculture governability has been secondary in this chapter, however, with the assumption that the argument for sustaining livelihoods in fisheries is largely self-evident. I have been much more concerned with understanding how livelihoods can be seen as a governability issue in fisheries. My basic argument in that regard has been that livelihoods and governability share a foundational complementarity. In epistemological terms, they are both approaches that fit with a post-equilibrium view of fisheries governance. This correspondence is brought out more strongly when they are coupled with the idea of wellbeing, which shows that livelihoods are diverse, complex and dynamic attempts to achieve a desirable way of living in particular social, cultural, economic, political and ecological contexts.

Translation of the insights arising from a wellbeing sensitive approach to livelihoods as a governability problem into practical tools for sustaining livelihoods is a wicked problem. Livelihoods are multiple and potentially competing; one person's pursuit of wellbeing may threaten another's. Moreover, configurations of livelihoods are dynamic and embedded within broader factors that transcend sectoral and regional governance, and, once undertaken, governance interventions stimulate their own remorseless consequences for good or bad. These complexities must not, however, be reason to abandon the effort to integrate livelihoods considerations into fisheries and aquaculture governance. Even if the analytical and methodological tools of livelihoods, wellbeing and governability are still in development and remain a tough sell in fisheries management circles, the fact that they have made inroads at WorldFish and the FAO (Andrew et al. 2007; Béné et al. 2007) indicates the shortcomings of alternative standard approaches and the dire straits of many populations and regions that depend upon fisheries and aquaculture for their economic and social wellbeing.

References

Allison, E.H. (2005). The fisheries sector, livelihoods and poverty reduction in eastern and southern Africa. In F. Ellis, & H.A. Freeman (Eds.), Rural livelihoods and poverty reduction policies (pp. 256–273). London: Routledge.

Allison, E.H., & Ellis, F. (2001). The livelihoods approach and management of small-scale fisheries. Marine Policy, 25, 377–388.

Allison, E.H., & Horemans, B. (2006). Putting the principles of the Sustainable Livelihoods Approach into fisheries development policy and practice. Marine Policy, 30(6), 757–766.

Andrew, N.L., Béné, C., Hall, S.J., Allison, E.H., Heck, S., Ratner, B.D. (2007). Diagnosis and management of small-scale fisheries in developing countries. Fish and Fisheries, 8(3), 227–240.

Armitage, D.R., Berkes, F., Doubleday, N.C. (Eds.). (2007). Adaptive co-management: Collaboration, learning, and multi-level governance. Vancouver: UBC Press.

Bavinck, M. (2001). Marine resource management: Conflict and regulation in the fisheries of the Coromandel Coast (Vol. 5, Livelihood and environment). New Delhi: Sage.

Bebbington, A. (1999). Capitals and capabilities: A framework for analyzing peasant viability, rural livelihoods and poverty. World Development, 27(12), 2021–2044.

Bebbington, A., Hinojosa-Valencia, L., Muñoz, D., Lizarazú, R.E.R. (2007). Livelihoods and resource accessing in the Andes: *Desencuentros* in theory and practice. In I. Gough, & J.A. McGregor (Eds.), Wellbeing in developing countries: From theory to research (pp. 176–198). Cambridge: Cambridge University Press.

Béné, C. (2003). When fishery rhymes with poverty: A first step beyond the old paradigm on poverty in small-scale fisheries. World Development, 31(6), 949–975.

Béné, C. (2006). Small-scale fisheries: Assessing their contribution to rural livelihoods in developing countries. (Fisheries Circular No. 1008, p. 46). Rome: FAO.

Béné, C., McFadyen, G., Allison, E.H. (2007). Increasing the contribution of small-scale fisheries to poverty alleviation and food security. FAO Fisheries Technical Papers (p. 125). Rome: FAO.

Berkes, F., Mahon, R., McConney, P., Pollnac, R., Pomeroy, R. (Eds.). (2001). Managing small-scale fisheries: Alternative directions and methods. Ottawa: International Development Research Centre.

Biswas-Diener, R., Diener, E. (2001). Making the best of a bad situation: Satisfaction in the slums of Calcutta. Social Indicators Research, 55(3), 329–352.

Boissevain, J. (2004). Hotels, tuna pens, and civil society: Contesting the foreshore in Malta. In J. Boissevain, & T. Selwyn (Eds.), Contesting the foreshore: Tourism, society, and politics on the coast (pp. 233–260). Amsterdam: Amsterdam University Press.

Chambers, R. (1983). Rural development: Putting the last first. Harlow: Longman Scientific & Technical.

Chambers, R. (1987). Sustainable livelihoods, environment and development: Putting poor rural people first. Brighton: Institute of Development Studies.

Chambers, R. (1997). Editorial: Responsible well-being – A personal agenda for development. World Development, 25(11), 1743–1754.

Charles, A. (2001). Sustainable fishery systems. Oxford: Blackwell Science.

Chuenpagdee, R., & Jentoft, S. (2009). Governability assessment for fisheries and coastal systems: A reality check. Human Ecology, 37(1), 109–120.

Coulthard, S. (2008). Adapting to environmental change in artisanal fisheries – Insights from a South Indian lagoon. Global Environmental Change, 18(3), 479–489.

Coulthard, S. (2011). More than just access to fish: The pros and cons of fisher participation in a customary marine tenure (Padu) system under pressure. Marine Policy, 35(3), 405–412.

Coulthard, S., Johnson, D.S., McGregor, J.A. (2011). Global fisheries in crisis: A social wellbeing approach to poverty and sustainability. Global Environmental Change, 21, 453–463.

Ellis, F. (2000). Rural livelihoods and diversity in developing countries. Oxford: Oxford University Press.

FAO. (2006). Topics factsheets: Development trends in fisheries. Retrieved August 9, 2010, from http://www.fao.org/fishery/topic/12344/en

FAO. (2009). The state of world fisheries and aquaculture 2008. Rome: FAO.

Gough, I.R., & McGregor, J.A. (2007). Wellbeing in developing countries: From theory to research. Cambridge: Cambridge University Press.

Gough, I.R., McGregor, J.A., Camfield, L. (2007). Theorising wellbeing in international development. In I. Gough, & J.A. McGregor (Eds.), Wellbeing in developing countries: From theory to research (pp. 3–43). Cambridge: Cambridge University Press.

Greenberg, J.B. (2006). The political ecology of fisheries in the upper Gulf of California. In A. Biersack, & J.B. Greenberg (Eds.), Reimagining political ecology (pp. 121–148). Durham: Duke University Press.

Gunderson, L.H., & Holling, C.S. (Eds.). (2002). Panarchy: understanding transformations in human and natural systems. Washington, DC: Island Press.

Jentoft, S., & Chuenpagdee, R. (2009). Fisheries and coastal governance as a wicked problem. Marine Policy, 33, 553–560.

Jentoft, S., & Eide, A. (Eds.). (2011). Poverty mosaics: Realities and prospects in small-scale fisheries. Dordrecht: Springer.

Jentoft, S., Bavinck, M., Johnson, D. (2009). Co-management and legal pluralism: How a methodological problem becomes and institutional one. Human Organization, 68(1), 27–38.

Johnson, D. (1999). Merchants, the state and the household: Continuity and change in a 20th century Acadian fishing village. Acadiensis, 29(1), 57–75.

Johnson, D. (2006). Category, narrative, and value in the governance of small-scale fisheries. Marine Policy, 30, 747–756.

Johnson, D. (2010). Institutional adaptation as a governability problem in fisheries: Patron-client relations in the Junagadh fishery, India. Fish and Fisheries 11(3), 264–277.

Johnson, D., & Sathyaplan, J. (2006). Legal pluralism in the marine fisheries of Junagadh District and the Union Territory of Diu. Working Papers Series 2006 (p. 80). The Hague: Indo-Dutch Program for Alternatives in Development.

Kooiman, J. (2002). Governing as governance. London: Sage.

Kooiman, J., Bavinck, M., Jentoft, S., Pullin, R. (Eds.). (2005). Fish for life: Interactive governance for fisheries. Amsterdam: Amsterdam University Press.

Marschke, M.J., & Berkes, F. (2006). Exploring strategies that build livelihood resilience: A case from Cambodia. Ecology and Society, 11(1), 42. Retrieved from http://www.ecologyandsociety.org/vol11/iss41/art42/

Ommer, R.E. (1989). Merchant credit and the informal economy: Newfoundland, 1919–1929. Historical Papers/Communications historiques, 24(1), 167–189.

Pascual, J. (2004). Littoral fishermen, aquaculture, and tourism in the Canary Islands: Attitudes and economic strategies. In J. Boissevain, & T. Selwyn (Eds.), Contesting the foreshore: tourism, society, and politics on the coast (pp. 61–82). Amsterdam: Amsterdam University Press.

Platteau, J.P. (1995). A framework for the analysis of evolving patron-client ties in agrarian economics. World Development, 23(5), 767–786.

Pollnac, R.B., & Poggie, J.J. (2008). Happiness, well-being and psychocultural adaptation to the stresses associated with marine fishing. Human Ecology Review, 15(2), 194–200.

Pollnac, R.B., Pomeroy, R.S., Harkes, I.H.T. (2001). Fishery policy and job satisfaction in three southeast Asian fisheries. Ocean and Coastal Management, 44(7–8), 531–544.

Ram, K. (1992). Mukkuvar women: Gender hegemony and capitalist transformation in a South Indian fishing community. New Delhi: Kali for Women.

Rubinoff, J. (1999). Pink gold: Globalization and the transformation of backwater aquaculture on Goa's Khazan (agricultural) lands. (Paper presented at the 28th Annual Conference on South Asia, Madison, Wisconsin).

Sathyapalan, J., & Johnson, D. (2008). Livelihood security in the marine fisheries sector of Gujarat State, India. In S.M. Dev, & K.S. Babu (Eds.), India's development: Even or uneven? Reflections on development theory and practice (pp. 277–304). New Delhi: Manohar Press.

Schwach, V., Bailly, D., Christensen, A.S., Delaney, A.E., Degnbol, P., van Densen, W.L.T., Holm, Petter, McLay, H. Anne, Nielsen, Kåre Nolde, Pastoors, Martin A., Reeves, Stuart A., Wilson, D.C. (2007). Policy and knowledge in fisheries management: A policy brief. ICES Journal of Marine Science: Journal du Conseil, 64(4), 798–803.

Scoones, I. (1999). Sustainable rural livelihoods: A framework for analysis. IDS Working Paper 72 (p. 22). Brighton: Institute for Development Studies.

Sen, A.K. (1981). Poverty and famines: An essay on entitlement and deprivation. Oxford: Clarendon Press.

Sen, A.K. (1992). Inequality reexamined. Oxford: Oxford University Press.

Sen, A.K. (1999). Development as freedom. Oxford: Oxford University Press.

Sievanen, L., Crawford, B., Pollnac, R., Lowe, C. (2005). Weeding through assumptions of livelihood approaches in ICM: Seaweed farming in the Philippines and Indonesia. Ocean and Coastal Management, 48, 297–313.

Thorpe, A., Andrew, N., Allison, E. (2007). Fisheries and poverty reduction. CAB Reviews: Perspectives in Agriculture, Veterinary Science, Nutrition and Natural Resources, 2(85), 1–12.

White, S., & Ellison, M. (2007). Wellbeing, livelihoods and resources in social practice. In I. Gough, & J.A. McGregor (Eds.), Wellbeing in developing countries: From theory to research (pp. 157–175). Cambridge: Cambridge University Press.

Chapter 6
Food Security in the Context of Fisheries and Aquaculture – A Governability Challenge

Roger S.V. Pullin

Abstract Food security is the right of all people to have access to sufficient, safe and nutritious food. Food security is governed across scales from household and local to national and global and food security issues are typically diverse, complex and dynamic – perfect territory for an interactive governance approach. Fish make essential contributions to food security at all levels, by providing high quality protein, lipids for brain development and function, and micronutrients. Many capture fisheries are in crisis and aquaculture, although having huge potential, is facing difficult intersectorial and environmental problems. These problems and potentials are reviewed from the perspective of making interventions to improve governability along the fish chains from fishery and farm to consumer. A worked example is attempted for three aquaculture fish chains.

Keywords Aquaculture • Capture fisheries • Fish chains • Fisheries • Fish trade • Food security • Governability

Introduction

Food security is a never-ending problem. Everyone always needs to know that the next meal, its sufficiency and acceptability are guaranteed. Food security is arguably the wickedest of wicked problems, as defined by Jentoft and Chuenpagdee (2009), because its supply and demand scenarios are always changing and because

R.S.V. Pullin (✉)
Independent Consultant Aquatic Biologist, 7A Legaspi Park View, 134 Legaspi St., Makati City, Philippines
e-mail: pullin.roger@gmail.com

M. Bavinck et al. (eds.), *Governability of Fisheries and Aquaculture: Theory and Applications*, MARE Publication Series 7, DOI 10.1007/978-94-007-6107-0_6,

the consequences of not adapting to those changes are severe: conflict, sickness, death and, where ecosystems are damaged irreversibly, increased risks of the same for future generations. The world's food factories are ecosystems and the baselines of those ecosystems are always shifting, as Pauly (1995) showed for fisheries. Food security is won or lost on ecological battlegrounds.

Palaeolithic humans were opportunistic omnivores (Eaton et al. 1996). Omnivory was a major factor in the evolution and success of *Homo sapiens*, with fish as important providers of the fatty acids that are essential for the development and functioning of healthy nervous systems. Rich herring (*Clupea harengus*) and Atlantic cod (*Gadus morhua*) fisheries contributed much to the historical food security, prosperity and power of Europe and North America (e.g., Blaxter 1990; Kurlansky 1999). Many inland, coastal and open sea capture fisheries have declined, largely because of overfishing and ecosystem abuse, during massive institutional failures. Meanwhile, aquaculture has undergone rapid growth, contributing increasingly to food security but including some environmentally and socially unacceptable impacts.

Fish, whether from aquaculture and capture fisheries, contribute to human food chains through fish chains (Thorpe et al. 2005; Mahon et al. 2008; Chap. 2 by Kooiman and Bavinck, this volume). Sustaining those contributions is the wicked problem that requires continuous solution. Assessments of the governability (see Kooiman et al. 2008; Chap. 6 by Chuenpagdee and Jentoft, this volume) of entire fish chains and of the individual links along them might indicate entry points for interventions to improve governability.

This chapter begins by reviewing the importance of food security and the right-to-food approach. This is followed by discussion of some of the definitions and dimensions of food security, including food safety. The contributions of fish to food security are emphasized and the two supply subsectors, aquaculture and capture fisheries compared. Governability is then discussed in relation to food security, from the perspective of human behaviour. The case is made for an interactive governance approach to food security, and governability assessment is applied to evaluate three aquaculture fish chains. The chapter concludes with general discussion about food security and opportunities for improving governability along the chains.

Importance of Food Security: The Right-to-Food Approach

Food security is a non-negotiable fact of life for the wellbeing of every person and is essential for a stable society. Food insecurity causes strife and food deprivation and has long been a weapon of coercion and war. Haddad and Oshaug (1998) summarized the progress of a human right to food, from its implicit inclusion in freedom from want – one of the four freedoms called for by President Roosevelt in 1941 – and found that an explicit linking of democracy and food security had not yet been developed, compared to that achieved for democracy and economic growth, but that elements of human rights were becoming interwoven in approaches to food policy.

After a long history of UN deliberations and communications on the inclusion of food as a human right among others, a clear exposition of the right to food was agreed in 2002 under the International Covenant on Economic, Social and Cultural Rights: "the right to adequate food is realized when every man, woman and child, alone or in community with others, has the physical and economic access at all times to adequate food or means for its procurement". However, as Haddad and Oshaug (1998) warned, the human rights approach to food security can be a blunt set of tools. Human behaviour sometimes ignores human rights, as well as the environmental uncertainties and risks that attend food production.

The difficulties of governing and funding food production and distribution often make it very difficult to fulfill all rights to food. One cannot eat rights, just as one cannot eat money. It is little use having a right to food, if the basis for its provision, a healthy food-producing ecosystem, has been irreversibly compromised. Parties to International Conventions that should have helped to sustain the contributions of fish to food security (e.g., the United Nations Convention on the Law of the Sea and the Convention on Biological Diversity) emphasize their rights under those conventions rather than their obligations, especially those for the conservation of fisheries ecosystems.

Moore reviewed progress of the Food and Agriculture Organization of the United Nations (FAO) in following a right-to-food and rights-based approach to food security, defining the latter as: "...recognition that all people have a legal right to adequate food and to be free from hunger, and (taking) this right as a focus for actions" (2005, 141). FAO (2002) has continued to make huge contributions towards achieving food security; for example, publishing standard methods for assessing food deprivation, and establishing the Right to Food Forum and information service (www.fao.org/righttofood/).

FAO regards food insecurity as: "A situation that exists when people lack secure access to sufficient amounts of safe and nutritious food for normal growth and development and an active healthy life. It may be caused by the unavailability of food, insufficient purchasing power or the inappropriate distribution or inadequate use of food at the household level. Food insecurity, poor conditions of health and sanitation and inappropriate care and feeding practices are the major causes of poor nutritional status. Food insecurity may be chronic, seasonal or transitory." Within the United Nations Millennium Development Goal 1. Eradicate Extreme Poverty and Hunger, Target 1.c. is to "halve, between 1990 and 2015, the proportion of people who suffer from hunger" (www.un.org/millenniumgoals).

Definitions of Food Security

The FAO definition of food security is used widely and is accepted here: "Food security exists when all people, at all times, have physical and economic access to sufficient, safe and nutritious food to meet their dietary needs and food preferences for a healthy and active life" (FAO 1996, 2). However, applying this broad definition to food fish is difficult, because most mainstream literature on food security and most food security policies emphasize, mainly or exclusively, food energy security, with hunger defined on the basis of inadequate per caput food energy intake per day. FAO defines the per caput hunger threshold as 1,600–2,000 kcal/day. Food security policy emphasizes the foods that provide the most energy: staple cereals, oils and fats, and sugar and related products. From that food energy security perspective, animal products (meat, offal, dairy products and fish) and other foods (fruit and vegetables) are often lumped together as minor contributors. FAO (2009a) gave average dietary energy contributions for high (H) and low (L) income countries as: cereals – H, 48%, L, 55%; oils and fats – H, 13%, L, 9%; sugar and related products – H, 11%, L, 9%; and totals for all these components – H, 72%, L, 73%.

Good nutrition, or comprehensive food security, involves much more than food energy security. This is where fish and fish products become important. Comprehensive food security requires adequate availability and affordability of all human nutritional requirements: proteins that provide all essential amino acids; lipids that provide for normal development and functioning of tissues and organs, especially cell membranes and the central nervous system; and micronutrients (essential vitamins and minerals). Emergency interventions by governments and by organizations such as the World Food Programme (www.wfp.org) and Oxfam (www.oxfam.org) recognize this. For the purposes of this chapter, food security means that same sufficiency in all human nutritional requirements, including assurance of food safety.

The very word security implies that the resources upon which food production and distribution depend be managed for their sustainability. The sustainability of food production is therefore synonymous with food security. In this context, sustainability does not necessarily mean reliance on the same foods and food producing ecosystems, as long as others are accessible and affordable. This is inevitable because of seasonal variations in the availability of many food products, short-term climatic uncertainties and long-term climate change, and changing dietary preferences. However, ringing the changes on human food items and their production and distribution systems should, as far as possible, avoid constraining the options of future generations. This was made explicit, as follows, in the following definition of sustainable development in the 1987 Report of the World Commission on Environment and Development, Our Common Future: "Sustainable development is development that meets the needs of the present without compromising the ability of future generations to meet their own needs" (www.un-documents.net/wced-ocf.htm).

Dimensions of Food Security

Food security is required at household level, at sub-national (village, district or province) level (for which the present author prefers the term local), and at national level (Alamgir and Arora 1991). All household members must avail of a minimum necessary food intake, relative to their age and size, gender, type of work and, for women, pregnancy or lactation status. Local food security requires that all in the locality meet their minimum food needs, to which Alamgir and Arora add the requirement that the available foods "reflect, at least partially, tastes and preferences and household status" (1991, 7). Kawarazuka and Béné (2010) identified three pathways by which those who work in small-scale fisheries and aquaculture improve their household nutritional security: direct consumption of fish; increased purchasing power to buy good food; and empowerment of the women who work in the fish chains to effect better management of family income. Moving up from the household to the local level already weakens the link to the food intakes of different types of individuals and this link largely disappears at national and global levels, where the main or exclusive target becomes average minimum food requirements per caput. FAO is developing voluntary, cross-sectoral guidelines, the purpose of which is stated as follows in the opening to its Preface as follows: "To improve the governance of tenure of land, fisheries and forests with the overarching goal of achieving food security for all to support the progressive realization of the right to adequate food in the context of national food security" (FAO 2012).

There are large differences within and among the vulnerabilities of rural and urban populations to food insecurity. Research on food system vulnerability, in the face of societal change and global environmental change, is increasing: see, for example: www.gecafs.org. Urban food security is particularly challenging because of burgeoning city populations and the logistics and economics of food distribution. Low-income urban households typically spend 60–80% of their budgets on food (FAO 2000). From almost 6,000 household surveys in Asian developing countries, Dey et al. (2005) found that fish consumption and the choice of fish species and products varied greatly with income class and location, with rural people consuming more fish than urban people and poor people consuming more low value, mostly freshwater, fish. Globalization is bringing many other pressures to bear on food security; for example, recent and still emerging intellectual property arrangements for food biotechnology and genetic resources have huge implications for ownership of and access to the means for food security (e.g., Tansey 2002).

For all of the reasons mentioned above *inter alia*, the realities and risks of food insecurity are widespread and persistent. FAO (2009a) estimated the world's population of hungry persons at 923 million in 2007, 80 million more than at a 1990–1992 baseline. ADB (2008) reported that 2008 global food prices had increased by over 50%, with food stocks at their lowest levels since the early 1980s. The concept of food sovereignty is increasingly promoted, largely by non-governmental organizations, citing the rights of all individuals to produce food, to access resources for food production, and to unite in using "*people power*" to realize their right to food; for examples, see www.peoplesfoodsovereignty.org and www.ibon.org.

Specific micronutrient deficiencies are a health problem for many millions of poor persons and are targetted by food fortification (e.g., Hardianti 2005) and by the breeding of vitamin-rich staple foods (e.g., the vitamin A-enriched golden rice; www.goldenrice.org). In 2008, the Copenhagen Consensus ranked micronutrient (Vitamin A and zinc) supplements for children as the most cost-effective among 30 forms of aid (www.copenhagenconsensus.com). Though less well supported to date by technophile donors and politicians, simple diet diversification is also potentially a very cost-effective and powerful contributor to the relief of protein, essential lipid, and micronutrient insecurity, especially at household level. In Ghana, for example, small homestead ponds have enabled family production of vegetables, condiments such as chillies, and small quantities of fish, that together not only improved household income by 229–679% but also transformed household micronutrient security (Ruddle 1996).

Food Safety

Food fish, meaning here all finfish and aquatic invertebrates and their products in the human food chain, are prone to rapid spoilage unless well stored and preserved. This and a wide variety of other human health hazards that can accompany fish production and consumption make food safety an integral requirement for food fish security. The principle here is that all human food must be safe to produce and to eat. In practice, however, absolute safety can never be guaranteed, because potential risks are always possible from factors that cannot be completely and continuously monitored; including microorganisms, allergens, and wind- and water-borne pollutants.

The Codex Alimentarius Commission (CAC) (www.codexalimentarius.net), established in 1963 by FAO and the World Health Organization, provides a near-comprehensive set of safety standards for human foods, including a Code of Practice for Fish and Fishery Products, with standards covering about 60 general or specific categories. Food safety has undergone substantial development, largely through the CAC, with widening implementation of Hazard Analysis Critical Control Point (HACCP) principles (FAO 1995a). Fish producers and processors who fail to adopt HACCP principles become greatly disadvantaged in the markets, especially for exports. The result has been a steady transition to greater safety in the production and consumption of fish and fish products. From the traditional focus on hygiene and on removal of unsafe foods from the human food chain, food safety risks are now analyzed and managed proactively along the human food chain and, in the case of fish and fish products, "from farm or sea to plate" (FAO 2006a, para. 25). FAO/NACA/WHO (1999) reviewed food safety issues associated with aquaculture fish chains, including risks from water-borne parasites and pathogens. Huss et al. (2004), for FAO, published a comprehensive review on seafood safety. In aquaculture, food fish safety issues are being linked to production site quality and environmental impacts (Focardi et al. 2005). Research on food fish safety has included analyses of consumer attitudes (e.g., Jussaume and Higgins 1998) and private sector understanding of compliance with regulations (e.g., Henson and Heasman 1998).

Despite these advances, food safety problems from fish and fish products remain, especially in some developing regions. Poor coastal dwellers are at risk from unsafe seafood that is self-caught or purchased from local markets, including bivalve molluscs contaminated with red tide organisms and pathogenic microorganisms, ciguatera toxin in reef fish, and histamine from poor quality, inadequately stored tuna (e.g., see Chamberlain 2000). Global warming is increasing the risks of toxic algal blooms, and Hales et al. (1999) have predicted that it will increase risks of ciguatera poisoning. On a wider front, in inland and coastal waters, government agencies do not always communicate effectively to the public the potential risks of eating self-caught fish from polluted waters (Chess et al. 2005). On the other hand, the mass media sometimes publish eye-catching but misleading items concerning food fish safety; for example, in the U.S.A, gross exaggeration of the risks of mercury poisoning from eating tuna, as opposed to its health benefits.

The debate over the potential benefits and risks of farming and consuming so-called Genetically Modified Organisms (GMOs) has long been highly polarized between those who see them as essential for future world food security and those who see them as products that are dangerous and that will enable rich corporations to dominate world food security (e.g., FAO 2001). This debate continues with widespread ignorance and misconceptions as to how GMOs might or might not be hazardous. Every captive-bred farmed organism is to some extent genetically changed (i.e., modified in the broad sense), whether it has been selectively bred, or produced by hybridization or by any biotechnology. Genetic engineering is just one type of biotechnology, which produces transgenic organisms, otherwise known as GMOs (narrow sense) or Living Modified Organisms (another confusing term used in international conventions and protocols). Hybrids that are crosses between two species are more genetically modified (broad sense) then transgenic organisms. The same applies to some other forms of genetic manipulation, including multiplication of chromosome numbers (polyploidy). The pros and cons of farming GM fish are being hotly debated, but a broader view is needed, encompassing the use of all biotechnology in aquaculture.

Contributions of Fish to Food Security

Fish and fish products make essential contributions to human food security worldwide (e.g., Elvevoll and James 2000; ADB 2005; www.seafood.net.au), both directly as dietary components and indirectly in feeds for farmed animals. FAO (2009b) summarized the latest (2006) average regional annual per caput fish consumptions in kg as: China, 26.1; Oceania, 24.5; Europe; 20.8; North and Central America, 18.9; Asia (excluding China), 13.9; South America, 8.4; Africa, 8.3. Gupta (2006) concluded that fish contributes over 20% of the animal protein intake of more than 2.6 billion people. FAO (2006b) gave per caput annual fish consumption in traditional Asia-Oceania fish eating countries as mostly above 25 kg/year (above 50 kg/year in some and 190 kg/year in the Maldives) and estimated that fish was providing 22% of total protein intake in sub-Saharan Africa and approaching or exceeding 50% in some poor countries: e.g., Sénégal, 47%; Gambia, 62%; Ghana and Sierra Leone, 63%. In the 1990s, the average annual fish consumption of Pacific islanders was 34 to 37 kg per caput, compared to the world average of 16.5 kg per caput, and provided 50% of their recommended protein intake. The value of total fish exports annually from Pacific island nations and territories has increased from US$48 million in 1999 to US$101 million in 2007 (Gillett 2009).

Fish as food provide humans with comparatively little dietary energy: a per caput average of 20–30 kcal/day, and up to 180 kcal/day where there is a lack of alternative food items (FAO 2009b). The main role of fish in food security is to provide the following: high quality animal protein containing all essential amino acids, especially those such as lysine that are low in other protein sources; lipids, especially the omega-3 polyunsaturated fatty acids required for brain development and function and energy- and vitamin-rich oils; and micronutrients (vitamins and minerals; especially calcium, iodine, iron, phosphorus, zinc, and Vitamins A, B1, B2 and D). Small indigenous fish species supply vitamins and minerals to poor consumers in Bangladesh; for example, 43–90% of Vitamin A and 31–36% of calcium recommended dietary allowances for children of 4–6 years (Thilsted et al. 1997). Subsistence inland and coastal fishing provides huge, though largely undocumented, contributions to the protein, lipid and micronutrient security of poor persons; for example, rice farmers (e.g., Halwart and Bartley 2005). Fish are particularly important providers of docosahexaenoic acid (DHA). DHA deficiency in humans has very serious implications for mental health and stable societies (e.g., Crawford et al. 2008; and other publications from the Institute of Brain Chemistry and Human Nutrition; www.londonmet.ac.uk).

Kurien (2004) recognized the complex relationships between fish trade, as imports and exports, and food security and endorsed the four entitlements, recognized by Sen (1981), that provide direct or indirect food security to an individual: production-based, growing one's own food; trade-based, selling or bartering other possessions for food; labor-based, selling labor for food; and transfer-based, receiving transfers or gifts of food. Kurien (2004) found that although the entry of a fish species into international trade was correlated with depletion of that species, there was little evidence that exporting fish results in food fish insecurity in the country of origin. For 11 widely differing countries, the impact of fish trade on domestic food security was as follows: positive and large, 1 (Nambia); positive and significant, 5 (Chile, Fiji, Nicaragua, Sri Lanka, Thailand); positive and small, 2 (Brazil, Sénégal); negative and small, 2 (Kenya, Philippines); negative and large, 1 (Ghana). In addition, many Low Income Food Deficit Countries earn from fish exports very significant foreign exchange, which is used for many purposes, including the financing of food, especially grains, and fuel imports.

Contributions of fish to the food security of poor persons have been enhanced greatly by imports of cheap and durable products such as canned fish and by coastal states exporting from seasonal gluts of wild caught fish to their land-locked neighbors (e.g., Hara 2001), though this brings with it a dependency on foreign fish stocks that might decline and can undercut progress towards more self sufficiency through development of sustainable aquaculture (present author's observations in Southern and West Africa).

Aquaculture

Aquaculture, the farming of aquatic organisms, is still widely regarded as a separate sector within food production and/or as a separate subsector of capture fisheries, despite the fact that farming domesticated fish and hunting wild fish have little in common. Unfortunately, monosectorial and subsectorial policies and institutions have typified many sectors – agriculture, fisheries, forestry, mining, tourism, transport, water supply and waste disposal that depend upon the very same ecosystems. This is a major impediment to equitable and sustainable use of natural resources and therefore a threat to the health of ecosystems and to food security. Institutional change to remedy this is likely to be slow.

The main shifts that are needed concerning the role of farmed aquatic organisms in food security are: (i) recognition that aquaculture is the farming of aquatic plants and animals and therefore requires subsectorial policies and institutions within the agriculture sector, like those for crops and livestock agriculture; (ii) capture fisheries, including those enhanced by artificial stocking, also require sectorial policies and institutions; and (iii) policies and institutions for both the aquaculture subsector and the capture fisheries sector must be integrated with those for all other sectors and subsectors that use the same resources, especially ecosystems, so that they can proceed in partnership rather than conflict.

In 1970, aquaculture provided only 3.9% of world fish supply. From 1990 to 2002, aquaculture production grew on average by 10.2% per year, more any other animal protein food sources: beef, 0.8%; mutton, 1.5%; pork, 2.5%; eggs, 3.6%; and poultry, 4.8% (Moffit 2005), and now contributes about 50% by volume to world fish supply. Traditional aquaculture, especially in China and adjacent countries, was closely integrated with agriculture, water management and waste recycling, but those highly resource-efficient systems, have now been largely replaced by non-integrated systems, including intensive pellet-fed fishponds (Edwards 2004). Similarly, in India, a long-established polyculture of six carp species, using the different natural feeding niches in fertilized ponds, has been intensified for higher yields, through supplemental feeds (Nandeesha 2001).

Intensive aquaculture, like all forms of intensive food production, has large ecological footprints. In 2003, at least 41.6% of production of farmed finfish and crustaceans was derived from feeding them with farm-based and/or industrially manufactured feeds (Tacon et al. 2006) who also reported, citing Gill (2005), that in 2004 aquaculture took 3% of the global total of industrially manufactured animal feeds, compared to: cattle, 24%; pigs, 32%; poultry, 38%: pigs, 32%, and other animals, 3%. Aquaculture's share of those feeds must now be much higher and increasing, though its reliance on fishmeal and fish oil is being substantially reduced through highly innovative feed formulations. Pullin et al. (2007) reviewed possible biological, ecological and intersectorial indicators for the sustainability of aquaculture and Pullin (2011) has further reviewed the paramount issue of choice of species and feeds, for responsible and sustainable aquaculture.

Aquaculture, like agriculture and forestry and in concert with them and other natural resources sectors, must become more responsible and sustainable, and less environmentally damaging. Some forms of aquaculture have had a bad history of booms and busts and environmental damage: over-abstraction of surface and ground waters; destroying mangrove and other wetlands to establish production units, some of which then fail anyway; exposing acid sulphate soils; increasing salinization of lands and aquifers; causing eutrophication of inland and coastal waters; introducing invasive alien species; spreading parasites and diseases; changing wild genetic resources by interbreeding etc.

Many of the same charges can be laid against other sectors; for example, shrimp aquaculture was found to be contributing only 1.5 and 0.9% respectively of the total anthropogenic sources of nitrogen and phosphorus entering Mexican coastal waters (Páez-Osuna et al. 1998). Aquaculture has a particularly bad image and gets a bad press where it is blamed, whether entirely correctly or not, for adverse impacts on nature and natural resources; for example, salmon farming on wild salmon stocks and fisheries (e.g., Ford and Myers 2008). Costa-Pierce summarized the solutions to this problem as follows: "the aquaculture world community needs to focus its attention on a new paradigm, in order to evolve an 'aquaculture revolution' that is technically sophisticated, knowledge-based, and ecologically and socially responsible" (2002, 364–365).

Aquaculture is indeed changing for the better, following the provisions of the FAO CCRF (FAO 1995b) and its Technical Guidelines. There are some excellent commentaries on what this change implies (e.g., Consensus 2006). In particular, there is a huge effort underway to replace with cheaper and more sustainable sources of lipids the fish oils that are used in farmed fish feeds (Turchini et al. 2009). Similarly, the use of vegetable proteins in farmed fish feeds is increasing, with a view to making large reductions in fishmeal and trash fish requirements. According to Finley and Fry (2009), soy protein will provide half of the protein requirements of farmed fish feeds by 2020. Aquaculture products are included in the organic food movement (e.g., www.ifoam.org) and the criteria for them being accredited as organically farmed often include broad assessments of the earth-friendliness of their production systems and not just the avoidance of use of chemicals etc.

Ahmed and Lorica (2002) and FAO (2009b) pointed to the high importance of aquaculture for food security, especially in Asia. Moreover, inland aquaculture is an obvious way to add value to scarce water resources, through their multipurpose use. It is therefore certain that the contributions of aquaculture to food fish security will continue to increase and will soon exceed those of capture fisheries. Aquaculture of plants and of herbivorous or omnivorous aquatic animals (mainly finfish, molluscs and crustaceans) is more feed- and energy-efficient than other ways of producing animal protein. As Brown put it: "The big winner in the animal protein stakes has been aquaculture, largely because herbivorous fish convert feed into protein so efficiently" (2006, 171).

Subasinghe et al. (2009) reviewed positively the future prospects for expansion of aquaculture, in spite of its many challenges, especially climate change. They concluded that aquaculture was expected to:

> Contribute more effectively to food security, nutritional well-being, poverty reduction by producing…with minimum impact on the environment and maximum benefit to society, 85

million tonnes of aquatic food by 2030, an increase of 37 million tonnes over... 2005 (Subasinghe et al. 2009, 7).

Contributions from aquaculture to fish supply have been increasing rapidly and will have to increase further. FAO (2006b) noted an increased contribution (1994–2003) of aquaculture to fish supply in the Near East/North Africa as 4.5–18.7%, but also forecast that fish supply in sub-Saharan Africa would have to increase by 28% to maintain even its poor concurrent annual average fish consumption of 6.6 kg per caput, and commented that aquaculture has much scope to provide more than its 2% contribution.

Capture Fisheries

The historical and ongoing crisis in capture fisheries is well documented in the primary scientific literature (e.g., Pauly et al. 1998; Jackson et al. 2001; Pauly et al. 2005) and in many books and other products that are accessible to the public (e.g., Pauly and Maclean 2003; Sadovy et al. 2003; Clover 2004; Roberts 2007; and, most recently, the film *The End of the Line*, www.endoftheline.com). The picture is one of rampant overfishing and destruction of the ecosystems that produce fish, and of massive institutional failures, including management based on lies (e.g., see Bromley 2009). Marine and inland capture fisheries are often typified by untruthful and incomplete statistics. About 20% of the world fish catch is derived from what is officially called Illegal, Unregulated and Unreported fisheries. About 50% of the fishing by vessels off the West African coast is probably illegal. Open ocean catches, such as tunas, are often transferred at sea to factory ships, without being landed and properly recorded. Sometimes, the truth is told about how bad things are. For example, the European Commission has admitted that 88% of its fish stocks are overfished, that 93% of North Sea cod are caught before they have any chance to spawn, and that fishing is becoming unprofitable, despite large subsidies (European Commission 2009). Even so, it remains politically difficult or impossible to set sensible catch quotas and fishermen "exceed quotas with impunity" (Anon 2009, 52).

What then are the prospects for turning at least some capture fisheries around, so as to restore or to increase their contributions to food security? Hutchings (2000) was generally pessimistic, especially in terms of the time needed, citing little or no recovery in gadoid (cod family) and flatfish stocks after 15 years. Enhanced or culture-based fisheries and fish ranching are also not likely to work well for most open water fisheries, except in some special cases such as: regular stocking and harvesting of lakes and reservoirs; stocking rivers with highly migratory fish, such as salmon; and stocking semi-enclosed coastal waters, such as bays and lagoons, with species that have limited or no movements, especially molluscs.

The environmental impacts of stocking large numbers of captive-bred organisms, genetically different from wild types and feeding at one specific level in the trophic pyramids of an open water ecosystem are open to obvious criticisms. But the more telling argument, against many operations that attempt to enhance capture by stocking, is that they simply have "limited or no demonstrable success" (Molony et al. 2003, 409). Nevertheless, once started, they are hard to close down because they appeal to the public, the mass media, fishers, fisheries managers and politicians, who see them, even ineffective or unproven, as measures to restore fisheries. Such efforts are sometimes accompanied by other expensive and unproven measures, such as artificial reefs. Governors and the governed seem to be happily complicit in funding these questionable developments, and lots of money flows for the supply of fish seed and associated structures and services.

However, if better governance can be achieved, the future for capture fisheries is not all doom and gloom. In addition to the FAO CCRF (FAO 1995b) and its Technical Guidelines, there are many sources of free advice on how to improve and sustain capture fisheries; e.g., see www.seaaroundus.org. Pauly et al. (2002) pointed out that most capture fisheries have been unsustainable, but found that reducing fishing efforts and subsidies, together with management of the oceans for sustainable fishing and fish conservation in marine reserves, can be the way forward. Ecosystem-based management is the key. There are some other specific tools that will help: for example, the application of eco-labeling to inform more discerning consumers (e.g., FAO 2005).

Governability and Human Behavior in Food Security Scenarios

Kurien emphasized that: "Preserving the resource base and the integrity of the aquatic ecosystem is a *sine qua non* for food security – with or without international trade" (2004, 153). Throughout history, however, human behaviour with respect to food security has typically been selfish and irresponsible, especially in terms of the annexation, degradation and conversion for other uses of the aquatic and terrestrial ecosystems that produce food, often spurred by unfettered population growth. Diamond (2005) described the Easter Islanders' complete destruction of the ecosystem that supported them, including all sources of wood to make canoes. He speculated on what the person who felled the last tree might have thought or said: "Jobs, not trees…Technology will solve our problems, never fear we'll find a substitute for wood…We don't have proof that there aren't palms somewhere else on Easter, we need more research, your proposed ban on logging is premature and driven by fear-mongering" (Diamond 2005, 114)? He went on to point out the obvious and chilling parallels between this Easter Island debacle and ecosystem abuse in the modern, now thoroughly interconnected and interdependent world. That applies especially to the mismanagement of world fisheries. Faced with the opportunity to catch the last few fish or whales, many otherwise responsible persons will do just that, reasoning that if they do not, someone else will.

To what extent then might human behaviour become more responsible (i.e., more governable) in terms of caring for the ecosystems that produce fish and other human food? There are some grounds for optimism. Institutions for the support of more responsible behaviour in aquaculture and capture fisheries are increasing in number and influence, especially those concerned with implementing the FAO Code of

Conduct for Responsible Fisheries (CCRF) and its many Technical Guidelines (FAO 1995b; available at www.fao.org); for example, the Marine Stewardship Council (MSC; www.msc.org). There is a strong relationship between increasing economic welfare and increasing consumption of animal protein. This requires that more and more grain equivalents be used to feed livestock (including farmed fish) rather than to feed people directly (Fresco and Rabbinge 1997). There are, however, strong global, national and local movements towards making responsible dietary choices for food security and for the natural environment.

It can hardly be called irresponsible to eat well and to enjoy eating some animal protein, but a more vegetarian diet is indeed more earth-friendly, more sustainable, and therefore more conducive to food security at all levels. Goodland proposed a food conversion efficiency tax,

> In order to reduce food wastage and to improve health and food availability...The least efficient converters (pork, beef) would be highly taxed; more efficient converters (poultry, eggs, dairy) would be moderately taxed. Most efficient converters (ocean fish) would be taxed lowest. Grains for human food would not be taxed (1997, 189)

Goodland's tax-friendly stance on ocean fish was too simplistic, as many of them feed at high trophic levels, as carnivores. He omitted the products of aquaculture from his tax rankings, and considered aquaculture as having two extremes: low productivity/low impact, fed with autotrophs; and high productivity/high input energy/high impact, fed with manufactured feeds. He concluded that even the latter "can be more productive and at much less environmental cost than its competitor, livestock if grain inputs only are counted. If fossil energy and water costs are included, (high productivity) aquaculture is not competitive" (Goodland 1997, 193). Many, including the present author, would now find this to be too simplistic a view of the huge diversity and potentials, for good and ill, of responsible and irresponsible aquaculture (see below). On the broader front, whereas Goodland's tax proposals would surely have been politically impossible in 1997, some of them might be possible now, given the increasing public and private acceptance of painful earth-friendly measures such greenhouse gas emissions.

Dietary choices are major issues in attempting to balance sovereign preferences and sustainability. Norton et al. (1998) discussed this, considering three rank ordered goals of ecological economics (ecological sustainability, fair distribution and allocative efficiency), relationships between preference change and price change, and the extents to which attempts to influence individual preferences might be undemocratic. They concluded that: "Actively seeking to influence preferences is not inconsistent with a democratic society (and that)...in order to operationalize real democracy, a two-tiered decision structure must be used... in order to eliminate 'preference inconsistencies' between the short term and long term and between local and global goals" (Norton et al. 1998, 209). Their main point was that individual sovereignty is exercised in achieving democratic consensus on the broad and long term goals of society and that those broad goals can then, still democratically, "limit and direct preferences at lower levels" (Norton et al. 1998, 209). For example, overconsumption of animal fats and sugars and under-consumption of fruit, vegetables, and fish, pose huge and costly health problems, not only in affluent western and northern societies, but also in many developing countries. Increasing affluence and sometimes irresponsible, corporate behaviour and marketing, have led to nutritional transitions into over-nutrition, a form of malnutrition (e.g., Gillespie and Haddad 2001; Popkin et al. 2001).

Is there more cause for optimism or for pessimism concerning the future dietary behaviour of humans? Again, the present author sees grounds for optimism. Education and free information on the health aspects of food choices are increasingly available worldwide. Religious and cultural determinants of food choices are usually non-negotiable, but most tend to enhance global food security. Persuasion of vegetarians to eat more meat and fish would have the opposite effect. The world's livestock produce 18% of greenhouse gas emissions, measured as carbon dioxide equivalent, and occupy, including that needed for growing their feeds, 70% of all agricultural land (Steinfeld et al. 2006). In a Rockhopper TV film, *Taking the Credit* (www.rockhopper.tv), broadcast by BBC World News on October 23, 2009, the representative of a leading Swedish hamburger restaurant chain stated that over 70% of it carbon offsets purchased in Africa were accounted for by the production of hamburger beef itself.

Concerning responsible dietary choices with respect to fish, there is an increasing availability of free advice about which fish to purchase or to avoid, from wild harvests and from farms. For example, the Environmental Defense Fund's recommendations target the USA and include farmed arctic char, farmed rainbow trout and sablefish among its "Eco-best" buys, "tilapia (Latin America)" and "lobster, American/Maine" among the "Eco-OK", and farmed Atlantic salmon (for some reasons considered far worse than other farmed salmonids) and imported shrimps and prawns among the "Eco-worst" (www.edf.org). FishBase (www.fishbase.org) and Sealifebase (www.sealifebase.org) provide links to the increasing online and mobile phone-accessible sources that provide advice to earth-friendly purchasers of fish in markets and consumers of fish in restaurants, including, where possible, recommended minimum size limits and eco-friendly fishing and farming methods.

Food Security: The Case for an Interactive Governance Approach

Food security requires not only responsible human behaviour, but also effective institutions and tenable economics for all parties along human food chains, from producers through processors and vendors to consumers. Farmers, fishers, processors and vendors must be able to earn acceptable returns to their investments and the prices of food products must be affordable to consumers. But all of those *negotiable* scenarios always face a *non-negotiable*, three-fold reality: (a) all food is produced in natural or artificial ecosystems; (b) most of those ecosystems are not entirely under human control; and (c) all food production is accompanied by uncertainty and risk, especially concerning the weather, pests and diseases. Food security, though itself a non-negotiable fact of life, is achieved or lost by mixes and interactions of those negotiable socioeconomic factors (mainly human behaviour, institutions and markets) and those non-negotiable ecological factors (sound ecosystem management, which means use *and* conservation). Responsible human behaviour, effective institutions and sound ecosystem management maximize and sustain food security in the face of climatic and other risks. Irresponsible behaviour, ineffective institutions and the degradation of ecosystems diminish food security and amplify risks.

Ineffective institutions for food security derive largely from the persistence of monosectorial perspectives on food supply. In reality, the agriculture, aquaculture, fisheries and livestock sectors share the same ecosystems, but they usually compete for resources (land, water, investment, feeds, fertilizers, labor, research support, etc.) and accept little or no accountability for their adverse impacts on others. Most food is produced amidst sectorial land, water, energy and labor wars, not intersectorial partnerships. Irrigated crop agriculture receives about 70% of world freshwater withdrawals. Non-food sectors (e.g., forestry, industry, tourism and waste disposal) also have huge impacts on the resources and ecosystems required for food production. Sound ecosystem management for food production requires a rapid transition from unrealistic and unsustainable sector-specific policies and institutions to intersectorial ones. This would help to hedge against sectorial failures and risks, as extreme weather events, inexorable climate change and unpredictable pandemics of diseases common to farm animals and humans have increasingly large and negative influences on food security. The same intersectorial approach is needed when forecasting the pros and cons of future reliance on any given food producing sector or subsector.

Delgado et al. (2003) compared different production and price scenarios (slower and faster aquaculture expansion, lower production by China, efficiency of use of fishmeal and fish oil, and ecological collapse), from 1997 to 2020 for low- and high-value finfish, crustaceans, molluscs, fishmeal, fish oil, beef, pork, mutton, poultry meat, eggs, milk, and vegetable meals. They showed well how fruitless it is for any one food sector or subsector to ignore the rest and for any policymaker to miss the big picture when seeking to identify best bets. Some of the standardized scenarios of Delgado et al. (2003) can be criticized; for example, their ecological collapse scenario was more gradual and milder than some real world examples are likely to be. Under their ecological collapse scenario, by 2020 all the aquatic food prices would increase by 26–70% and fishmeal and fish oil prices by 134 and 128% respectively, while milk prices would reduce by 5% and egg and meat prices increase by 1–7%. That seems over-optimistic on the livestock front and probably over-pessimistic on what can be achieved in expansion of the more responsible aquaculture that is less dependent on fishmeal and fish oil. Accounting for the large influences of China's demand for and production of food would also need some recalculation.

Countries that operate distant water fishing fleets have long fished the rich waters of other countries, through agreements of varying degrees of equity and sometimes illegally. As food demand increases and food production and distribution have become globalized, foreign acquisitions and use of lands and inland waters have recently increased (Anon 2009). The People's Republic of China (PRC) is but one among an increasing number of Asian, Middle Eastern and North African countries (e.g., South Korea, Saudi Arabia and Libya) that are acquiring lands and waters in the developing countries of sub-Saharan Africa, Central and Southeast Asia, and Latin America, for producing food, fibre and biofuels. Cotula et al. (2009) found that 2,492,684 ha of such lands (excluding deals less than 1,000 ha) had been thus acquired in Ethiopia, Ghana, Madagascar, Mali and Sudan. Such arrangements are usually styled as cooperation in "*agribusiness*". They can include attempted and established cooperation in coastal and inland aquaculture; for example, PRC-Philippines and PRC-Ghana (www.ibon.org; present author's observations). From 2004, the PRC has been implementing a so-called "Going Out" policy to develop collaboration in business, including agribusiness, but announced in 2008 that its new 20-year Food Security Policy explicitly excludes any foreign land acquisition (Xinhua News Agency 2008, cited in Cotula et al. 2009). This situation will probably remain highly dynamic and somewhat opaque.

Kurien (2004) saw food security for all as a guided outcome and found that trade-enhanced food security inevitably requires cooperation between market, state and civil society. Kurien also described the so-called chains of custody of internationally traded fish products as "long and varied" (2004, 17). He saw the end of such chains that were closest to fishers and fish workers, as well as "first sale transactions and first product transformations," as most important for the food security of the poor, with the prospect of "people power" at these chain positions to "match 'market power'" (Kurien 2004, 17).

Food security is therefore highly diverse, complex and dynamic, and is typically defined across a range of scales, from household and local to national and global. These attributes make the case for an interactive governance approach (Kooiman 2003; Kooiman et al. 2005), but the result would have to be more than a re-description of well-published scenarios in a new jargon and there would have to be practical applications. At present, the best way forward seems to be assessing governability along fish chains and seeking critical entry points for improvement. To explore this, the following three examples of entire aquaculture fish chains were chosen as systems-to-be-governed, from ecosystem, through production cycles, harvests, post harvest processing, marketing, wholesale, and retail to consumers: A. pond farming of carps by community groups, in Bangladesh; B. pond farming of tilapia, in the Philippines; and C. lake-based cage farming of tilapia, also in the Philippines (ADB 2005). Their governing systems included actors and institutions that deal with the negotiable (man-made arrangements) and the non-negotiable (climate and ecosystem function). The governance interactions between these systems-to-be-governed and their governing systems were seen as the main determinants of governance success.

Governability Assessments for Three Aquaculture Fish Chains

Two methods were used to assess the governability of each link in the three chains, from prerequisites for farming, through seed procurement and growout, to harvesting to post harvest operations. With both methods, governabilities were scored as low (L), moderate, (M) or high (H). For the first method (I), a rapid, empirical and intuitive quick guess was made, without reference to any of the detailed parameters and relationships used in the interactive governance approach, of the likelihood of success (L, M, or H) of attempts to improve each link through better governance. With that approach, governability was taken as that score for amenability to better governance in general. With the second method (II), a systematic, though still entirely subjectively scored application of the interactive governance approach was attempted, to score governability (L, M, or H). Following Chuenpagdee et al. (2008), two of the main descriptors and criteria specified in the interactive governance approach were considered, i.e., prevalence of system properties and presence/absence of governing interactions (see Table 6.1).

The two methods used for estimating governability showed similar results. The types of links (rows) where there seemed to be scope for exploring improvement of governability (from L-, or M-) were the same for the two methods in many cases. The numbers of cells indicating scope for improvement of governability were also broadly similar for the three fish chains. Although no firm conclusions can be drawn from these governability assessments, there are indications that governability assessment can indicate weak links in the chains and potential scope for improvement. It also seems possible that a simple approach can be as good a method for this highly subjective, qualitative exercise, as can use of the full gamut of interactive governance descriptors and criteria.

General Discussion

Fish chains start with ecosystems and end with consumers. Fish as food come, directly or indirectly, from open water ecosystems that are exploited by capture fisheries and agro-ecosystems that host aquaculture. The health of those ecosystems is the non-negotiable bottom-line for optimizing and sustaining contributions of fish to food security. Some of the world's historical fish chains no longer exist because of ecosystem change or collapse. Folke et al. took the view that "a resilient social-ecological system may make use of a crisis to transform into a more desired state" (2005, 441). Fish chains have plenty of ongoing crises, and food security is more than just a desired state, it is an essential need. But it is usually specific links in fish chains that are seen as the systems-to-be-governed, not the whole chain and not whole social-ecological systems, resilient or not. Chain-long policies and their effective implementation, through chain-long governance are still rare. Any chain is only as strong as its weakest link.

At the institutional level, the most important entry points along a fish chain are those where institutional visions, roles and responsibilities can be broadened to recognize and to respond to the needs of the rest of the chain and of other interdependent sectors. Changes in human behaviour, as well as new and reformed institutions, are sorely needed, at international, national and local levels, to learn and to tell the truth about what is possible and sustainable in terms of the contributions of wild caught and farmed fish to food security, and then to act accordingly. Such

Table 6.1 Governabilities of three aquaculture fish chains, estimated by two methods: I. empirically; II. by an interactive governance approach (Chuenpagdee et al. 2008)

				Method II: estimates using an interactive governance approach					
	Method I: empirical estimates			Prevalence of system properties			Presence/absence of governing interactions		
Fish chain links	A	B	C	A	B	C	A	B	C
Prerequisites									
Sites	H	**M**	**L**	**L**	M	H	H	**M**	**L**
Permits	H	M	M	**L**	M	**L**	M	M	**L**
Skills,	**L**	M	M	M	M	M	**L**	M	**L**
Information	**L**	**M**	**L**	**L**	M	M	**L**	**M**	**M**
Policies	M	M	**L**	M	M	M	**L**	M	M
Financing	**L**	M	M	M	M	M	**M**	M	M
Markets	M	H	H	H	H	H	**L**	M	M
Engineering	H	H	**M**	**L**	M	M	M	M	M
Seed procurement									
Wild	**L**	–	–	–	L	–	L	–	–
Hatchery (own)	M	M	M	L	M	M	L	M	M
Hatchery (bought)	**L**	M	M	**L**	M	M	**L**	M	M
Growout									
Feed, fertilizers	**L**	M	M	M	M	M	**L**	M	M
Labour (family)	H	H	H	M	M	M	H	M	M
Labour (hired)	H	H	H	H	M	M	H	M	M
Fuel etc.	–	M	M	–	M	M	L	L	L
Equipment	**L**	M	M	**L**	M	M	**L**	M	M
Harvesting									
Labour (family)	H	H	H	H	M	M	M	M	M
Labour (hired)	M	M	M	M	M	M	M	M	M
Equipment	**L**	M	M	H	M	M	**L**	M	M
Fuel, ice etc.	**L**	M	M	H	M	M	M	M	M
Post harvest									
First handling	M	M	M	H	M	M	H	M	M
First sales (farm)	H	H	H	H	M	M	H	M	M
Processing	–	M	M	–	M	M	–	M	M
Domestic markets	**L**	H	H	M	M	M	M	M	M
Export markets	–	M	**L**	–	**M**	**L**	–	**L**	**L**
Certification	–	**L**	**L**	–	**L**	**L**	–	**L**	**L**
Promotion	–	M	M	–	M	M	–	L	L

A: Group carp ponds, Bangladesh; B: Tilapia ponds, Philippines; C: Tilapia cages, Philippines (ADB 2005)

Links where improvement of governability (from L to M, or from M to H) might be explored are in *bold font*

Abbreviations used are: L low governability, M moderate governability, H high governability

behavioural and institutional changes will be possible only if individuals, households and those in authority pursue them pro-actively, not just reactively in response to food and environmental crises. Food fish security will be achieved only if humans can agree to care, indefinitely, for the ecosystems that produce their food fish.

That would be somewhat analogous to the process of domestication. Animals that are amenable to domestication (i.e., governable) strike a great bargain with their governors, receiving food security, shelter, health care, and mating rights, in exchange for providing a wide range of goods and services: meat, milk, eggs, hunting, winning races and fights, companionship etc. If humans behaved more as domesticates of nature, their governor, and less as its wild exploiters, their food security and many other benefits would be maximized. The necessary change from irresponsibility to responsibility in food production is just that: a change from wild behaviour to governed behaviour.

Can humans become sufficiently governable to achieve that relationship with nature, or will human nature always preclude it? Time will tell. Irresponsible behaviour in food fish production, from aquaculture and capture fisheries, is still currently threatening food fish security. Economic growth itself can be at odds with food security. For example, the announcement of a forum on the conflicts between economic growth and the recovery of wild salmon populations contained the following quote from its author's son: "Dad, get a life. Most people out here in the real world just don't care that much about restoring wild salmon. They have other things to worry about!" (Lackey 2005, 21; see also Czech et al. 2006 for a summary of that forum). But food insecurity is a very powerful motivator for change. If planet Earth is indeed to support a population of over nine billion people, before reductions to levels more appropriate to its available resources, then human governability *will* have to improve throughout all food chains, including fish chains.

Further work is needed to explore whether governability estimates can indeed help to identify the links in human food chains in general, and links in fish chains in particular, at which improvements are most needed and possible. This will require robust, objective, quantitative methods. Existing methods that could be explored include the Delphi Method, which has been used for a wide range of purposes, including fish conservation (Barrett 2009), and Environmental Damage Schedules from community judgments (Chuenpagdee et al. 2001).

References

ADB. (2005). An evaluation of small-scale freshwater rural aquaculture development for poverty reduction. Manila: Asian Development Bank.

ADB. (2008). Soaring food prices – Response to the crisis evaluation. Manila: Asian Development Bank.

Ahmed, M., & Lorica, M.H. (2002). Improving developing country food security through aquaculture development – Lessons from Asia. Food Policy, 27, 125–141.

Alamgir, A., & Arora, P. (1991). Providing food security for all. London: Intermediate Technology Publications.

Anon. (2009). Buying farmland abroad: Outsourcing's third wave. The Economist, 391(8632), 55–57.

Barrett, P.J. (2009). Estimating Devil's Hole pupfish lifestage ratios using the Delphi Method. Fisheries, 34(2), 73–79.

Blaxter, J.H.S. (1990). The herring. Biologist, 37(1), 27–31.

Bromley, D.W. (2009). Abdicating responsibility: The deceits of fisheries policy. Fisheries, 34(6), 280–290.

Brown, L.R. (2006). Plan B 2.0: Rescuing a planet under stress and a civilization in trouble. New York: W.W. Norton & Company.
Chamberlain, T. (2000). Histamine in artisanal tuna fisheries of the Pacific Islands. Technical Report 2000/06. Suva: University of the South Pacific.
Chess, C., Burger, J., McDermott, M.H. (2005). Speaking like a state: Environmental justice and fish consumption advisories. Society and Natural Resources, 18, 267–278.
Chuenpagdee, R., Knetsch, J.L., Brown, T.C. (2001). Environmental damage schedules: Community judgments of importance and assessments of losses. Land Economics, 77(1), 1–11.
Chuenpagdee, R., Kooiman, J., Pullin R.S.V. (2008). Assessing governability of capture fisheries, aquaculture, and coastal zones. Transdisciplinary Environmental Studies, 7(1), 20 pp.
Clover, C. (2004). The End of the line: How overfishing is changing the world and what we eat. New York: Ebury Press.
Consensus. (2006). *Sustainable aquaculture in Europe*. CD-ROM; http://www.euraquaculture.info
Costa-Pierce, B.A. (2002). Ecology as the paradigm for the future of aquaculture. In B.A. Costa-Pierce (Ed.), Ecological aquaculture (pp. 339–372). Oxford: Blackwell Science.
Cotula, L., Vermuelen, S., Leonard, R., Keeley, J. (2009). Land grab or development opportunity? Agricultural investment and international land deals in Africa. Rome: FAO/International Institute for Environment and Development/International Fund for Agricultural Development.
Crawford, M.A., Broadhurst, C.L., Galli, C., Ghebremeskel, K., Holmsen, H., Saugstad, L.F., Schmidt, W.F., Sinclair, A.J., Cunnane, S.C. (2008). The role of docosahexaenoic and arachidonic acids as determinants of evolution and hominid brain development. In K. Tsukamoto, T. Kawamura, T.D. Beard and M.J. Kaiser (Eds.), Fisheries for global welfare and environment. Memorial book of the proceedings of the 5th World Fisheries Congress (pp. 57–76). Tokyo: TERRAPUB. Available at: http://www.terrapub.co.jp/onlineproceedings/fs/index.html
Czech, B., Alam, S.K. et al. (2006). Economic growth, fish conservation, and the AFS: Conclusion to a forum, beginning of a movement? Fisheries, 31(1), 40–43.
Delgado, C.L., Wada, N., Rosegrant, M.W., Meijer, S., Ahmed, M. (2003). Fish to 2020: Supply and demand in changing global markets. Washington, DC: International Food Policy Research Institute.
Dey, M.M., Rab, M.A., Paraguas, F.J., Piumsombum, S., Bhatta, R., Alam, M.F., Ahmed, M. (2005). Fish consumption and food security: A disaggregated analysis by types of fish and classes of consumers in selected Asian countries. Aquaculture Economics and Management, 9, 89–111.
Diamond, J. (2005). Collapse: How societies choose to fail or to survive. London: Penguin Books Ltd.
Eaton, S.B., Eaton, S.B. III, Konner, M.J., Shostak, M. (1996). An evolutionary perspective enhances understanding of human nutritional requirements. Journal of Nutrition (American Institute of Nutrition), 126, 1732–1740.
Edwards, P. (2004). Traditional Chinese aquaculture and its impact outside China. World Aquaculture, 35(1), 24–27.
Elvevoll, E.O., & James, D.G. (2000). Potential benefits of fish for maternal, foetal and neonatal nutrition: A review of the literature. Food, Nutrition and Agriculture, 27, 28–39.
European Commission. (2009, April 22). Green paper: Reform of the Common Fisheries Policy. Brussels: European Commission.
FAO (1995a). The use of Hazard Analysis Critical Control Point (HACCP) principles in food control. Rome: FAO.
FAO (1995b). Code of conduct for responsible fisheries. Rome: FAO.
FAO. (1996, November). Rome declaration on World Food Security and World Food Summit Plan of Action. (Paper presented at the World Food Security and World Food Summit of the FAO, Rome).
FAO. (2000). Food for the cities: Food supply and distribution policies to reduce urban food insecurity. Rome: FAO.
FAO. (2001). Genetically modified organisms, consumers, food safety and the environment. FAO Ethics Series 2. Rome: FAO.
FAO. (2002, June). Summary of proceedings: Measurement and assessment of food deprivation and undernutrition. (Paper presented at the International Scientific Symposium of the FAO, Rome).
FAO. (2005). Guidelines for the ecolabelling of fish and fishery products from marine capture fisheries. Rome: FAO.

FAO. (2006a). FAO's strategy for a food chain approach to food safety and quality: A framework document for the development of future strategic directions. (Paper presented at the 17th Session of the Committee on Agriculture, FAO, Rome). Available at http://www.fao.org/DOCREP/MEETING/006/Y8350e.HTM

FAO. (2006b). State of world aquaculture 2006. Rome: FAO.

FAO. (2009a). The state of food insecurity in the world. Rome: FAO.

FAO. (2009b). The state of world fisheries and aquaculture 2008. Rome: FAO.

FAO. (2012). Voluntary guidelines in the responsible governance of tenure of land, fisheries and forests in the context of national food security. Rome: FAO.

FAO/NACA/WHO. (1999). Food safety issues associated with products from aquaculture. Geneva: WHO.

Finley, W., & Fry, J. (2009). Expanding opportunities for vegetable proteins in aquaculture feeds. World Aquaculture, 40(2), 29–32.

Focardi, S., Corsi, I., Franchi, E. (2005). Safety issues and sustainable development of European aquaculture: New tools for environmentally sound aquaculture. Aquaculture International, 13, 3–17.

Folke, C., Hahn, T., Olsson, P., Norberg, J. (2005). Adaptive governance of social-ecological systems. Annual Review of Environment and Resources, 30, 441–473.

Ford, J.S., & Myers, R.A. (2008). A global assessment of salmon aquaculture impacts on wild salmonids. PloS Biology, 6(2), e33.

Fresco, L.O., & Rabbinge, R. (1997). Keeping world food security on the agenda: Implications for the United Nations and the CGIAR. Issues in Agriculture 11. Washington, DC: Consultative Group on International Agricultural Research.

Gill, C. (2005). World feed panorama: Disease takes toll, but feed output bounces back. Feed International, 26(1), 4–9.

Gillespie, S., & Haddad, L. (2001). Attacking the double burden of malnutrition nutrition transition in Asia and the Pacific. Manila: Asian Development Bank.

Gillett, R. (2009). Fisheries in the economics of the Pacific Island countries and territories. Manila: Asian Development Bank.

Goodland, R. (1997). Environmental sustainability in agriculture: Diet matters. Ecological Economics, 23, 189–200.

Gupta, M.V. (2006). Challenges in sustaining and increasing fish production to combat hunger and poverty in Asia. Naga, WorldFish Center Quarterly, 29(1 and 2), 4–10.

Haddad, L., Oshaug, A. (1998). How does the human rights perspective help to shape the food and nutrition policy research agenda? Food Policy 23(5), 329–346.

Hales, S., Weinstein, P., Woodward, A. (1999). Ciguatera (fish poisoning), El Niño, and Pacific sea surface temperatures. Ecosystem Health, 5(1), 20–25.

Halwart, M., & Bartley, D.M. (Eds.). (2005). Aquatic biodiversity in rice-based ecosystems. studies and reports from Cambodia, China, Lao People's Democratic Republic and Viet Nam. CD-ROM. Rome: FAO.

Hara, M.M. (2001). Could marine resources provide a short-term solution to declining fish supply in SADC inland countries? The case of horse mackerel. Food Policy, 26, 11–34.

Hardianti, K. (2005). Special evaluation study of selected ADB interventions on nutrition and food fortification. Manila: Asian Development Bank.

Henson, S., & Heasman, M. (1998). Food safety regulation and the firm: understanding the compliance process. Food Policy, 23(1), 9–23.

Huss, H.H., Ababouch, L., Gram, L. (2004) . Assessment and management of seafood safety and quality. Rome: FAO.

Hutchings, J.A. (2000). Collapse and recovery of marine fishes. Nature, 406, 882–885.

Jackson, J.B.C., Kirby, M.X. et al. (2001). Historical overfishing and the recent collapse of coastal ecosystems. Science, 293, 629–637.

Jentoft, S., & Chuenpagdee, R. (2009). Fisheries and coastal governance as a wicked problem. Marine Policy, 33, 553–560.

Jussaume, R.A. Jr., & Higgins, L. (1998). Attitudes towards food safety and the environment: A comparison of consumers in Japan and the U.S. Rural Sociology, 63(3), 394–411.

Kawarazuka, N., & Béné, C. (2010). Linking small-scale fisheries and aquaculture to household nutritional security. Food Security, 2(4), 343–357.
Kooiman, J. (2003). Governing as governance. London: Sage.
Kooiman, J., Bavinck, M., Jentoft, J., Pullin, R. (2005). Fish for life: Interactive governance for fisheries. Amsterdam: Amsterdam University Press.
Kooiman, J., Bavinck, M., Chuenpagdee, R., Mahon, R., Pullin, R. (2008). Interactive governance and governability – An introduction. The Journal of Transdisciplinary Environmental Studies, 7(1).
Kurien, J. (2004). Fish trade for the people: Toward understanding the relationship between international fish trade and food security. Report of the Study on the Impact of International Trade in Fishery Products on Food Security. Rome: FAO.
Kurlansky, M. (1999). Cod. London: Vintage Books.
Lackey, R.T. (2005). Economic growth and salmon recovery: an irreconcilable conflict? Fisheries, 30(3), 30–32.
Mahon, R., McConney, P., Roy, R.N. (2008). Governing fisheries as complex systems. Marine Policy, 32, 104–112.
Moffit, C.M. (2005). Environmental, economic and social aspects of animal protein production and the opportunities for aquaculture. Fisheries, 30(9), 36–38.
Molony, B.W., Lenanton, R., Jackson, G., Noriss, J. (2003). Stock enhancement as a fisheries management tool. Reviews in Fish Biology and Fisheries, 13, 409–432.
Nandeesha, M.C. (2001). Farmers as scientists: Andhra Pradesh fish farmers go into revolutionary carp research. Aquaculture Asia, VI(4), 29–32.
Norton, B., Costanza, R., Bishop, R.C. (1998). The evolution of preferences: Why ‘sovereign’ preferences may not lead to sustainable policies and what to do about it. Ecological Economics, 24, 193–211.
Páez-Osuna, P., Gurrero-Galvan, S.R., Ruiz-Fernández, A. (1998). The environmental impact of shrimp aquaculture and the coastal pollution in Mexico. Marine Pollution Bulletin, 36(1), 65–75.
Pauly, D. (1995). Anecdotes and the shifting baseline syndrome of fisheries. Trends in Ecology and Evolution, 10(10), 430.
Pauly, D., & Maclean, J.L. (Eds.). (2003). In a perfect ocean: The state of fisheries and ecosystems in the North Atlantic Ocean. Washington, DC: Island Press.
Pauly, D., Christensen, V., Dalsgaard, J.P.T., Froese, R., Torres, F. Jr. (1998). Fishing down marine food webs. Science, 279, 860–863.
Pauly, D., Christensen, V., Guénette, S., Pitcher, T., Sumaila, U.R., Walters, C.J., Watson, R., Zeller, D. (2002). Towards sustainability in world fisheries. Nature, 418, 689–695.
Pauly, D., Watson, R., Alder, J. (2005). Global trends in world fisheries: Impacts on marine ecosystems and food security. Philosophical Transactions of the Royal Society B, 360, 5–12.
Popkin, B.M., Horton, S.H., Kim, S. (2001). The nutrition transition and prevention of diet-related diseases in Asia and the Pacific. Manila: Asian Development Bank.
Pullin, R.S.V. (2011). Aquaculture up and down the food web. In V. Christensen, J.L. Maclean (Eds.), Ecosystem approaches to fisheries: A global perspective (pp. 89–119). Cambridge: Cambridge University Press.
Pullin, R.S.V., Froese, R., Pauly, D. (2007). Indicators for the sustainability of aquaculture. In T.M. Bert (Ed.), Ecological and genetic implications of aquaculture (pp. 53–72). Dordrecht: Springer.
Roberts, C. (2007). The unnatural history of the sea. Washington, DC: Island Press.
Ruddle, K. (1996). The potential role of integrated management of natural resources in improving the nutritional and economic status of resource poor farm households in Ghana. In M. Prein, J.K. Ofori, C. Lightfoot (Eds.), Research for the future development of aquaculture in Ghana (pp. 53–72). ICLARM conference proceedings 42. Manila: International Center for Living Aquatic Resources Management.
Sadovy, Y.J., Donaldson, T.J. et al. (2003). While stocks last: The live reef food fish trade. Manila: Asian Development Bank.

Sen, A. (1981). Poverty and famines: An essay on entitlement and deprivation. Oxford: Clarendon.

Steinfeld, H., Gerber, P., Wassenbar, T., Castel, V., Rosales, M., de Haan, C. (2006). Livestock's long shadow. Rome: FAO.

Subasinghe, R., Soto, S., Jiansan, J. (2009). Global aquaculture and its role in sustainable development. Reviews in Aquaculture, 1, 2–9.

Tacon, A.G.J., Hasan, M. R., Subasinghe, R.P. (2006). Use of fishery resources as feed inputs for aquaculture development: trends and policy implications. FAO Fisheries Circular, 1018.

Tansey, G. (2002). Food security, biotechnology and intellectual property: Unpacking some issues around TRIPS. A discussion paper. Geneva: Quaker United Nations Office.

Thilsted, S.H., Roos, N., Hassan, N. (1997). The role of small indigenous fish species in food and nutrition security in Bangladesh. Naga, 13–15.

Thorpe, A., Johnson, D., Bavinck, M. (2005). Part II: The system to be governed. Introduction. In J. Kooiman, M. Bavinck, S. Jentoft, R. Pullin (Eds.), Fish for life: Interactive governance for fisheries (pp. 41–44). Amsterdam: Amsterdam University Press.

Turchini, G.M., Torstensen, B.E., Ng, W-K. (2009). Fish oil replacement in finfish nutrition. Reviews in Aquaculture, 1, 10–57.

Xinhua News Agency. (2008). Faganwai: Wu haiwai duntian jihua (NDRC China has no plans to acquire land overseas) (In Chinese). Retrieved from http://news.xinhuanet/com/fortune/2008-11/13/content_10351772.htm

Chapter 7
Ecosystem Health in the Context of Fisheries and Aquaculture – A Governability Challenge

José J. Pascual-Fernández and Ratana Chuenpagdee

Abstract The importance of marine and ocean ecosystems to the livelihoods and wellbeing of fishing and coastal communities around the world is well recognized. Global efforts have been made to prevent these ecosystems from deteriorating, but the challenges are huge, with ongoing pressures and stresses driven largely by a wide range of human activities. In this chapter, we first employ the governability concept to examine these stressors in terms of their diversity, complexity, dynamics and scales in relation to the natural and social systems-to-be-governed, the governing systems and the governing interactions. Recognizing that the health of the ecosystem is an outcome of governing efforts and interactions between governing institutions and social actors, we apply the governability perspective to assess factors affecting the ability of the social system-to-be-governed and the governing system to cope with the present state of the marine and ocean ecosystems, and draw policy implications based on that analysis.

Keywords Human ecosystems • Ecosystem health • Fisheries • Coastal zone • Aquaculture • Global warming • Governability

J.J. Pascual-Fernández (✉)
Institute of Social and Political Sciences, University of La Laguna, Tenerife, Spain
e-mail: jpascual@ull.es

R. Chuenpagdee
Department of Geography, Memorial University of Newfoundland, St. John's, NL, Canada
e-mail: ratanac@mun.ca

M. Bavinck et al. (eds.), *Governability of Fisheries and Aquaculture: Theory and Applications*, MARE Publication Series 7, DOI 10.1007/978-94-007-6107-0_7,

Stresses and Challenges to Ecosystem Health[1]

Humans have been related to and dependent on coastal areas and oceanic ecosystems since hunter and gatherers' societies. This dependency has increased in the last centuries with the growing and changing demographics in coastal areas around the world. The diversity of human uses in these areas has also increased with expansions in the exploitation of underwater mineral resources, such as petroleum and gas, and renewable resources like fisheries, as well as ongoing development of harbor and related infrastructure for maritime transport. Human pressures on the oceans are much higher today than in any other moment in history. Increased attention must therefore be paid to questions regarding the means by which we govern the ecosystems we depend on, and how to mediate the relationship between human societies and nature.

The continuous modification of coastal and marine ecosystems by human societies means their survival is at stake. They are vulnerable to changes that may end up transforming their functions to the point that they are no longer able to provide ecosystem services and goods. Past lessons show that some of these changes are not reversible and that their effects can be drastic. Cod fisheries collapse in Newfoundland is a good example of how anthropogenic changes may strike back on coastal populations (Finlayson and McCay 1998). Human-nature interactions are frequently inscribed in co-evolution processes that, for instance, allowed humans to develop seeds (for agriculture) that slowly changed with and because of human behavior. This consequently made the transformation from hunter and gatherer societies into agricultural adaptations possible (Rindos 1984). In other circumstances, human-nature interactions may be conducive to abrupt changes and unforeseeable consequences, such as when pressures on the ecosystems or human populations compromise their integrity and resilience. As described, the relationship between human societies and ecosystems creates increasing concerns about the health of coastal and marine ecosystems.

Several integrated management frameworks and ecosystem-based approaches have been employed to address these concerns. Some of them, such as the one promoted by Resilience Alliance (Berkes and Folke 2000; Armitage et al. 2007), emphasize linkages between social and ecological systems. While the interactive governance approach has a similar focus, it is explicit in the examination of all related systems, i.e., those that are being governed, the governing system and the interactions between the two, throughout the entire 'fish chain' (Kooiman et al. 2005). These systems receive even attention because it is understood that factors fostering or inhibiting governance can be found in any of them. In addition, looking at ecosystem health from the governability perspective, as suggested here, is a

[1] The research summarized in this chapter is related to the project *Governability analysis applied to the creation process of Marine Protected Areas* (GOBAMP, CSO2009-09802), supported by the Spanish Ministry of Economy and Competitiveness. We also acknowledge the support from the Social Sciences and Humanities Research Council of Canada.

systematic way of analyzing what makes the system more or less governable, and what governing interventions may therefore be required.

Before delving in any further, we would like to first acknowledge that the word 'ecosystem' evokes multiple connotations, even within the scientific community. It generally refers to a complex system that relates living organisms and physical factors in an environment. One of the most widely accepted definitions of the term is provided by Convention on Biological Diversity. Accordingly, 'ecosystem' means a dynamic complex of plant, animal and micro-organism communities and their non-living environment interacting as a functional unit (CBD 1993 [1992]).[2] However, some elements of the concept may be understood from different perspectives. For instance, where or how the limits of an environment can be located may constitute a controversial issue, because natural systems are globally interrelated. Another relevant issue, perhaps the most important for our aims, is how the role of human societies in ecosystems can be conceptualized. For some authors it is evident that humans have a clear, dominating role in the recent evolution of ecosystems in our planet (Vitousek et al. 1997). Others even argue for the use of a 'human ecosystems' concept for those ecosystems where humans constitute a central agent (Stepp et al. 2003). As posited by Vitousek et al., "It is clear that we control much of Earth, and that our activities affect the rest. In a very real sense, the world is in our hands - and how we handle it will determine its composition and dynamics, and our fate" (1997, 498–499). This position looks increasingly clear after considering all of the accumulated evidence on how humans induce ecosystem transformations in the world, as is the case with climate change.

Next, defining what a healthy ecosystem is may be subject to debate. System evaluators may have different opinions about the baseline, because ecosystems are continuously changing. Moreover, based on their experiences, each generation may have a very different perspective of the state of a healthy ecosystem. Finding pristine ecosystems without clear human impact, whether it is at sea or on land, is no longer possible. Even remote areas are being exploited directly by humans or indirectly by other activities (Kulbicki 2005), including through processes like climatic change.

Well-functioning ecosystems are crucial not only for fisheries productivity (Chuenpagdee et al. 2005), but also for other functions. The capacity of the sea to absorb CO_2, for instance, is a key factor in assimilating climate change. Ocean currents are also essential for climate stability in many areas of the world. The extent to which human societies depend on ecosystem services, specifically on marine ecosystems, may not be completely known. Yet, humans may very well be the only species on earth to have unprecedentedly impacted ecosystem health far beyond recovery. This is perhaps due to confidence in our ability to develop technologies to control the natural realms. However, such ability is increasingly contested by

[2] The relevance of this concept even conduces to detailed legal definitions, for instance as "a dynamic complex of plant, animal and micro-organism communities and their non-living environment interacting as a functional unit" (Commonwealth of Australian Law 1999, 466).

obstinate facts. Climate change is a good example of this global challenge, with major technological or political efforts providing meager results.

Maintaining marine ecosystem health and its functions is a governability issue that requires the involvement and commitment from actors across the board and at all levels of governance. Most programs and initiatives related to ecosystem health have both local and global dimensions. Marine protected areas (MPAs), for instance, are one of the most popular ecosystem management tools promoted around the world, but they are also a classic example of a wicked problem that needs to be examined through the governability lens. Because MPA proposals involve large areas, achieving agreements on what needs to be done, as well as on who and how to do them, raises significant challenges. Globally, nations struggle to meet the various targets set and adopted to increase the level and extent of marine protection, including the latest recommendation at the 2003 World Parks Congress to establish networks of MPAs (including strictly protected areas) covering at least 20–30% of each habitat by 2012.[3] Debates are ongoing about how the targets were set, whether they are realistic and whether these MPAs would indeed contribute to improving ecosystem health. At the local level, MPAs are often ridden with conflicts among user groups with incompatible practices and competing interests, making their creation problematic (Jentoft et al. 2011). Without local support, MPAs are unlikely to be successful or sustained. Considering all the challenges described above, serious consideration about our governing capacity and the understanding of the characteristics and qualities of the ecosystems that need to be governed is essential.

Following a governability perspective, we characterize challenges affecting fisheries and coastal ecosystem health in terms of the system-to-be-governed, the governing-system and the government interactions, taking into account their diversity, complexity, dynamics and scale/vulnerability (Table 7.1). The detailed analysis of these challenges is presented around three relevant global issues related to ecosystem health: coastal zones; fisheries and aquaculture; and climate change.

Coastal Zones: Diversity, Complexity and Dynamics on the Rise

Coastal Transformations

Coastal zones constitute a key element in many ecosystems. They are crucial for the growth, reproduction and vitality of many species, including humans. These areas play essential roles in the development and evolution of societies and cultures across

[3] WPC Recommendation 5.22, see http://www.internationalwildlifelaw.org/MPARecs.pdf

Table 7.1 Challenges related to fisheries and coastal development affecting ecosystem health

Characteristics	System-to-be-governed	Governing system	Governing interactions
Diversity	New economic opportunities like tourism, transportation, commerce, recreation, etc. conducive to coastal development; increasing number of ethnic communities due to coastal migration; increasing use of fishing gear types; new entrants into the fishery	Increasing actors in fisheries and coastal areas requiring a higher number – or broader scope – of responsible government agencies; new rules and regulations established to address the multiple demands and interests	Greater level, and variation in types, of interactions between government agencies and fisheries-coastal actors required to manage growth and to prevent negative consequences for fisheries resources and the coastal ecosystem
Complexity	Fishers of different gear types, coastal communities of different ethnic groups and multiple coastal stakeholders relying on the sharing of limited resources and space	Various governing agencies with different responsibilities requiring increasing coordination to achieve their individual mandates	Not all interactions are mutually supportive. For instance, two-way information sharing between stakeholders can generate trust, while one-way information flow can be seen as exploitative
Dynamics	User conflicts and social tension arising from resource and space competition – gear conflict may surge because risks associated to specific gears; cultural differences leading to an intensification of effort in fishing and coastal development and deterioration of social capital	Mandates and goals of the various agencies may not be compatible causing confusion and uncertainty among stakeholders	Interactions can result in positive outcomes such as compliance and partnership building, or negative outcomes such as conflict and disagreement among stakeholders
Scale	Impacts of fishing and coastal development are not contained within the areas where the activities take place; they also do not happen immediately but can be accumulated over time. Other general processes may deeply affect coastal areas (climate change)	Most governing agencies are restricted in their spatial jurisdiction, having thus difficulty coping with the different magnitude of the impacts and the consequences of phenomena like climate change, sea level rise and tsunamis	Interactions are difficult to coordinate when space is widened and when time sensitivity may be an issue, as in the case of adaptive management where learning and feedback are required. Also, contradictory interests, like those related to climatic change between powerful countries, may pose enormous challenges to reach agreements

the globe. As shown by archeological evidence, early complex cultures and civilizations emerged along the coasts. In addition, communication and commerce have depended on sea pathways at least since the appearance of pristine natural states. At present, these coastal areas are probably more important than ever, as population and economic growth have converged in the last decades, seen in particular in the "Pacific Rim" countries (Lundin and Linden 1993). In densely populated coastal areas, pollution due to waste disposal is one of the main concerns. Organic waste disposal may increase the relevance of certain algae, transforming the ecosystem with undue consequences. The increasing incidence of red tides exemplifies this. Shellfish gathering in the affected areas becomes extremely dangerous for human health. Other target fish species may suffer from the same problems. In areas where industrial development is prominent, chemical pollution may threaten the entire ecosystem, hindering many human uses of the coasts. In many tropical countries, coastal development involves conversion and clear-cutting of mangrove forests. An estimated 35% of the global mangrove area has been lost since 1980. Mangrove forests consequently constitute the most threatened coastal habitat in the world (Valiela 2006). Previously, mangroves had generally been considered 'wasteland,' with low ecosystem values except for their harvestable products such as wood for charcoal and the construction of houses and fishing gear. The conversion of mangrove areas for other uses was therefore deemed suitable. Loss of mangrove forests, however, has major negative consequences due to their important ecosystem functions and services, for example, as habitats and nursery grounds for juvenile fish and other living marine organisms. Mangroves also provide shoreline stabilization and help waste assimilation, thereby mitigating coastal pollution problems. Despite its vast destruction, the Indian Ocean tsunami in December 2004 has led to renewed appreciation for the values of mangroves. Coastal communities in the tsunami affected areas avowed the importance of mangrove forests in mitigating the devastation.

Coastal Processes Accelerating Changes

In the last decades, coastal areas have been modified by human activities to the point that the future of humankind may be threatened. Coastal erosion and displacement, water and air pollution, disappearance of living marine resources, and degradation of coastal habitats are all signs of harmful coastal changes brought about by anthropogenic and natural causes. These various coastal processes create tensions and compromise the previous social arrangement and relative ecosystem homeostasis. An increase in diversity and complexity are likely consequences of these changes, because the new system elements contain more actors who may not fit perfectly with each other. For example, new entrants into the coasts may mean more fishers with different gear types, coastal communities with different ethnic groups, and a host of other coastal stakeholders all relying on the sharing of limited resources and space. Depending on the relationship, conflicts among traditional users and new entrants may arise. Resource and space competition is going to be permanent source of rivalry among different coastal activities. Such is already the case when tourism development enter an area previously linked to fishing activities. The same human and economic capital, as well as political resources, is diverted from one to the other.

Population Growth and Displacement

More than half of the human population lives within 100 km from coasts, and a large percentage also inhabit regions that are less than 3 m above sea level (Ojeda Zújar et al. 2001). Economic forces and industrial development patterns have shaped these processes. For example, sea trade is still a cheap and efficient mode of transportation for industrial production. Industries tend to establish their factories in coastal areas for this reason. In some cases they also need the vicinity of the sea or large masses of water for refrigeration or waste disposal. Many of the largest cities in the world, including several 'megacities' with more than ten million inhabitants, are also located in coastal areas. The number of coastal megacities is expected to grow more than fourfold from 1975 to 2015 (Ojeda Zújar et al. 2001). One activity that perhaps has grown the most in the twentieth and twenty-first centuries, transforming what was a privilege of upper classes at the beginning of twentieth century into a mass activity, is tourism. The tourism industry has shaped coastal landscapes, transforming mostly uncultivated or unpopulated lands into densely inhabited resorts in only a few decades. This is certainly the case in many areas of the Mediterranean or Caribbean coasts, as well as Southeast Asia (Boissevain and Selwyn 2004). In many cases, coastal tourism has clearly contributed to the displacement of populations to the shore, increasing the complexity in the system because of the new mix of stakeholders and the power that some of these new entrants may exercise in policy-making.

Capture Fisheries and Aquaculture: Dynamics and Scale Issues

One of the most important activities directly impacting ecosystem health is fishing. Around the world, there is a long history of modernization of capture fisheries. Traditional small, wooden, non-mechanized boats are being replaced by motorized boats, equipped with mechanized and powerful gear and modern technology (such as radar and sonar for fish finding). Such modernization, along with the improvement in storage and onboard processing facilities, has enabled fishing to take place further offshore, in deeper areas and with longer duration. Post-harvesting technology has also been developed to accommodate the increasing amount of fish removed from the sea, changes in consumers' taste and preferences for seafood products, and globalization. These transformations affect the dynamics of the system, reinforcing user conflicts because not all the stakeholder groups have the same opportunities to access the new technology or means of production. This may even lead to expelling user groups from their traditional activities due to changes in the market or the overexploitation of resources. This happened to artisanal fishers in many coastal areas or inland waters in Africa after the entrance of more powerful industrial fleets. Two processes in this area deserve special attention. The introduction of trawling and the generalization of aquaculture have introduced systemic changes in the fish chain, as well as in the dynamics and scale of the ecosystem.

The Development of Trawling and Gear Conflicts

From a technological perspective, one of the most prominent changes in fisheries in the last century came with the progressive development of trawlers. This technology has a long history in the West, first appearing in the form of sailing trawlers in the latter Middle Ages, but its evolution during nineteenth and twentieth centuries is closely linked to the adoption of onboard steam and diesel engines. These advances increased trawler capacity to not only extract resources, but also to alter the ocean floor (Roberts 2007). Later in the twentieth century, this technology was extended all over the world. In Southeast Asia, for example, trawls were first used in Manila Bay at the end of the Second World War, before spreading throughout the whole region in the early 1960s (Butcher 2004). Trawling, a non-selective fishing method that results in a mixture of target and non-target species, including juveniles and young fish, is an intensive operation that cannot be sustained. In the Gulf of Thailand, for instance, catches from trawling started to decline from 300 kg of catch per hour in 1961 to only about 50 kg/h in the 1980s, and eventually to 20–30 kg/h in the 1990s (Eiamsa-Ard and Amornchairojkul 1997). The decline in catch per effort was accompanied by changes in catch composition (e.g., more small and short-lived species, including 'trash fish'), after which the 'fishing down the food web' phenomena followed (Pauly and Chuenpagdee 2003). In addition to causing ecosystem effects, including habitat destruction and bycatches, trawling competes directly (in terms of space and target species) with other small-scale and stationary fishing gear, such as pots and traps, creating conflicts among fishers. While many countries have regulations that prevent trawling from operating close to shore, gear conflicts between fishing sectors remain one of the key governability issues related to both ecosystem health and social justice. It is important to remember that the governments of many countries heavily subsidized the development of trawling fleets. Bearing in mind their significant contribution to the world's fisheries production, more support at international and national levels needs to be given to small-scale fisheries, (Chuenpagdee et al. 2006; Chuenpagdee 2011). While it is true that some small-scale fishing practices also damage the ecosystem, namely those that involve the use of cyanide and bomb blasting, the overall impact from this sector is still relatively low in comparison to the large-scale fisheries. Importantly, the contribution of small-scale fisheries to the society, in terms of jobs, income and livelihoods, is a lot greater than that of industrial fisheries (Pauly 2006).

Aquaculture Increasing Diversity, Complexity and Impacts on Ecosystem Health

Another major coastal transformation is induced by aquaculture. Enterprises of different types, sizes and technology have spread around the globe, generating impacts on ecosystem health. Aquaculture development increases system complexity with its presence in areas previously devoted to fishing or other activities. It brings new stakeholders with different images of reality and visions for the future of the coastal area, marginalizing traditional users who have little capital or may not be eager to join (Pascual 2004), and thus increasing the diversity of the system-to-be-governed and the governability challenges.

From an ecosystem perspective, great diversity and complexity can also be found in the aquaculture system, with huge differences between herbivorous, carnivorous and omnivorous species. The latter two depend, to some extent, on compound feed, made up of fishmeal and fish oil, among other ingredients. Carp and shrimp, for instance, consume more than 40 and 18% of the total world production of this aquafeed, respectively (FAO 2009). Problems with this are related to the fact that raw materials used in the production of fishmeal are comprised of small and juvenile fish, some of which may have little to no value but are often important sources of protein for the poor (Tacon and Hasan 2007). Furthermore, they are mostly caught with destructive gear such as bottom trawling and push net. Another concern resulting from aquaculture development is environmental health. For instance, the early development of intensive shrimp culture, mainly tiger prawn (*Penaeus monodon*) in Southeast Asia in the late 1980s involved clear-cutting of large areas of mangrove forests. Large amount of feed is input into ponds that are densely stocked with larvae from hatcheries. The high density culture system and the intensive feeding, coupled with use of fertilizers, biocides and antibiotics for disease control, has compromised their long term viability and created side effects such as high concentrations of nitrogen and organic wastes. Euthrophication due to the excess nutrients may also constitute a risk, because some algae species can be toxic to humans or marine organisms.

Another side effect of aquaculture is the introduction of alien species; typically those that have sufficient market value ensure the culture is profitable. However, all ecosystems are subject to the arrival of alien species, and recent increases in water temperature have fostered this process in the oceans. Ballast water in cargo ships also contributes to such introductions. No less relevant are the intentional or accidental releases from different sources, as in the case of a tropical green alga (*Caulerpa taxifolia*), which was released from the Monaco Oceanographic Museum before 1984 and colonized large areas of the Mediterranean Sea (Meinesz et al. 2001; Valiela 2006). Some of these introductions are controversial. For instance, zebra mussels (*Dreissena polymorpha*) brought to the Great Lakes by ballast water was considered by some to have helped clean up the water. This, however, means that they have filtered out most of the phytoplankton and small zooplankton in the waters, leaving larval and juvenile fish without any food. Debate about the benefits and risks of non-native oysters (*Crassostrea ariakensis*) to the Chesapeake Bay has been on-going since their introduction in the 1990s, as suggested for example by Graczyk et al. (2006).

According to the FAO (2010), aquaculture development since 2000 has mostly followed an ecosystem approach to management principles, and is in accord with the Code of Conduct for Responsible Fisheries. While such trends are not consistent in all regions of the world, it is certainly a step in the right direction that should lead to better overall environmental performance of the aquaculture sector. Such improvement can be attributed to various factors, including technological innovations, lead to improved feed conversion and reduction of fishmeal, and appropriate legislation and governance.

Sea Level Rise and Global Warming: Widening the Scale

Human impacts on ecosystem may be local, but increasingly scientific evidence suggests that they generate problems at a global scale. For instance, when humans transform coastal areas by developing tourist resorts and associated infrastructures, the local impact is coastal alteration, while at a distance they contribute to increasing CO_2 emissions. This results in global warming, which is an increasing challenge to coastal and ocean ecosystems, and may alter large areas in a process that is still largely uncertain. A high rate of ecosystem change is one of the effects of global warming, driving, for instance, the extinction of some species. Corals are among the most affected species, as the rise in temperature results in an incident referred to as 'coral bleaching,' threatening their survival (Hoegh-Guldberg et al. 2007).

Sea level rise, due to, among other factors, global warming and the melting of polar ice (Warrick et al. 1993), may increase the challenges for coastal populations, especially in some island states. Scientific evidence has clearly shown the capacity of humankind in altering natural cycles and accelerating changes that have previously occurred at lower rates, surpassing even the ability of science to predict the consequences. In fact, according to Church et al. (2001) several ice sheets (like the Antarctic and the Greenland) and other hundred thousand non-polar glaciers contain water sufficient to raise significant sea level if they were melted. In addition to islands like the Maldives, many highly populated estuaries may be inundated by these changes, although the rate of these processes is the subject of scientific debate.[4] Other effects of global warming include variations in the acidity of the ocean, wind patterns and hurricanes, all of which pose high risks for many coastal territories. In general, aquatic ecosystems may buffer climatic changes and reduce shocks, but not when the magnitude of change is high. The growing risks and scales of these processes increase governability challenges.

[4] For instance (Mörner 2007), criticizes the assumptions about the current process of sea level rise and their impacts in Maldives or Sri-Lanka.

The level of scale is perhaps the most problematic, especially when actions to alleviate some of the problems imply changes that many countries are not ready to bear. The struggle with the Kyoto protocol is a good illustration of the global governability challenges.

Governability Analyses

Pressure on coastal ecosystems has grown in developed and less developed countries. While it is frequently assumed that the developed world is better equipped with the technology and resources needed to care for the environment and minimize ecological degradation, such assumptions may be questioned. Even with the best available science, a complete understanding of these processes is still lacking. Disasters such as the collapse of the Northern cod stocks in Canada, despite large sums of funding and effort invested on research and fisheries management, can be attributed to insufficient knowledge of the fish species and their life history (Finlayson 1994; Hannesson 1996; Finlayson and McCay 1998). Restoring ecosystems may constitute a huge challenge in many cases, as the decades of attempts in oyster restoration in the Chesapeake Bay illustrate. In the Newfoundland cod fisheries, the ecosystem changes occurring after the collapse have driven some scientists to conclude that recovery would be impossible, and that current fishing practices may also hinder the possibility of this recovery (Davies and Rangeley 2010). It is very difficult and extremely costly, if not altogether impossible, to go back to the previous stages of ecosystems that have been hugely transformed by human activities. Precautionary approaches have been proposed since the 1990s as a general principle in order to avoid risks and irreversible processes (FAO 1996). Yet, practical applications of this principle are still rare (Punt 2006). The challenges are greater for developing and less developed countries coping with these environmental issues because of the scarcity of human and financial resources. Often, short-term goals to secure subsistence and livelihoods of coastal communities overcome long-term considerations for the ecosystem effects of human activities.

Ecosystem health can be considered an outcome of the governing interactions between governing actors and institutions and the social system-to-be-governed in the process of dealing with multiple stressors. It can also be treated as the existing ecosystem conditions, posing conditions and limitation for governability that the governance system needs to deal with. In the latter case, governability is related mainly to the ability of the social system-to-be-governed and the governing system to adapt to the present ecosystem state. The analysis of multiple stressors, particularly fisheries, aquaculture or coastal development, in terms of how they affect the system's governability has already been demonstrated by Chuenpagdee et al. (2008), and is therefore only presented in summary below. The discussion instead focuses on how to improve governability when faced with ecosystem health under stress.

Table 7.2 Characteristics of the fisheries, aquaculture and coastal ecosystems, and their relative governability

	Capture fisheries			Aquaculture			Coastal zones		
	SG	GS	GI	SG	GS	GI	SG	GS	GI
Diversity	M	M	M	M	L	L	H	H	H
Complexity	M	M	M	M	L	L	H	H	H
Dynamics	M	L	L	M	L	L	H	H	H
Scale	M	L	L	L	L	L	H	H	H
Governability	Moderate			High			Low		

Source: Adapted from Chuenpagdee et al. (2008)
SG system-to-be-governed, *GS* governing system, *GI* governing interactions

Governability of Multiple Stressors Affecting Ecosystem Health

In the examination of diversity, complexity, dynamics and scale in relation to capture fisheries, aquaculture and coastal zones, Chuenpagdee et al. (2008) assert that, relatively speaking, these properties are featured most prominently in coastal zones, followed by capture fisheries and aquaculture. In other words, the governability of the aquaculture industry is considered to be highest when compared to capture fisheries and coastal zones (Table 7.2).

As described above, diversity and complexity in capture fisheries arise in both natural and social systems, and more so in tropical areas than in temperate waters. Habitats such as mangroves and coral reefs are rich in biodiversity and high in productivity, and therefore support multi-species fisheries. In terms of trophic interaction, the relationships of these species, in addition to their dependency on the habitats, are generally complex and dynamic. When the balance is upset, the natural fisheries system generally shows signs of degenerating health. On the social side, fisheries stakeholders are numerous and diverse, each with their own complexity and dynamics. They also interact among themselves and with governing institutions in ways that are not easy to understand or predict. Regarding scale, the range and representation of the natural and social boundaries in capture fisheries give rise to governance challenges. In terms of the governing system, institutions dealing with capture fisheries, aquaculture and coastal areas need to acknowledge the intricate properties of the natural and social systems. Frequently, increasing the diversity of actors requires a higher number, or a broader scope, of responsible government agencies. New rules and regulations may be needed to cope with an increased diversity of demands and interests. Principles such as precaution for the natural system and social justice for the human system are the operational foundations for the governing system that lead to improvements in ecosystem health.

Diversity and complexity in coastal zones is generally higher than that found in fisheries. There are more actors involved in multiple livelihood opportunities, and there is a vast array of investment in development for urban and industrial purposes. Depending on the activities, changes in coastal areas take place on a daily basis and at varying scales. The thermodynamics of the ocean may result in long-term change observed through sea level rise. Temperature fluctuation is associated with seasonal variation or with phenomena such as El Niño and La Niña. Coastal development for tourism, and as part of coastal sprawl and urban development, often happens quickly and not necessarily with proper planning. Coastal infrastructure is primarily developed on an ad-hoc basis to support rising demands. They rarely account for possible changes brought about by coastal hazards or global warming. Conflicts among various user groups in coastal zones are largely due to direct competition for space, resources and economic gains. These can, however, also be accelerated by many causes, for example, when the various governing institutions dealing with coastal zone issues lack clarity in their vision and goals. The various governing agencies that deal with overlapping issues need increased coordination, potentially posing challenges to their traditional operations. The recent hype in stakeholder participation in fisheries and coastal management adds another complication to the governance of coastal areas. In addition, the incompatibility of the mandates, goals and concrete policies of the various agencies causes uncertainty and confusion among stakeholders, and reduces the governability of the coastal zone. Improving ecosystem health requires efforts from the governing system to foster interactions among coastal stakeholders through transparent and accountable processes. It also requires a rethinking of the governing institutions. They must reconsider, for instance, the scale issues affecting governability.

Compared to capture fisheries and coastal zones, the overall level of governability of aquaculture is the highest. Aquaculture can be diverse, complex, dynamic and of varying scales. The extent of each of these properties depends on species culture, types of operation and areas where farming takes place. The number of actors and their multiplicity is generally smaller in comparison to the other two systems. While coastal aquaculture is highly governable, there are some concerns about aquaculture and ecosystem health that qualify them as wicked problems requiring interventions at global and national levels. For example, the 'Good Aquaculture Practices' (or GAP) program promoted by the FAO aims to regulate and standardize farming operations worldwide for food safety and environmental sustainability. Ethical considerations are needed when discussing the contributions and threats to food security associated with farming issue, such as the use of low value fish as raw materials in fishmeal, the conversion of mangrove forests and other land areas, and the space competition between fishing and aquaculture that takes place in coastal areas. Finally, similar to large-scale capture fisheries, aquaculture operations often receive high levels of subsidies in comparison to the small-scale fishing sector, adding yet another layer of conflicts and justice issues among fisheries stakeholders.

Improving Governability of Ecosystem Under Stress

The focus on the 'poor' state of the ecosystem, in which way it is defined, does not suggest that governability of an ecosystem that is in good health can be neglected. Rather, it reflects two observations: that the majority of the world's marine ecosystem is under stress, and that governability of such systems is truly a wicked problem. Using the framework suggested by Jentoft and Chuenpagdee (2009), we begin by looking at the natural ecosystem under stress for ways to improve its governability. It has, however, been recognized that improving ecosystem health requires long-term planning and the commitment of human and financial resources. The examination of what can be done at the social system-to-be-governed is therefore critical. As has previously been mentioned, social, cultural and economic diversity is great among people whose livelihoods depend on healthy fisheries and coastal ecosystems (Pascual Fernández 1991; Pascual-Fernández et al. 2005). The level of complexity and dynamism in these social systems constitute both challenges and opportunities for governance. Communities that have strong social capital and a traditional network of support tend to cope well in stressful situations and would therefore likely be able to find adaptive mechanisms to deal with challenges. Communities with complex and multiple livelihood strategies that draw from a wide range of resources can rotate to using other resources or can acquire food and income by other means while waiting for resource recovery. The role of government agencies is then to develop policies that encourage livelihood diversification. These could include appropriate incentive schemes and capacity building and training programs. However, such policy developments have to be based on a thorough understanding of the dynamics of the social systems and the intra- and inter-sectoral relationships. The latter is particularly important as the tendency for conflicts among numerous fisheries and coastal stakeholders is high. The governing system needs to become familiar with the analysis of stakeholders, in terms of their resource dependency and power relationships, to devise suitable interventions. The increased diversity in the coastal zone begs for a greater level of interaction between stakeholders and governing institutions, as well as a variation in types of interactions between government agencies and fisheries-coastal actors.

Alternative livelihoods for small-scale fishing communities are among the most popular interventions, but are difficult to accomplish without an understanding of the diversity and complexity of fishers' socio-economic conditions and the cultural and traditional importance of fishing activities to their way of life. As many fishers would profess, fishing is not just a livelihood but a preferred lifestyle (Onyango 2011). Fishers have their reasons to be reluctant about leaving the fishing occupation or to contemplate doing other things when fishing is not profitable. Some argue, for example, that fishing is the family tradition, while others may prefer it for the freedom it offers. In these instances, non-fishing activities outside the fishing season or during area or seasonal closures should be promoted to enable food and income supplement. Those owning lands are likely to already be doing this by engaging in, for example, vegetable growing or small-scale animal farming. Some support should then be given to help these fishers maintain their ability to derive income and food sources from non-fishing activities. Examples of these efforts are securing market access or controlling prices for the sale of their crops, offering training in the making of value-added fisheries and non-fisheries products, and providing low interest loans to help them start small businesses.

Demands for the governing system to cope with the diversity, complexity, dynamics and scale issues affecting ecosystem health are high. For instance, adapting to the continuously evolving situation in the marine ecosystem and coastal zones requires the governing system to be capable of re-designing institutional arrangements appropriate for new circumstances. Increasing actors and stakeholders in the coastal areas means a higher number or a broader scope of government agencies to adjust existing rules or create new ones. The capacity of the agencies to cope with the new conditions may be compromised, however, by a mismatch in the scale of the problems they confront and their competencies. The nature, level and variation of interactions of fisheries and coastal actors with the governing institutions would be expected to expand as a consequence of the increased diversity. Similarly, the multiplicity of stakeholders and the increased complexity in the social system-to-be-governed requires reconsideration of the types of information and the methods of sharing and dissemination. Information sharing, for instance, can enhance trust when considered adequate. It can also induce suspicion if it is interpreted as exploitative. There are no recipes for perfect interactions, but some emphasis should be placed on partnership building as a way to enhance compliance and avoid conflict and disagreement among stakeholders. The way in which effective partnerships are built depends on the characteristics of the system, especially the scale extent. It may be difficult to encourage positive and direct interaction between groups that are spatially widespread. Time issues may also have an impact on the interactions, as the period required to address the challenges may discourage collective action or increase difficulties for effective learning and feedback processes.

Conclusion

Challenges posed by the current health of the marine and ocean ecosystems around the world are overwhelming. The human capacity to alter these systems through direct physical modification, as happened in coastal urbanization or bottom trawling is enormous. The consequences of these processes are immediate and long-term, both begging for effective policy measures and appropriate governing interventions.

Many of the ecosystem health concerns discussed in this chapter follow the wicked problem definition. They are difficult to define and delineate from other concerns. Their specificities are difficult to grasp, and there is no simple solution. This is due in large part to the fact that any intervention implies major changes in the social system-to-be-governed and the governing system. Responses to these problems need collective efforts that may be beyond our present capacity to interact adequately in reaching agreements and finding solutions. Nevertheless, the heightened awareness of the poor state of the world's marine and ocean ecosystems at global, national, and local levels, and the ongoing attempts to address it are encouraging signs. The analytical lens offered by the interactive governance and the governability approach, contribute to enhancing our understanding of the factors that affect ecosystem health, as well as the areas and types of interventions that may help address the problem. Our analysis shows that the diversity, complexity, dynamics and scale of the key challenges affecting ecosystem health, such as those generated from coastal zone development, intensifying fisheries and aquaculture, and climate change, can be found in the system-to-be-governed, the governing system and the governing interactions. Likewise, the ways to improve governability can also be found in all three systems.

References

Armitage, D. R., Berkes, F., Doubleday, N. (2007). Adaptive co-management: Collaboration, learning, and multi-level governance (Sustainability and the environment). Vancouver: UBC Press.

Berkes, F., & Folke, C. (Eds.). (2000). Linking social and ecological systems: management practices and social mechanisms for building resilience (1st paperback ed.). Cambridge: Cambridge University Press.

Boissevain, J., & Selwyn, T. (Eds.). (2004). Contesting the foreshore: Tourism, society and politics on the coast (MARE series). Amsterdam: Amsterdam University Press.

Butcher, J. G. (2004). The closing of the frontier: A history of the marine fisheries of Southeast Asia c. 1850–2000. Singapore: Institute of Southeast Asian Studies.

CBD. (1993 [1992]). Convention on biological diversity. United Nations — Treaty Series. Rio de Janeiro-New York: United Nations.

Chuenpagdee, R. (Ed.). (2011). World small-scale fisheries: Contemporary visions. Delft: Eburon.

Chuenpagdee, R., Degnbol, P., Bavinck, M., Jentoft, S., Johnson, D., Pullin, R. et al. (2005). Challenges and concerns in capture fisheries and aquaculture. In J. Kooiman, M. Bavinck, S. Jentoft, R. Pullin (Eds.), Fish for life: Interactive governance for fisheries (pp. 25–37). Amsterdam: Amsterdam University Press.

Chuenpagdee, R., Liguori, L., Palomares, M.L.D., Pauly, D. (2006). Bottom-up, global estimates of small-scale marine fisheries catches. Fisheries Centre Research Reports, 14(8), 1–94.

Chuenpagdee, R., Kooiman, J., Pullin, R.S.V. (2008). Assessing Governability in Capture Fisheries, Aquaculture and Coastal Zones. The Journal of Transdisciplinary Environmental Studies, 7(1), 1–20.

Church, J.A., Gregory, J.M., John, H.S., Karl, K.T., Steve, A.T. (2001). Sea level change. In J.H. Steele, S.A. Thorpe, K.K. Turekian (Eds.), Encyclopedia of ocean sciences (pp. 2599–2604). Oxford: Academic Press.

Commonwealth of Australian Law. (1999). Environment Protection and Biodiversity Conservation Act 1999. Act Compilation (current) – C2010C00505. Vol. 2. Australia: Australian Government, Attorney-General's Department. Retrieved from http://www.comlaw.gov.au/ComLaw/Legislation/ActCompilation1.nsf/0/93CECCC69FFF754BCA2577660 0173C09?OpenDocument

Davies, R.W.D., & Rangeley, R. (2010). Banking on cod: Exploring economic incentives for recovering Grand Banks and North Sea cod fisheries. Marine Policy, 34(1), 92–98.

Eiamsa-Ard, M., & Amornchairojkul, S. (1997). The marine fisheries of Thailand, with emphasis on the Gulf of Thailand trawl fishery. In G. Silvestre, & D. Pauly (Eds.), Status and management of tropical coastal fisheries in Asia (pp. 85–95). Philippines: Asian Development Bank & International Center for Living Aquatic Resources Management (ICLARM).

FAO. (1996). Precautionary approach to capture fisheries and species introductions. FAO Technical Guidelines for Responsible Fisheries 2. Rome: FAO.

FAO. (2009). The state of world fisheries and aquaculture 2008. Rome: FAO.

FAO. (2010). The state of world fisheries and aquaculture 2010. Rome: FAO.

Finlayson, A.C. (1994). Fishing for truth: A sociological analysis of Northern Cod stock assessments from 1977 to 1990. St. John's: Institute of Social and Economic Research, University of Newfoundland.

Finlayson, A.C., & McCay, B.J. (1998). Crossing the threshold of ecosystem resilience: The commercial extinction of northern cod. In F. Berkes, & C. Folke (Eds.), Linking social and ecological systems: Management practices and social mechanisms for building resilience (pp. 311–337). Cambridge: Cambridge University Press.

Graczyk, T.K., Girouard, A.S., Tamang, L., Napier, S.P., Schwab, K.J. (2006). Recovery, bioaccumulation, and inactivation of human waterborne pathogens by the Chesapeake Bay nonnative oyster, Crassostrea ariakensis. Applied and Environmental Microbiology, 72(5), 3390–3395.

Hannesson, R. (1996). Fisheries mismanagement: The case of North Atlantic Cod. Oxford: Fishing News Books.

Hoegh-Guldberg, O., Mumby, P.J., Hooten, A.J., Steneck, R.S., Greenfield, P., Gomez, E. et al. (2007). Coral reefs under rapid climate change and ocean acidification. Science, 318(5857), 1737–1742.

Jentoft, S., & Chuenpagdee, R. (2009). Fisheries and coastal governance as a wicked problem. Marine Policy, 33(4), 553–560.

Jentoft, S., Chuenpagdee, R., Pascual-Fernandez, J.J. (2011). What are MPAs for: On goal formation and displacement. Ocean & Coastal Management, 54, 75–83.

Kooiman, J., Bavinck, M., Jentoft, S., Pullin, R. (Eds.). (2005). Fish for life: Interactive governance for fisheries. Amsterdam: Amsterdam University Press.

Kulbicki, M. (2005). Aquatic ecologies. In J. Kooiman, M. Bavinck, S. Jentoft, R. Pullin (Eds.), Fish for life: Interactive governance for fisheries (pp. 45–69). Amsterdam: Amsterdam University Press.

Lundin, C.G., & Linden, O. (1993). Coastal ecosystems – Attempts to manage a threatened resource. Ambio, 22(7), 468–473.

Meinesz, A., Belsher, T., Thibaut, T., Antolic, B., Mustapha, K.B., Boudouresque, C.F. et al. (2001). The introduced green alga Caulerpa taxifolia continues to spread in the Mediterranean. Biological Invasions, 3(2), 201–210.

Mörner, N.A. (2007). Sea Level changes and tsunamis, environmental stress and migration overseas: The case of the Maldives and Sri Lanka. Internationales Asienforum, 38(3–4), 353–374.

Ojeda Zújar, J., Vallejo Villalta, I., González Romero, G. (2001). El acceso de los estados al mar: Áreas costeras y cuencas marinas. In J.L. Suárez de Vivero (Ed.), Los océanos: medio ambiente, recursos y políticas marinas (1ª ed., pp. 41–67, Vol. 25). Barcelona: Ediciones del Serbal.

Onyango, P.O. (2011). Occupation of last resort? Small-scale fishing in Lake Victoria, Tanzania. In S. Jentoft, & A.H. Eide (Eds.), Poverty mosaics: realities and prospects in small-scale fisheries (pp. 97–124). Dordrecht: Springer.

Pascual, J.J. (2004). Littoral fishermen, aquaculture and tourism in the Canary Islands: Attitudes and economic strategies. In J. Boissevain, & T. Selwyn (Eds.), Contesting the foreshore: Tourism, society and politics on the coast (pp. 61–82). Amsterdam: Amsterdam University Press.

Pascual Fernández, J. (1991). Antropología marítima: Historia, ecología, organización social y cambio económico entre los pescadores. Madrid: Ministerio de Agricultura, Pesca y Alimentación.

Pascual-Fernández, J.J., Frangoudes, K., Williams, S.B. (2005). Local institutions. In J. Kooiman, M. Bavinck, S. Jentoft, R. Pullin (Eds.), Fish for life: interactive governance for fisheries (pp. 153–172). Amsterdam: Amsterdam University Press.

Pauly, D. (2006). Major trends in small-scale marine fisheries, with emphasis on developing countries, and some implications for the social sciences. Maritime Studies (MAST), 4(2), 7–22.

Pauly, D., & Chuenpagdee, R. (2003). Development of fisheries in the Gulf of Thailand Large Marine Ecosystem: Analysis of an unplanned experiment. In G. Hempel, & K. Sherman (Eds.), Large marine ecosystems of the world: Trends in exploitation, protection, and research (pp. 337–354). Amsterdam: Elsevier.

Punt, A.E. (2006). The FAO precautionary approach after almost 10 years: Have we progressed towards implementing simulation-tested feedback-control management systems for fisheries management? Natural Resource Modeling, 19(4), 441–464.

Rindos, D. (1984). The origins of agriculture: An evolutionary perspective. Orlando: Academic Press.

Roberts, C. (2007). The unnatural history of the sea. Washington, DC: Island Press/Shearwater Books.
Stepp, J.R., Jones, E.C., Pavao-Zuckerman, M., Casagrande, D., Zarger, R.K. (2003). Remarkable properties of human ecosystems. Conservation Ecology, 7(3), 11.
Tacon, A.G.J., & Hasan, M.R. (2007). Global synthesis of feeds and nutrients for sustainable aquaculture development. In M.R. Hasan, T. Hecht, S.S. De Silva & A.G.J. Tacon (Eds.), Study and analysis of feeds and fertilizers for sustainable aquaculture development (pp. 3–17). Rome: FAO.
Valiela, I. (2006). Global coastal change. Malden/Oxford: Blackwell Publications.
Vitousek, P.M., Mooney, H.A., Lubchenco, J., Melillo, J.M. (1997). Human domination of Earth's ecosystems. Science, 277(5325), 494–499.
Warrick, R.A., Barrow, E.M. & Wigley, T.M.L. (Eds.). (1993). Climate sea level change: Observations, projections and implications. Cambridge: Cambridge University.

Part III
Thematic Applications

Chapter 8
Applying the Governability Concept in Fisheries – Explorations from South Asia

Maarten Bavinck and Jan Kooiman

Abstract This chapter investigates the variations in governability that occur in fisheries systems. It builds upon the notion that diversity, complexity, dynamics, and scale affect the performance of societal systems profoundly, and that these effects emerge at the level of their three components. Variations in the governability of systems-to-be-governed are examined through a chain approach. Theories of legal pluralism, institutional nestedness and adaptability are subsequently applied to assess the governability of governing systems. Finally, governing interactions are considered through the lens of three ideal-typical modes – self-governance, hierarchical governance and co-governance. The many variations of governability are illustrated with cases from the capture fisheries of South India. The central message is that connections between the components of a fisheries system matter and that a better match may result in higher levels of governability.

Keywords Governability • Fish chain • Legal pluralism • Nestedness • Institutional dynamics

M. Bavinck (✉)
Department of Human Geography, Planning and International Development Studies, University of Amsterdam, Amsterdam, The Netherlands
e-mail: j.m.bavinck@uva.nl

J. Kooiman
Centre for Maritime Research, University of Amsterdam, Amsterdam, The Netherlands
e-mail: jkooiman@xs4all.nl

M. Bavinck et al. (eds.), *Governability of Fisheries and Aquaculture: Theory and Applications*, MARE Publication Series 7, DOI 10.1007/978-94-007-6107-0_8,

Introduction

Fisheries differ in many respects, natural and social, between and within countries. It is therefore quite logical to expect that these differences affect the opportunities available for successful steering, or governance. Similarly, some styles of governance may be more conducive to addressing particular circumstances than others. The challenge, as Ostrom (2007) points out, is to move beyond policy panaceas and develop a fine-tuned diagnostic approach for problems and potentialities in linked social-ecological systems. Such systems are invariably complex, nonlinear, multi-scale and changing (cf. Berkes et al. 2003). They require appropriate governance solutions. After all, "many variables affect the patterns of interactions and outcomes observed in empirical studies" (Ostrom 2007, 15181).

Interactive governance theory suggests that the governability of any societal system, such as fisheries, depends on the condition of its three parts: the system-to-be-governed, the governing system, and the realm of governing interactions (Kooiman 2008). The system-to-be-governed consists of a social and a natural sub-system, which are joined together by the act of fishing. After all, each fisher is part of an economic and social world, which drives him to fish the way he does, and also a natural world (including an ecosystem) that affects and is affected by him. Cumulatively, a collection of fishers within a particular sub-sector, or *métier*, fills up a specific niche in the natural sub-system to make up a system-to-be-governed. Various systems-to-be-governed possess different governance requirements as well as different levels of governability. Interactive governance theory ascribes a special influence to four features that affect both natural and social sub-systems and are present in different degrees – diversity, complexity, dynamics, and scale.

This chapter explores the consequences of diversity, complexity and dynamics for the governability of fisheries systems, and also considers the implications of geographical scale.[1] To enliven and illustrate what might easily become an abstract discussion, we provide examples at each step of the analysis. Almost all of these examples derive from field studies undertaken by the first author in the state of Tamil Nadu, India over a period of 15 years (1995–2010). The fact that one geographical region, and one 'collection' of fisheries, is capable of delivering a full set of examples of governability is noteworthy and indicative of diversity at a national scale level.[2] We also note that the examples connect different

[1] Gibson et al. (1998) provide a useful overview of the use of scale in the social sciences. They define scale simply as "the dimension used in any effort to measure a phenomenon", and distinguish three basic types: space, time and quantity (Gibson et al. 1998, 6). In this chapter we highlight the implications of spatial scale. Johnson (2006) discusses facets of scale with regard to fisheries.

[2] The fact that one setting is capable of delivering illustrations of almost the full set of governability options raises important questions regarding societal coherence. Is Tamil Nadu a particularly chaotic fisheries setting? Could other settings provide a similarly broad range of examples? We are inclined to believe that most fisheries settings, if investigated carefully, actually contain a large variety of governability arrangements, not dissimilar to Tamil Nadu. We must also point out, however, that the fisheries of Tamil Nadu may actually be more structured than this array of examples suggest.

scales internal to and beyond the case, and highlight more social and economic characteristics than natural science features. Moreover, all of these instances can actually be supplemented and debated. This fits in the interactive governance approach of not treating systems as objective facts lying 'out there', but instead as heuristic tools (Jentoft 2007, 361). In this line of thought, Kooiman (2008, 174) points out that: "What a system looks like, how it can be broken down, and what its boundaries and other qualities might be, depends on the perspectives of the observers."

Governability and Systems-to-Be-Governed

There are two basic ways of considering what constitutes a system-to-be-governed in fisheries studies. The first takes a spatial, or territorial, approach, and is concerned with defining system identities and boundaries. The ecosystem approach is a prominent exponent of this spatial orientation (Garcia et al. 2003). The second focuses on what have become known as value chains (Gereffi et al. 2005), or, in our field of interest, fish chains. The primary interest here is in identifying and tracing connections as they occur from an ecosystem to the consumer. The following discussion is based on the fish chain approach.

Diversity, Complexity, Dynamics and Scale of Fish Chains

The *Fish for Life* volume (Kooiman et al. 2005) presents the system-to-be-governed as a bundle of fish chains running from an ecosystem to the consumer's plate. Fish chains constitute the basic unit of any fisheries system and are therefore the object of governance activity. Within any fisheries system there are generally numerous fish chains, organized around various target species or seafood markets. These chains involve a variety of primary actors, such as the fish, fishers, processors, traders, transporters, wholesalers and retailers. Each category can of course be subdivided further (for example, the category 'fish' consists of numerous marketable species, whereas fishers can be divided into owners, skippers and crew members). Moreover, each fish chain involves a range of secondary actors – other species in an ecosystem and people who create the conditions upon which the fish chain operates. The latter include net makers, engine shops, and ship builders, right down to the road workers paving the routes that transport fish from landing centers to markets.

Individual fish chains vary with regard to their diversity, complexity, dynamics, and scale. The same can be said for the bundles of fish chains that constitute a fisheries system. Although diversity, complexity and dynamics exist in a continuum, we choose to distinguish between two basic values: high and low. We also discuss some of the implications of spatial scale.

Kooiman argues that *diversity* "calls attention to the specific and varying qualities of actors and other entities. [...] It is a source of creation and innovation, but also carries the danger of disintegration" (2008, 176). For the natural system, diversity concerns features such as biological heterogeneity, species abundance, habitats and ecosystem health. Notably, the composition of the species assemblage in these ecosystems, the species richness and the status as rare, endangered, endemic or keystone species have to be identified. For the socio-economic system, the issues of diversity pertain to the composition of stakeholder groups in terms of their demographic profile, their organization, vested interests, property and access rights, and political orientation. Their capacity to influence the way the system works is determined by the power they exercise, which may or may not enable governability. Thus, what constitutes power (the capacity to ignore, exclude, marginalize or mobilize other groups) is a central issue.

According to scholars in the field of interactive governance, it is only through replication that systems-to-be-governed of high diversity can be adequately governed. They follow Ashby's Law of Requisite Variety, which argues that only variety in governance can 'destroy' societal mimesis. Wilden (1987, 192) has expanded on this idea, adding a principle of 'requisite diversity in representation'. He argues that it is not the diversity of the whole system-as-a-whole that must be represented; rather a governing system must be able to represent the basic types of variety found in the system-to-be-governed. Low diversity of a system-to-be-governed can be met by simple, unitarian styles of governance, whereas high diversity necessitates varied approaches.

Complexity is related first of all to how the natural system interconnects, how species interact, and how particular habitats (such as mangroves and coral reefs) contribute to the productivity of the system as a whole. This is the field of systems ecology. The complexity of the socio-economic system on the interactions among the actors involved, their interdependency, collaboration and integration, which rests partly on their "organic solidarity" (Durkheim 1964), and the normative ties that exist among them. Governance of the combined socio-ecological system is the topic of much recent scholarly activity (cf. Berkes et al. 2003; Folke 2006). According to Kooiman (2008, 176), "complexity invites examination of societal structures, interdependencies and interrelations and is a condition for combining interdependencies. The difficulty is how to reduce it in an effective and responsible manner." Kooiman (2008) highlights the problems involved in understanding complexity – in building an appropriate image – and thereby also in finding a suitable response.

Generally speaking, Kooiman (2008, 176) argues that *dynamics* "create the potential for change, but can have disruptive consequences." The focus is on alterations that occur over time, in the short as well as in the long term. They may involve both internal and external factors. Climate change (e.g. temperature, weather patterns, storm frequency and intensity) and invasive species are examples of natural system dynamics. Socio-economic system dynamics include alterations in stakeholder compositions, relationships, and interactions. Conflict and power differentials play an important role. The relationship between the human and the natural sub-system includes changes in use patterns or variations in the levels of extraction.

In acknowledging that uncertainty and risk are primary characteristics of human-in-nature systems, it becomes necessary to revise our understanding of their governability. As Mahon et al. (2008, 104) point out, human-in-nature systems are not easily controllable. In other words, they face important limits to their governability. Jentoft (2007, 363) notes that the system-to-be-governed sometimes "alters rapidly, unpredictably, irreversibly," and that "for those affected, the result is uncertainty and surprise." In terms of governability, the system-to-be-governed may alter rapidly between one state and the next, sometimes appearing highly governable and subsequently reverting to a state of low governability. Jentoft (2007, 364) therefore argues for a response that emphasizes 'flexibility', whereby "flexibility is [...] about the governing system's aptitude in adapting promptly to system dynamics and change." Governors' inability to respond flexibly to the dynamics of the system-to-be-governed in question will affect the appropriateness of their responses and, in turn, negatively influence governability.

Fish chains clearly operate at various *spatial scales* (see Kooiman et al. 2005, Part 2). While some ecosystems are geographically small, and some species have only limited spatial range, others are extremely large (see, for example, the global migratory patterns of high-seas tuna and the large marine ecosystems distinguished by the United Nations Environment Programme and others). The same is true of fisher operations and of markets. Whereas some fish chains are based on local fisheries catering to local markets, others reach out and operate on various scale levels. In linking the local to the national and international, fish chains often span the jurisdictions of territorially based governing systems, creating multitudinous interactions as well as tensions and ambiguities. We will highlight some scale effects below, without pretending to be comprehensive.

Feature Effects on Systems-to-Be-Governed: Examples from South Asia

The six fish chains included in the cells of Table 8.1 exemplify lower or higher values with regard to diversity, complexity and dynamics. In the following section, we briefly discuss each example, taking along scale effects, and providing hints of possible implications for governance.

Case 1: Lower diversity. In the mid-1990s, fish merchants introduced a simple hoop net for gathering *sea snails* (species unknown) to the small-scale fisheries of the Coromandel Coast in Tamil Nadu (Bavinck 1996, 1998). This métier required few fishing skills and could be carried out by small inshore fishing units, all of which utilize the same fishing methods. Produce was shipped to markets in the Middle East, thereby integrating the *métier* into a global value chain. The region's fishers, however, protested against the use of the net, arguing that it interfered in the food web and would contribute to resource depletion. The operation of this fishery thereby became a governance issue at the regional level.

In governance terms the challenge was relatively simple. As the same technique was being introduced to a larger region, the governing system could have decided on a common measure to be implemented centrally, prohibiting, regulating or leaving free the use of the net. As it turned out, however, the Fisheries Department, not

Table 8.1 Case studies governability and system-to-be-governed

Intensity/features	Diversity	Complexity	Dynamics
Lower	1. Snail net fishery	3. Mackerel fishery	5. Chank fishery
Higher	2. Mixed fishery	4. Tuna fishery	6. Shrimp fishery

being convinced of the danger of the fishery, did nothing at all. A series of fisher village councils prohibited the use of the net in their waters, while other councils refused to take measures. This resulted in conflicts between fishing villages, which could only be solved through police action (Bavinck 1998).

Case 2: Higher diversity. The Palk Bay in southern Tamil Nadu is recognized internationally as a *biodiversity hotspot* (Bavinck and Vivekanandan 2011). However, the area also harbors a large and socially diverse fishing population. Not only are there many castes and religions involved, but fishers also engage in a large variety of different métiers, varying from beach seining to trawling, gillnetting and diving, as well as the manual collection of, for instance, sea grass. These fisheries cater to local, national and international markets. The government of India, recognizing the ecological importance of the region, decided to implement a National Park (1985) and a Biosphere Reserve (1989) in the Gulf of Mannar. Fishers, however, have been protesting the various restrictions that are being implemented, creating new governance challenges.

Jentoft (2007) argues that a system-to-be-governed that is characterized by high diversity requires a governance approach that is 'sensitive' to difference. One possibility would be to opt for maximum decentralization and self-governance.

Case 3: Lower complexity. Indian mackerel (*Rastrelliger kanagurta spp.*) is one of the mainstays of the small-scale fishing sector (Bavinck 2001, 94ff). Most fishing households in the Coromandel Coast region possess several mackerel nets and operate them throughout the year. Although pelagic species such as mackerel have a large geographical range, the human dimensions of the fish chain in this case are short and straightforward: fisher women or small traders generally sell mackerel to consumers on the local market at relatively low prices. The mackerel fishery therefore plays a key role in the food security of the local agricultural population, and presents a societal issue for governance.

The fish chain is compact and of low complexity. No urgent resource problems manifest themselves at present, and the coordination issues that do arise are generally solved by individual market actors. Nevertheless, fisher councils keep watch over the fairness of the auctions in which fishers sell their catches. With limited economic interests and well-established procedures, governance activity is typically low key.

Case 4: Higher complexity. The *tuna fishery*, which has arisen in southern India since 1995, is linked into a complex international fish chain with a large number of agents and institutions playing a role. The Indian government has been promoting the conversion of shrimp trawlers into gillnetters and long liners, and 25 boat owners in Chennai have actually moved into the tuna fishery (field notes MB 2008). Their catches are destined for export, mainly to the Far East, and quality control is an important consideration.

Contrary to the mackerel fishery discussed in Case 3, the tuna fishery chain is long, of high monetary value, and very intricate. The governance challenges are diffuse and of a serious nature. On the one hand, international agencies are pushing for a management regime that prevents overfishing of tuna stock. On the other, agencies at various scale levels keep watch over hygiene and food quality. Coordination of governance effort between actors is an important concern.

Case 5: Lower dynamics. The chank (*Turbinella pyrum*) fishery of the Gulf of Mannar dates back to pre-colonial times (Hornell 1914), and is linked mainly to North Indian ornamental markets. It is one of the few marine fisheries to have been regulated (and taxed) by the colonial government. Nowadays, divers, who own no more than a mask and a set of fins, operate this fishery from small boats, offering their catches to traders waiting on the shore. The main dynamic in this fishery is market demand and the resulting pressure on stocks.

This fish chain requires governance efforts of a stable kind, primarily directed at curbing fishing effort within ecological limits. In parallel to the pearl fishery that took place along this coast, the Tamil Nadu government historically issued licenses for chank diving and monitored the fishing closely. This practice was, however, recently discontinued for, what were in all likelihood, financial reasons. Some village councils are currently putting limits on diving technology, such as by prohibiting the use of scuba diving equipment (van Haastrecht and Schaap 2003).

Case 6: Higher dynamics. The shrimp fishery of India is an example of high dynamics, precipitated by a sudden integration, from the 1960s onwards, into a high-value, international market (Kurien 1978). The trawl fishery now counts almost 30,000 craft (CMFRI 2005) and still focuses largely on shrimp. The movement of trawler fleets from one Indian state to another has caused significant social tensions, as has their incursion into the fishing grounds operated by small-scale fisher populations. These dynamics are also influenced by rising fuel prices and the advent of shrimp aquaculture.

Governance theory argues that a system-to-be-governed characterized by high dynamics and unpredictability can be successfully addressed only when the governing system is flexible and organized for learning. As Mahon et al. (2005, 369–70) point out, "Dealing with unpredictable systems is like moving through uncharted territory. The only way to function in such systems is to constantly monitor where one has been and where one has reached, and then to reflect on the progress and to move forward guided by the learning." In reality, the Fisheries Department has followed a reactive approach by trying, for the most part, to exercise damage control. Learning still has a low priority in departmental policy.

Implications for Governability

Each of thc cases discussed above includes a comment about what constitutes an urgent governance issue or challenge, and what a 'successful governance intervention' might therefore look like. Biodiversity concerns (Case 2 above) first emerged in the realms of international science and environmental politics, and then impinged on the fisheries of the Gulf of Mannar. Concerns about the snail net (Case 1), however, arose within the localized fishing societies of the Coromandel Coast, resulting in fisher (not governmental) action.

The mackerel fishery of the Coromandel Coast (Case 3) does not present serious governability challenges to any governor at present and is largely taken care of by regular market mechanisms. The simplicity of that fish chain stands in contrast to the complex tuna fishery (Case 4), which is the scene of globalized governance efforts, ranging in location from the high seas to harbors and long liners, processing industries, trade and quality control. Finally, the slow dynamics of the localized chank fishery (Case 5) stand out against the hectic of shrimp trawling along the Indian coasts (Case 6). The latter can be handled only at a national or state level, such as through the imposition of a uniform closed season (Bavinck et al. 2008). Furthermore, due to its volatility and many externalities, the trawl fishery also requires constant vigilance.

The intention of the illustrative examples in this section of the paper has been to persuade the reader that the four features – diversity, complexity, dynamics, and spatial scale – do permeate and affect systems-to-be-governed, and, secondly, pose varying challenges for governance. Not only does each empirical situation provide different combinations of features, but the intensities by which these features manifest themselves also vary. Not every fishery is equally shaped by dynamics, nor is it equally diverse or complex. For governance action to be successful, governing systems and governing interactions must take on different shapes and styles of functioning.

Before closing this section, one point still needs addressing. We mentioned above that fisheries systems operating in a specific setting are frequently made up not of single, but of multiple fish chains, or 'bundles'. The individual chains that make up a bundle intersect – and interconnect – in various ways. The most obvious intersection occurs in the persons of the fishers (or traders, processors, etc.) who participate in multiple fisheries (and are affected by multiple governance arrangements). The composite 'bundle' is characterized by greater diversity and complexity than each single chain, with interferences between chains affecting their dynamics in, at times, unpredictable ways. This creates substantial challenges for governors aiming to create a holistic approach, because the various chains function at different scale levels. Their goals may only be achieved through linking with governing actors at other scales.

Governability and the Governing System

Governing activity involves more than finding an appropriate response to a particular societal issue. It is also about getting the structure and performance of a governing system right. This section considers governability of a fisheries system from the viewpoint of a governing system, highlighting again the effects of diversity, complexity, dynamics and scale. Kooiman and Chuenpagdee (2005, 328) explain: "The governing system and the framework of actors engaged in governing are often as diverse, complex, and dynamic as the system to be governed." These features greatly influence its potential performance.

Theorists argue that a governing system makes use of elements – images, instruments and willpower, or action – for its operation (Kooiman 2003). Ideally, the governing system also possesses various 'levels' or 'orders' of functioning. Managers at the ground level make decisions about routine and urgent issues, applying the tools that they have available in accordance with their estimation of risks and opportunities. At a secondary, institutional level, governors reflect on and adapt the rules and organizational structures at their disposal. Thirdly, at a meta-governance level, discussions take place on the basic values, principles and norms that should drive governance activity (Kooiman and Jentoft 2009).

In the pages below, we discuss the diversity, complexity, dynamics and scale of governing system under three headings. Diversity is considered from the perspective of legal pluralism, which argues that societal systems often possess multiple sources of law. The degree of legal pluralism in a fisheries system impacts its governability. We then present complexity and scale from the viewpoint of institutional nestedness. Finally, we consider the dynamics of a governing system through the notion of institutional adaptability and learning.

Diversities in Legal Pluralism

Governors are the people or organizations that address societal problems and opportunities. Our approach allows for the inclusion of part-time or incidental governors in addition to full-time ones. It suggests the existence of informal governors – without uniforms, offices and business cards – operating next to official governors with formal mandates. Additionally, it notes that community and market institutions coexist with governmental frameworks. All these governors have opinions about and undertake action with regard to societal problems and opportunities. Together, these actors constitute the diversity of the governing system.

Governors are of course not completely free in their choice of action. As pointed out before, governing actors are constrained or enabled by conditions in their social and natural environments. These conditions include a range of institutions and, more specifically, the rules of the game (North 1991). Such rules generally do not come singly. Instead they are coupled in sets that can be analyzed as legal systems. Contrary to a formal definition of law, such legal systems may be anchored outside of government in realms such as religion, ethnicity or professional life.

Scholars in the field of legal pluralism argue that most societies are characterized by some form of legal pluralism (Merry 1988; von Benda-Beckmann 2002), i.e. the incidence of multiple legal systems applying to similar situations (Vanderlinden 1971). This is certainly the case for fisheries. There are numerous descriptions of strong, non-state legal systems governing fisheries that are connected to specific ethnic groups, communities, and professions.[3] Collective action theory (Ostrom 1990) inquires how such legal systems can emerge in the daily practice of fishing. A rich body of literature has emerged on institutional arrangements, which have developed from the bottom up, both in the North and in the South.

Adding a new institutional layer on top of pre-existing non-state legal systems, state agencies in many countries have extended their jurisdiction over fisheries in response to their increasing profitability, environmental problems and associated social conflicts (Bavinck 2011a). Governments have expanded their range of instruments, including law, to deal with these new concerns. The result is the co-existence of multiple legal systems in fisheries.

Legal pluralism is essentially about diversity, or difference. However, not all situations of legal pluralism are equally 'different'. In some cases the state has come to dominate other legal systems, whereas in others it is met with substantial opposition. There are cases too that have culminated in institutional 'bridges' – forms of co-management that link state with non-state legal systems. The meeting of different bodies of law has often also resulted in 'institutional bricolage' (Cleaver 2002; Kraan 2009) and in the development of hybrid law. In short, one can distinguish between a variety of legal pluralist situations in fisheries, which can be posited to occur on a scale running from monopoly to strong differentiation. Jentoft et al. (2009, note 3) refers to these as instances of 'weak' versus 'strong' legal pluralism.

Legal pluralism has important implications for governance. Where pluralism is extreme, governors can be expected to be at loggerheads. They disagree about the substance of law, about procedures, as well as about the question who is ultimately in charge. Where differences are small and governors agree about most matters of substance and procedure, on the other hand, the path to governance is smoothened.

Complexities of Nestedness and Scale

We owe the term nestedness to Simon (see Hill and Fujita 2003), who introduced the image of the Chinese boxes. He observed, "Opening any given box discloses not just a new box within but a whole set of boxes; and opening any of the component boxes discloses a new set in turn." Nestedness is now considered an important feature of complex systems, both human and natural. It has been applied to cities (Hill and Fujita 2003), forests (Beckley 1998), fisheries (Jentoft 2004), environmental governance (Young 2002; Marshall 2008), and more generally to the study of institutions (Ostrom 1990, 2005; Hollingworth and Boyer 1998).

Important for our purposes is the conceptual relation between scale, level, nestedness and hierarchy. We view the analysis of governance and governability of a specific system as a choice of level in a hierarchy of nested systems. In exclusive hierarchies levels are not nested within one another – the higher level does not contain the

[3] See, for example, Bavinck (2005), Jentoft et al. (2009), Wiber and Kearney (1996), and Wylie (1989).

objects of a lower one (e.g. a food chain). In contrast, higher levels in inclusive, or aggregational, hierarchies such as taxonomic classifications, do embrace lower levels. There is, however, no interaction between them. Constitutive hierarchies are the most interesting type for our purposes, because the lower levels are not only included in higher ones, but all these units have functional relationships and continuously affect one another (Gibson et al. 2000). Bureaucracy provides a good example. Each unit in, for example, a department of fisheries has a specialized task that encompasses or is encompassed by other administrative units. However, the incidence of substantial interaction between units at different levels results in new governance forms and processes. In the literature constitutive hierarchies are often connected with complex systems theories and the way these systems emerge (Gibson et al. 2000).

As governors and governing issues are situated at various geographical scale levels, nestedness is also an issue of scale. Not only do perspectives change as one moves up from one level to the next, so too do mandates, tools and action potential. Increasing scale coincides with what is called multi-level governance, or "the dispersion of authoritative decision making across multiple territorial levels" (Hooghe and Marks 2001, xi). Young (2002 in Marshall 2008, 79) points out that multi-level governance generates two important challenges: (1) how to assign governance tasks across different levels, and (2) how to manage cross-level interactions arising in governing activity. Institutional nesting, or smaller organizations becoming "part of a more inclusive system without giving up their essential autonomy" (Marshall 2005, 47 in Marshall 2008), is one of the approaches used to address such challenges. We argue that multi-level governance in general, and nested governance in particular, generates different levels of complexity in the governing system with concomitant issues of governability.

In the examples that follow this section, we take nestedness to refer to the relationship between institutions at different levels. Lower complexity refers to the situation where local institutions are not nested in institutions at higher scale levels and governance efforts are channelled separately. Higher complexity, on the contrary, exists when institutions at various scale levels have been designed to interconnect. First, however, we consider issues pertaining to the dynamics of governing systems.

Institutional Dynamics

Governing is a spatial but also a temporal phenomenon – it is a process rather than a stable condition. The people and organizations involved in governing, and the rules that they apply, change with time. We know that some governing institutions are characterized by extreme flux, while others stagnate, atrophy or display robustness (Jentoft 2004). Institutional dynamics sometimes derive from challenges in the fisheries themselves. Managers, for example, reflect on the needs of fishers and others in the fish chain and take corrective action. In many instances, however, institutional dynamics have other origins. Scholars use the term 'path dependence' to indicate that present governing efforts are shaped by past choices (Page 2006). In reality, managers often build upon existing structures and policies, and frequently have little space for independent action. In many coastal nations, forces outside of the fisheries frequently drive institutional change (Taylor et al. 2007). A lack of institutional innovation may reflect a general disinterest in the future of the industry and other governmental priorities. Moreover, if institutional changes do take place, they may have other inspirations, such as the need to cut budgets or a policy favoring coastal tourism or industry over fisheries. In all these instances, it is likely that governing structures and processes do not match with the priorities of those engaged in the sector.

The rate, or speed, of institutional change is another concern. Rapid institutional changes put high pressure not only on those who are involved in the governing process, but also on people partaking in the system-to-be-governed. This, in turn, may lead to confusion and resistance. Slow institutional change creates other problems, particularly if the fisheries in question are highly dynamic and require quick action.

Scholars concerned with improving the governability of fisheries currently champion the notion of 'learning organizations', because "fish chains are by their very nature unpredictable" (Mahon et al. 2005, 369; also see Armitage et al. 2007; Berkes 2009). Mahon et al. point out that "most organizations in the fisheries sector can improve the extent to which they 'learn' from experience as well as from their surroundings" (2005, 370). The dynamics of a governing system includes the extent to and rate at which institutions and institutional systems adapt to changing circumstances.

Feature Effects on Governing Systems: Examples from South Asia

Case 7: Lower institutional diversity (weak legal pluralism). The government of India ratified the United Nations Convention on Law of the Sea in 1995. This law delineated Exclusive Economic Zones (EEZ) with a width of 200 nautical miles (NM) and transferred responsibility over large, hitherto unregulated ocean territories to the governments of coastal nations. The government of India is now formally in charge of an EEZ encompassing 1.6 million km^2. Although governmental prerogatives over territorial waters (up to 12 NM) are regularly challenged by non-state actors (see Case 8 below), save for a number of border regions, its jurisdiction over the EEZ is not in doubt.

The term 'weak legal pluralism' refers to a lack of legal contention and is equivalent to hegemony. If legal prerogatives are not backed up by the exercise of power, however, the governability of fisheries is dictated not by the governing system, but instead by events and processes in the fisheries. Although the Indian Navy and Coast Guard patrol the borders of the EEZ, control efforts apparently do not dissuade foreign fishing vessels from regularly transgressing Indian waters (Pramod 2010). In this case, weak legal pluralism coincides with relatively weak enforcement of existing rules and regulations.

Case 8: Higher institutional diversity (strong legal pluralism). In the context of the so-called blue revolution, the government of Tamil Nadu introduced trawling as a new fisheries technique in the 1950s. The new class of trawler fishermen, operating from newly founded harbors, is safeguarded by the constitution of the Republic of India, which not only gave every Indian the right to enter any occupation, but also precluded any other law than that of the government from being applied to the fishery. However, the small-scale fishers of the Coromandel, who inhabit villages governed through traditional caste law, protested vociferously against trawlers fishing 'their' waters. After all, according to their law, each village controls adjacent waters and decrees the technology that is to be applied. Government law thus juxtaposed caste law. The violent conflicts between the two groups of fishers that hit the streets of Chennai in the late 1970s were arguably a result of the non-compatibility of legal values (Bavinck 2001). These conflicts have continued to a lesser extent until the present (Bavinck 2011b).

Strong divergences in a governing system have a negative impact on the governability of the societal system, which is being tugged in various directions. As long as neither of the contesting legal systems gains the upper hand, or a sufficient body of hybrid law has not developed, governability of the fisheries is impaired and fishers suffer.

Case 9: Lower institutional complexity (nestedness). Fishers in the village of Valinookkam, along the Gulf of Mannar in Ramnad District, have a long and successful history of beach seine fishing. Although the number of beach seine companies has gone down in recent years, company owners still wield considerable power at the village and regional levels. They have informed trawler fishers in the region that trawling is prohibited in village waters during the beach seine season (Hopewell 2004). This local rule (it is unknown in other fishing villages in the region) has no basis in governmental law, which claims a monopoly on fisheries regulation. In line with current international opinion, officers of the Fisheries Department also look on the beach seine industry with disfavor, criticizing in particular the small mesh sizes used, and are not inclined to offer it protection.

The village rule protecting beach seine fishing in Valinookkam is not nested in wider institutional frameworks – on the contrary, this industry has acquired a negative reputation with government and international organizations alike. Although customary law in the region still provides beachseining with shelter, its long-term governability prospectives are bleak.

Case 10: Higher institutional complexity (nestedness). With declining catches and increasing evidence of overfishing, the sizeable small-scale fishing population of India put pressure on governments of coastal states to start restricting the trawl industry. The instrument deemed most suitable for this purpose was a closed season, which would coincide with the spawning season of important species (Bavinck et al. 2008). As coastal fisheries are a prerogative of state governments (and not the national government), the first closed seasons for trawling in India, which emerged after 1988, were not coordinated with regard to time. The drawbacks of this lack of synchronization soon emerged: trawl fleets started moving up and down the coast, making use of the variable timings of the closed seasons. It was only when the central government instigated a coordinated policy for east and west coasts in 1996 that closed seasons for trawling became adequately enforced.

Governability of the trawl fisheries in this case clearly improved after the central government brokered a national arrangement. The fact that small-scale fishers were in favor provided political backing. In time trawl fishers too have come to perceive the benefits of a closed season, as catches have gone up in the post-closure period. A long-drawn process of institutional nesting has thereby increased the effectiveness of the measure.

Case 11: Lower institutional dynamics (change). Trawler owners in Tamil Nadu maintain strong, long-term connections with a range of middlemen, such as export agents and female *varattavechchis,* or auctioneers (Bavinck 2001). Johnson (2010) analyzed the manifold dealings between such fishers and traders in terms of patron-client relations, whereby the latter provide services to the former in exchange for a regular supply of produce. The 'services' provided by middlemen in Tamil Nadu are mainly financial in nature: trawler fishers require a flexible and, frequently, immediate supply of credit for a variety of purposes. As Johnson (2010, 272) points out, "the patron-client institution at the heart of the […] fishery is an adaptive system with its own governance logic." Its adaptiveness is evidenced by its response to changing economic circumstances and needs. Johnson's conclusion that patron-client relations have been non-responsive to urgent ecological realities, such as the evidence of overfishing, and tend to "reinforce the fishery's path dependency" (Johnson 2010, 273), appears to be valid for Tamil Nadu as well. Consumer campaigns for responsible fisheries have not reached this part of the world, and middlemen experience no incentives to green their image.

This case illustrates the paradoxical nature of governance perspectives. A very successful governing arrangement may, from a different perspective, be analyzed as playing an obstructing role. While patron-client relations provide the proverbial grease to keep the trawl fishery going, they do not provide an answer to the emergence of larger governance challenges. From a governability viewpoint, their role therefore retains an ambivalent characteristic.

Case 12: Higher institutional dynamics (change). As India exports much of its marine produce, including shrimp, to markets in Europe, Japan and North America, its fortunes depend on the conditions imposed by institutions in those regions. Since the mid-1990s, an array of demands, varying from the use of turtle-excluder devices to measures for food safety, has been inflicted on Indian exporters, forcing many changes in the structure of the marine industry. Contrary to governments in other coastal states, Thailand for example, the Indian government has left the industry to essentially fend for itself. In the eyes of those that perform governing roles in the value chain, such as the export companies, the indifference of the state with regard to rapidly changing international regulations has resulted in numerous bankruptcies (Van der Pijl 2010).

Governing arrangements for international food chains are currently in flux (Swinnen and Maertens 2007), and their movements, which largely originate in the demand markets, reverberate at all levels. As a consequence, participants require vigilance as well as adaptive capacity in order to endure. Although many of these new governance efforts aim at improving, from one point of view or another, the governability of the system-to-be-governed, the net results of varying and sometimes contradictory arrangements are not easily evaluated.

Implications for Governability

In section "Governability and Systems-to-Be-Governed" above we considered the challenges created by the diversity, complexity, dynamics and scale of a system-to-be-governed for a governing system, and the implications for the governability of the societal system-as-a-whole. In this section we have attempted to do the reverse. That is, investigate how the diversity, complexity, dynamics and scale of a governing system can create impediments or opportunities for a system-to-be-governed. We have noted first of all that high institutional diversity (legal pluralism) can generate conceptual confusion and opportunities for the contradictory interpretation of situations and events. In the most extreme instance, such diversity can result in institutional paralysis or social conflict.

Whether institutional diversity is a good or a bad thing depends, however, on context and perspective. It also depends on one's perception of the reverse phenomenon, namely institutional hegemony. Whereas institutional hegemony can help to create a road map and promote governing effectiveness, it may also bury dissent. In this vein, one could argue that legal pluralism is an expression of countervailing opinions, which different segments of the population hold to be more relevant and fair.

Higher complexity (nestedness) makes for increased policy coherence. It would also seem to offer opportunities for expanding the governing scope to multiple scale levels. This is useful particularly when problems and opportunities are not localized, but are instead cut across geographical, sectorial, social and administrative spaces. This is frequently the case in fisheries. Institutional complexity also correlates with specialization and well-developed divisions of labor. As a result, governance capacities can reach new heights. The removal of many governing actors and their activities from the actual scene of fisheries is, however, a possible disadvantage. It may – as we shall see in the following section – be corrected through co-management.

Lower institutional complexity is best exemplified by the fisher-cum-governor: the fisher who, together with a group of compeers, creates and enforces his own fishing regulations. The people engaged in the system-to-be-governed here are also its governors. The division of labor is limited, as is the scope of governing, except through institutional replication along a coastline. The advantages of the fisher-cum-governor, however, are multiple. Chief among them is the fact that the fisher-cum-governor has an immediate, albeit localized, knowledge of the fishery.

Institutional dynamics can affect governability in various ways. If the pace of institutional change is too slow, it results in an accumulation of problems and unused opportunities in the industry, thereby lowering the system's governability. If, on the other hand, the pace of institutional change is very high, it may surpass the system-to-be-governed's capacity to adapt. Here again governability may suffer.

Governability and Governing Interactions

In previous sections we discussed the relations between units, or actors, in the fish chain, as well as those between governing individuals and organizations. All these relations are expressed through interactions, which involve the to-and-fro movement of information, goods and services. As noted by Kooiman (2003, 8): "Interactions shape actors and actors shape interactions; they are 'equal' as basic units of analysis and theory development." Interactions between governors are often purposive, as is their governing activity. The net effect of their activities, however, may be at odds with original intentions. This is because it is rare to find a single governor in charge of a fishery. Instead, there are often many governors involved, and their manifold actions and interactions have numerous impacts, including those not intended by any single actor. Just as a game of chess is never wholly determined by one of the two players, however strong he or she may be, governing activities have divergent and often unpredictable outcomes (Elias 1970). It is therefore only by taking a broad perspective – including the full range of governors and governing activities – and coordinating activities that the possibility of steerage increases.

Interactions also characterize the relations between the governing system and the system-to-be-governed. Whereas traditional governance theory viewed governing activity as 'one-way traffic' between the governing system and system-to-be-governed, contemporary analysts prefer a 'two-way traffic' model, with streams of interaction moving and having impact in both directions (Kooiman 2008, 2010). Interactive governance theory identifies three types, or modes, of interaction. Each differs in terms of its distribution of power. These modes are: hierarchical governance, co-governance, and self-governance. The first hierarchical mode of interaction takes place through 'interventions', or authoritative intrusions from above. In this mode, relations are fundamentally unequal, with superiors (the governors) being distinguished from inferiors (those being governed). The second set, involving 'interplays', aimed to reach goals by engaging actors in collective, rather than independent action. Interplays connect people or organizations of relatively equal status, and result in processes of co-governance. The third and most spontaneous set of interactions coincides with acts of self-governance. In this mode, the people or organizations directly involved in the fish chain steer their own activities. There are no 'outside' governors present.

As diversity, complexity, dynamics, and scale are ubiquitous, they permeate the realm of governing interactions too. We suggest that the *diversity* of governing interactions is likely to be highest in the self-governing mode, because actors in the various segments of systems-to-be-governed make their own choices. This situation is mirrored by a relative lack of diversity in the hierarchical mode. Here governors have chosen to employ a pyramid-like and, hence, repetitive structure of interaction (cf. Bundy et al. 2008). Co-governance takes an intermediate position.

When we turn to *complexity*, the situation described above is reversed. Complexity of interactions may be largest in the hierarchical mode. This is namely due to the scale of operations and the number of partaking units is frequently larger. The self-governing mode, which involves a smaller number of units at a lower scale level, reveals a less complex set of interactions. Here again, the co-governance mode is likely to hold the middle ground.

Finally, *dynamics*, or the propensity towards change, run parallel to the expectations of diversity. They are generally largest in the self-governance mode, where positions and interaction styles are less formalized and therefore more flexible. The hierarchical mode leans on bureaucracy and standardized procedures, both of which tend to resist change. Interactions in the co-governance mode are definitely more dynamic than in the hierarchical mode, connecting governors from a host of different origins. The varying values, principles and objectives available, and the different styles of operation, make for an ongoing process of public reasoning (Sen 2009) and negotiated decision-making. We suggest therefore that co-governance will generally take a medium position as far as dynamics are concerned.

Effects of Governing Modes on Interactions: Examples from South Asia

In order to account for the three distinguished modes, Table 8.3 has a slightly different setup than Tables 8.1 and 8.2 above. A third row capturing intermediate intensity has been included. In addition, rather than identifying a separate example for each of the

Table 8.2 Case studies governability and governing system

Intensity/features	Diversity (legal pluralism)	Complexity (nestedness)	Dynamics (change, learning)
Lower	7. Governing EEZ waters	9. Community management	11. Patron-client relations
Higher	8. Government vs. caste law	10. Closed season	12. International standards

Table 8.3 Case studies governability and governing interactions

Intensity/features	Diversity	Complexity	Dynamics
Lower	Implementation CITES	Gear regulations	Implementation CITES
Medium	Time-zoning	Time-zoning	Time-zoning
Higher	Gear regulations	Implementation CITES	Gear regulations

nine cells (which would be theoretically feasible, but impractical in terms of length), we have chosen to limit ourselves to illustrating the three interaction modes.

Case 13: Implementation of the CITES-List (hierarchical mode: high complexity, low diversity and low dynamics). The government of India ratified the Convention on International Trade in Endangered Species of Wild Fauna and Flora (CITES) in 1976 and added a list of species – including a significant number of marine flora and fauna – to the Wildlife Protection Act (1972). Wildlife wardens were put in charge of local enforcement; these officers make use of the machinery and hierarchical governance style of the Tamil Nadu Forest Department, by whom they are employed. The implementation of regulations prohibiting the harvest and trade of endangered species, such as sea cucumbers in the ecologically-rich Gulf of Mannar, is, however, resented by fishers and traders alike. Not only are these rules therefore frequently evaded, but Forest Department personnel are occasionally threatened with physical violence (Bavinck and Vivekanandan 2011).

Being nested at international, national and local levels, the interactions over CITES rulings are necessarily complex. A hierarchical – top-down – mode of interaction between decision makers at the national level and local people partaking in the system-to-be-governed has resulted in low legitimacy and substantial levels of illegal fishing.

Case 14: Time-Zoning in the Palk Bay (co-governing mode: medium-level diversity, complexity and dynamics). The large-scale violence that erupted between small-scale and trawler fishers in the Palk Bay from the 1970s onwards resulted in a special governance arrangement found nowhere else along the Tamil Nadu coastline. Rather than introducing a spatial zoning system, recommended by the Tamil Nadu Fishing Regulation Act (1983) and nominally implemented (see Case 8 above), the administrators of adjoining districts decided to proceed on the basis of a time-zoning agreement reached by the conflicting parties themselves. According to the terms of this agreement, trawler fishers from adjacent harbor locations are allowed to fish in the Palk Bay for three days a week, with the remaining days being reserved for small-scale fishing. This rule is monitored through concerted action by the Fisheries Department (which only provides trawler fishers with subsidized fuel on their fishing days), the Indian Coast Guard and local fisher organizations. It has survived, with minor modifications, for over 30 years (Bavinck 2003).

In this case, governing interactions are centered at the district level, and involve governmental administrators and fisher leaders from both sides. With the issue deemed both urgent and, there was great pressure to find a mutually satisfactory solution. The result – the endorsement by government of an agreement between the two fisher parties – was legitimate and long-lasting. It could not, however, be easily transferred to other geographical regions, where different conditions prevail.

Case 15: Gear regulations (self-governing mode: high diversity and dynamics, low complexity). The trammel net, which was introduced to Tamil Nadu's Coromandel Coast in the 1980s, triggered excitement as well as trepidation. Many fishers were enthused by the sudden increase of high-value shrimp catches, while others, like 35 year old Siva who, looking back on the arrival of the trammel net in his village, expressed worry about the long-term consequences: Would some fishers, who procured the net first, benefit at the expense of others? And if trammel net fishing became popular, might fish flee from inshore areas? What would then happen to older members of the community, who no longer had the capacity to undertake long fishing voyages? As a result of such anxieties, many village councils gathered and passed rules structuring the introduction of the trammel net (for example, setting a date before which no one was allowed to use the gear) as well as its use.

This case is centered at the local level. Such governing efforts, which are initiated in response to events deemed to negatively affect fishing practices (Bavinck and Karunaharan 2006), presume a high density of diverse interactions between village fishers, as well as between fishers and the marine ecology. Interactions of this kind commence with the sharing of observations from fishing practice, and the recounting of stories about experiences elsewhere. They then move on to the more formal dealings of the village councils. Enforcement, which is carried out by the body of village fishers, brings about its own body of interaction – sometimes verbal and other times involving physical force.

Implications for Governability

The interactive governance perspective assumes that no style of interaction and no governing mode is *a priori* preferable to another. Instead, various governability situations are said to call for different modes of governing. Hierarchical governance may thus be appropriate in some situations, while other situations call for self- or co-governance. The opposite condition also holds true: a particular mode of governing may be quite inappropriate for the situation at hand. For example, in Case 13, which discusses the implementation of the CITES list, the hierarchical governance mode used is in some ways counterproductive, because it provoked hostility among actors in the system-to-be-governed. Self-governing modes of interaction like the one presented in Case 15, on the other hand, possess great capacity for generating legitimacy. Their participants, however, may lack a broad view and the decisiveness to deal sufficiently with the governance issues that arise.

In practice, we rarely encounter the three governing modes in pure form. Instead, societal systems display a large variety of 'mixes', or hybrid varieties, of governing interaction (von Benda-Beckmann et al. 2009, 9–11). These modes are only malleable to an extent. While governing actors can deliberately change some aspects of their interactions with others, they also have to make use of, and adjust to, the patterns that have developed in every societal system over time.

Concluding Remarks

In this chapter we have discussed the ways in which the characteristics of fisheries systems may impact upon their governability. We conceptualized the system-to-be-governed as possessing both a natural and human dimension –typified by varying extents of diversity, complexity, dynamics, and scale effects. Subsequently, the attention shifted to governors and the nature of the governing system, which sometimes evolves in reaction to demands posed by the system-to-be-governed, but may also have other origins. Like systems-to-be-governed, governing systems also prove to be diverse, complex and dynamic. Finally we considered the manifold interactions that occur within and especially between a governing system and a system-to-be-governed, distinguishing between three relationship modes.

The central message of this chapter has been that the connections between the three parts of a fisheries system matter for governability and that a better match may result in higher levels of governability. Conversely, where the fit between the system-to-be-governed, governing system and governing interactions is lacking and mutual responsiveness is low, governability is likely to suffer. In this case, the fisheries in question are not appropriately steered and have unwanted societal consequences. These can vary – depending on perspectives of the society in question – from environmental destruction to economic hardship, social injustice and political strife. In such cases, remedial action is required. The focus of such action, however, depends on the nature of the situation at hand. Governors must determine the contours of the societal system and the issues to be addressed. They may then choose to either focus on adapting the governing system to the needs of the system-to-be-governed, or – the other way around – modeling the system-to-be-governed to better match governing system's capacities and values. They may also try to change the style of governing interactions, thereby influencing the legitimacy of rulings. Whatever the case may be, the overall governability of the system in question is affected, for better or for worse.

Acknowledgements This chapter has benefitted greatly from the comments provided by Derek Johnson, to whom thanks are due.

References

Armitage, D., Berkes, F., Doubleday, N. (2007). Adaptive co-management – Collaboration, learning, and multi-level governance. Vancouver: UB Press.

Bavinck, M. (1996). Fisher regulations along the Coromandel coast: A case of collective control of common pool resources. Marine Policy, 20(6), 475–482.

Bavinck, M. (1998). 'A matter of maintaining the peace.' State accommodation to subordinate legal systems: the case of fisheries along the Coromandel Coast of Tamil Nadu, India. Journal of Legal Pluralism, 40, 151–170.

Bavinck, M. (2001). Marine resource management. Conflict and regulation in the fisheries of the Coromandel Coast. New Delhi: Sage.

Bavinck, M. (2003). The spatially splintered state: Myths and realities in the regulation of marine fisheries in Tamil Nadu, India. Development and Change, 34(4), 633–657.
Bavinck, M. (2005). Understanding fisheries conflicts in the South – A legal pluralist perspective. Society and Natural Resources, 18(9), 805–820.
Bavinck, M. (2011a). The megaengineering of ocean fisheries: A century of expansion and rapidly closing frontiers. In S.D. Brunn (Ed.), Engineering earth: The impacts of megaengineering projects (pp. 257–273). Dordrecht: Kluwer.
Bavinck, M. (2011b). Governance, poverty and social justice in the coastal fisheries of India. In N.R.M. Pouw, & I.S.A. Baud (Eds.), Local governance and poverty in developing nations (pp. 115–136). New York: Routledge.
Bavinck, M., & Karunaharan, K. (2006). Legal pluralism in the marine fisheries of Ramnad District, Tamil Nadu, India. IDPAD Working Paper Nr 2. New Delhi/The Hague: Indo-Dutch Program for Alternatives in Development.
Bavinck, M., & Vivekanandan, V. (2011). Conservation, conflict and the governance of fisher wellbeing – Analysis of the establishment of the Gulf of Mannar National Park and Biosphere Reserve. Environmental Management, 47, 593–602.
Bavinck, M., de Klerk, L., van Dijk, D., Rothuizen, J.V., Blok, A.N., Bokhorst, J.R., van Haastrecht, E.K., van de Loo, T.J.C., Quaedvlieg, J.G.J., Scholtens J. (2008). Time-zoning for the safe-guarding of capture fisheries: A closed season in Tamil Nadu, India. Marine Policy, 32, 369–378.
Beckley T.M. (1998). The nestedness of forest dependence. Society and Natural Resources, 11(2), 101–120.
Berkes, F. (2009). Evolution of co-management: Role of knowledge generation, bridging organizations, and social learning. Journal of Environmental Management, 90(5), 1692–1702.
Berkes, F., Colding, J., Folke, C. (2003). Navigating social-ecological systems: Building resilience for complexity and change. Cambridge: Cambridge University Press.
Bundy, A., Chuenpagdee, R., Jentoft, S., Mahon, R. (2008). If science is not the answer, what is? An alternative governance model for the world's fisheries. Frontiers in Ecology and the Environment, 6. doi:10.1890/060112.
Central Marine Fisheries Research Institute. (2005). Marine fisheries census 2005. Cochin: CMFRI.
Cleaver, F. (2002). Reinventing institutions: Bricolage and the social embeddedness of natural resource management. European Journal of Development Research, 14(2), 11–30.
Durkheim, E. (1964). The division of labor in society (trans: Simpson, G.). New York: The Free Press.
Elias, N. (1970). Was ist Soziologie? Weinheim: Juventa Verlag.
Folke, C. (2006). Resilience: The emergence of a perspective for social-ecological systems analyses. Global Environmental Change, 16, 253–276.
Garcia, S.M., Zerbi, A., Aliaume, C., Do Chi, T., Lasserre, G. (2003). The ecosystem approach to fisheries: Issues, terminology, principles, institutional foundations, implementation and outlook. FAO Fisheries Technical Paper, No. 443. Rome: FAO.
Gereffi, G., Humphrey, J., Sturgeon, T. (2005). The governance of global value chains. Review of International Political Economy, 12, 78–104.
Gibson, C., Ostrom, E., Ahn, T.K. (1998) Scaling Issues in the social sciences. Working Paper No. 1. Bonn: International Human Dimensions Programme on Global Environmental Change (IHDP).
Gibson, C.C., Ostrom, E., Ahn, T.K. (2000). The concept of scale and the human dimensions of global change: A survey. Ecological Economics, 32, 217–239.
Hill, R.C., & Fujita, K. (2003). The nested city. Urban Studies, 40(2), 207–217.
Hollingworth, J.R., & Boyer, R. (Eds.). (1998). Contemporary capitalism. Cambridge: Cambridge University Press.
Hooghe, L., & Marks, G. (2001). Types of multi-level governance. European Integration Online Papers, 5,11. doi:10.2139/ssrn.302786.
Hopewell, J. (2004). 'When the shore seine is shot, the whole village eats!' The change in shore seine organization in Valinokkam village and the decline of shore seining in southern Ramnad District, Tamil Nadu, India. Dissertation, Master thesis, University of Amsterdam.

Hornell, J. (1914). The sacred chank of India, a monograph of the Indian conch (Turbinella pyrum). Madras: Government Press.
Jentoft, S. (2004). Institutions in fisheries. Marine Policy, 28, 137–149.
Jentoft, S. (2007). Limits to governability? Institutional implications for ocean and coastal governance. Marine Policy, 4, 360–370.
Jentoft, S., Bavinck, M., Johnson, D.S., Thomson, K.T. (2009). Fisheries co-management and legal pluralism: How an analytical problem becomes an institutional one. Human Organization, 68(1), 27–38.
Johnson, D. (2006). Category, narrative and value in the governance of small-scale fisheries. Marine Policy, 30, 747–756.
Johnson, D. (2010). Institutional adaptation as a governability problem in fisheries: Patron–client relations in the Junagadh fishery, India. Fish and Fisheries, 11, 264–277.
Kooiman, J. (2003). Governing as governance. London: Sage.
Kooiman, J. (2008). Governability: Exploring a concept. Journal of Comparative Policy Analysis, 10(2), 171–190.
Kooiman, J. (2010). Governance and governability. In S. Osborne (Ed.), The new public governance (pp. 87–104). London: Routledge.
Kooiman, J., & Chuenpagdee, R. (2005). Governance and governability. In J. Kooiman, M. Bavinck, S. Jentoft, R. Pullin (Eds.), Fish for life – Interactive governance for fisheries (pp. 325–349). Amsterdam: Amsterdam University Press.
Kooiman, J., & Jentoft, S. (2009). Meta-governance: Values, norms and principles, and the making of hard choices. Public Administration, 87(4), 818–836.
Kooiman, J., Bavinck, M., Jentoft, S., Pullin, R. (Eds). (2005). Fish for life. Amsterdam: Amsterdam University Press.
Kraan, M. (2009). Creating space for fishermen's livelihoods – Anlo-Ewe beach seine fishermen's negotiations for livelihood space within multiple governance structures in Ghana. African Studies Collection, 19. Leiden: African Studies Centre.
Kurien, J. (1978). Entry of big business into fishing, its impact on fish economy. Economic and Political Weekly, 13(36), 1557–1565.
Mahon, R., Bavinck, M., Roy, R.N. (2005). Governance in action. In J. Kooiman, M. Bavinck, S. Jentoft, R. Pullin (Eds), Fish for life (pp. 351–373). Amsterdam: Amsterdam University Press.
Mahon, R., McConney, P., Roy, R.N. (2008). Governing fisheries as complex adaptive systems. Marine Policy, 32, 104–112.
Marshall, G.R. (2005). Economics for collaborative environmental management: Renegotiating the Commons. London: Earthscan.
Marshall, G.R. (2008). Nesting, subsidiarity, and community-based environmental governance beyond the local level. International Journal of the Commons, 2(1), 75–97.
Merry, S.E. (1988). Legal pluralism. Law & Society Review, 22(5), 869–896.
North, D.C. (1991). Institutions. The Journal of Economic Perspectives, 5(1), 97–112.
Ostrom, E. (1990). Governing the commons: The evolution of institutions for collective action. Cambridge: Cambridge University Press.
Ostrom, E. (2005). Understanding institutional diversity. Princeton: Princeton University Press.
Ostrom E. (2007). A diagnostic approach for going beyond panaceas. Proceedings of the National Academy of Sciences, 104(39), 15181–15187.
Page, S.E. (2006). Path dependence. Quarterly Journal of Political Science, 1, 87–115.
Pramod, G. (2010). Illegal, unreported and unregulated marine fish catches in the Indian exclusive economic zone. Field report, policy and ecosystem restoration in fisheries. Vancouver: Fisheries Centre, University of British Columbia.
Sen, A. (2009). The idea of justice. London: Allen Lane.
Swinnen, J.F.M., & Maertens, M. (2007). Globalization, privatization, and vertical coordination in food value chains in developing and transition countries. Agricultural Economics, 37(s1), 89–102.

Taylor, W.W., Schechter, M.G. et al. (Eds.). (2007). Globalization: Effects on fisheries resources. Cambridge: Cambridge University Press.

Van der Pijl, W. (2010). The implications of international food standards and regulations for stakeholders in the export oriented shrimp trade value-chain: A case study in Chennai, India. Unpublished Master Thesis, University of Amsterdam.

van Haastrecht, E., & Schaap, M. (2003). A critical look at fisheries management practices: The 45-day ban in Tuticorin District, Tamil Nadu, India. Unpublished Master Thesis, University of Amsterdam.

Vanderlinden, J. (1971). Le pluralisme juridique, essai de synthese. In J. Gilissen (Ed.), Le pluralisme juridique (pp. 19–36). Brussels: l'Université de Bruxelles.

von Benda-Beckmann, F. (2002). Who's afraid of legal pluralism? Journal of Legal Pluralism, 47, 37–82.

von Benda-Beckmann, F., von Benda-Beckmann, K., Eckert, J. (2009). Rules of law and laws of ruling: Law and governance between past and future. In F. von Benda-Beckmann, K. von Benda-Beckmann, J. Eckert (Eds.), Rules of law and law of ruling – On the governance of law (pp. 1–30). Farnham: Ashgate Publishing Ltd.

Wiber, M., & Kearney, J.F. (1996). Stinting the commons: Property, policy or power struggle? Comparing quota in the Canadian dairy and fisheries sectors. In J. Spiertz, & M.G. Wiber (Eds.), The role of law in natural resource management (pp. 145–165). The Hague: VUGA Uitgeverij B.V.

Wilden, A. (1987). The rules of the game. London: Routledge and Kagan.

Wylie, J. (1989). The law of the streets, the law of the courts, and the law of the sea in a Dominican fishing village. In J. Cordell (Ed.), A sea of small boats (pp. 152–176). Cambridge, MA: Cultural Survival Inc.

Young, O.R. (2002). Institutional interplay: The environmental consequences of cross-scale interactions. In E. Ostrom, T. Dietz, N. Dolšak, P.C. Stern, S. Stonich, E.U. Weber (Eds.), The drama of the commons (pp. 263–291). Washington, DC: National Academy Press.

Chapter 9
Poverty in Lake Victorian Fisheries – Understanding Governability

Paul Onyango and Svein Jentoft

Abstract Why is poverty such a challenging problem in small-scale fishing communities, despite noble efforts to eradicate it? This chapter argues that poverty requires a governability lens to be understood and effectively alleviated. More specifically, the chapter uses the interactive governance approach's three systems model, i.e., the system-to-be-governed, the governing system and governing interactions, to discuss poverty as a "wicked problem" that involves challenges pertaining to governability. We explore the limits of, and conditions for, governability and how these affect poverty. Our point of view is shaped by the poverty conditions prevalent in the fishing communities of Lake Victoria in Tanzania.

Keywords Governability • Interactive governance • Lake Victoria • Poverty • Small-scale fisheries • 'Wicked problems'

Introduction

Poverty continues to be an issue of global concern. In small-scale fisheries, it is often linked to the overexploitation and degradation of fisheries resources. Subsequently, the relationship between small-scale fisheries management and poverty has been a subject of debate and a concern for countries in which these fisheries

P. Onyango (✉)
Department of Aquatic Sciences and Fisheries,
University of Dar es Salaam, Dar es Salaam, Tanzania
e-mail: onyango.paul@udsm.ac.tz

S. Jentoft
Faculty of Biosciences, Fisheries and Economics,
University of Tromsø, Tromsø, Norway
e-mail: svein.jentoft@uit.no

M. Bavinck et al. (eds.), *Governability of Fisheries and Aquaculture: Theory and Applications*, MARE Publication Series 7, DOI 10.1007/978-94-007-6107-0_9,

operate, their development partners and academics (Smith 1979; Copes 1989; World Bank 1992; Cunningham 1993; Béné 2003, 2004; FAO 2005; Jentoft and Eide 2011). For instance, poverty has been perceived to occur in fishing communities as a result of a decline in fisheries resources brought about by overexploitation. This is an argument that rests on a conceptualization of sustainable fisheries in which over-exploitation occurs when the rate of fishing exceeds the capacity of the natural environment to replenish (Gordon 1954; Beverton and Holt 1957). An alternative explanation is that poverty is in itself a cause of overexploitation (Béné 2003): Small-scale fishers, who are generally poor, will overexploit fisheries resources as they seek to meet their livelihood needs. Similarly, small-scale fisheries may be "an occupation of last resort" for people who cannot feed themselves in other sectors and are thus attracted by the opportunity to take up a new livelihood in fisheries (Onyango 2011).

To confront poverty among small-scale fishers, the adoption of a sustainable management system for the harvesting of fish resources is perceived by many as paramount (WCED 1987; Jalal 1993). Although it may be a necessary thing to do, management alone is hardly sufficient given the multiple factors that are at play. Small-scale fisheries are too diverse, complex and dynamic for simple solutions, such as restricting access or banning certain gear types. Poverty in small-scale fisheries requires an understanding that is embedded within a social, economic and political framework, and must be addressed accordingly. Notably, poverty is experienced by small-scale fishers within their communities, and should therefore also be addressed at that level as well as higher scales. It could be part of a problem that links to how fishers relate with themselves and with government. It could also be part of a larger socio-cultural issue within these communities. Poverty is envisioned here as multifaceted, extending beyond the income/expenditure nexus. It involves more than fisheries *per se*, including issues such as health, literacy, access to markets, political participation and the like. Poverty therefore requires efforts beyond technical approaches. In order to understand how it can be alleviated in small-scale fisheries, it is important to note that poverty does not have technical or scientific solutions, as Garrett Hardin (1968) pointed out. This also means that there is no single way to recognize and address the problem.

This chapter discusses how the poverty problem requires an interactive governance approach. While we agree that poverty in small-scale fisheries is attributable to multiple variables, including those mentioned above (Alkire 2008; Alkire and Foster 2008), we argue that these variables are related to each other. We note that poverty is a result of systemic causes, and that governance may be part of both the problem and the solution. Consequently, we need to focus on how poverty alleviation functions as a governance process and within governance settings. We perceive governance beyond the characteristics of accountability and transparent government, free and fair elections, and governance according to the rule of law. In particular, we view governance as "the whole of interactions taken to solve societal problems and to create societal opportunities, including the formulation and application of principles guiding those interactions and care for institutions that enable them" (Kooiman et al. 2005, 17).

The chapter begins by discussing existing perceptions of poverty. It shows how these views of the poverty problem are linked to the approaches that are taken to alleviate it, and how these perspectives have changed over time. We then present some background information about the study area. This is followed by discussion about the "wickedness" of the poverty problem (Conklin 2006) within the three component systems of the interactive governance approach. The section discusses how the wickedness of poverty is an inhibiting factor for its governability. Based on this discussion, we present a framework for assessing poverty as a governability problem.

Perceptions About Poverty

What is poverty and how is it experienced by those who suffer from it? In responding to questions such as these, it is necessary to understand the nature of the questions and, further, the context in which an answer may be given. The question 'What is poverty?' is one whose nature and the context available for its resolution are generally quite varied. Poverty cannot be isolated from the poor; this raises questions with respect to whether its alleviation should really start from defining it from an objective, detached position or simply begin from how it is understood, lived and experienced by the poor themselves. This indicates the complication that poverty tends to mean different things to different people in different contexts (Narayan et al. 2000; Jentoft and Midré 2011). Historically, its understanding has ranged from being attributed to those who are free commoners (in the medieval period), to people with an individual disability (in the thirteenth and fourteenth centuries), and to a systemic failure (in the eighteenth and nineteenth centuries) (Hanson 1997). In recent years, however, three analytical perspectives of poverty have emerged (see Onyango 2009).

The first is the individual perspective, which contends that the poor are themselves largely responsible for their pathological habits and ways of life (Hyatt 2001). This is a line of thinking that led to what Lewis (1963, 17) called, the "culture of poverty". That is, poverty as a way of life sustained by the poor within their own community. The second is the structural perspective, which contests the individualistic notion by arguing that the poor are led into such a life by external factors and that poverty alleviation, therefore, requires intervention by the government or some other external agent (Hyatt 2001). The third is the multifaceted dimension perspective. It argues that poverty is neither an individual nor a structural issue alone, but instead a combination of both.

Inspired by the work of Amartya Sen (1981, 1999), Alkire and Foster (2008) discuss the multidimensional nature of poverty in terms of physical safety and security, employment and quality of work, empowerment and agency, the ability to go about without shame, meaning, and psychological and subjective well-being. Physical safety and security imply that one is not a victim of property, physical or lethal violent crime or conflict. Employment and quality of work mean that one is neither underemployed nor working in an unsafe environment and receiving low pay. One's empowerment and agency means not being forced or compelled against will in one or more domains. The ability to go without shame implies that one does not suffer from stigma, humiliation, isolation, indignity and discrimination. Lastly, meaning and psychological and subjective well-being regard deprivation, alienation, anomie and dissatisfaction.

We take this multidimensional view a step further by arguing that these variables are interrelated and are affected by a change in any of them. This relationship and dynamism provides a rich ground for the governability lens. For instance, a person's physical safety and security is a prerequisite for quality employment, his empowerment and agency, ability to go without shame and to hold a perception on his subjective well-being. Similarly employment and quality of work will be determined by psychological and subjective well-being, the ability to feel dignified and live with physical safety and security. This is to say that any of the listed variables will be affected and affect all the other variables at the personal as well as the systemic level.

Thus in confronting poverty in small-scale fisheries, its governors should not only understand the multiple variables, but also explore how the relationship among the variables is experienced, understood and shared between and among the governing actors and fishers/stakeholders. This, we argue, requires an interactive governance mechanism where the poor, the government and civil society partners act together in developing a shared understanding in confronting poverty. The governability lens fits the aforementioned criteria. It is therefore argued in this chapter that an interaction mechanism that affords actors influence on the decisions or activities that are geared toward poverty alleviation is likely to promote the overall capacity to alleviate poverty.

The Study Area

Our study was undertaken in two communities on the Tanzanian side of Lake Victoria. One community is located on the western side of the Lake (Kasheno), while the other is on the southeastern side (Nyakasenge) (Fig. 9.1). Poverty is widespread, deeply entrenched, but predominantly a rural phenomenon. It is reported that about 87% of the poor population are located in the rural areas (RAWG 2007). Based on the Household Baseline Survey (HBS) of 2000/01, poverty is highest among households who depend on agriculture, a statistical category that includes fisheries.

The Lake's fisheries have undergone a rapid expansion of the fish processing industry and seen the volume of exported processed fish increase significantly (Lukunga 2005). On the face of it, this is a welcome development. Fish exports earn the country valuable foreign exchange. The processing industry offers employment opportunities to the populous lake region and beyond, and also provides a tax base for the government. The export market should therefore increase the price of fish and eventually improve the income of the fishers. The vast economic opportunities the

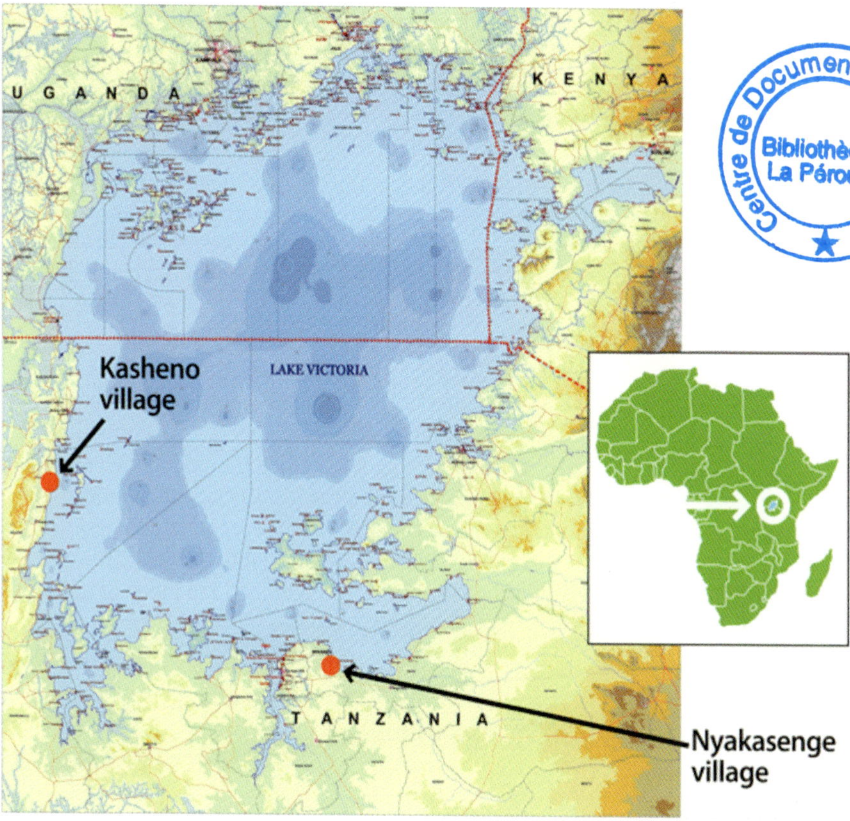

Fig. 9.1 Lake Victoria map showing Nyakasenge and Kasheno villages (Credit: P. Onyango)

lake affords its proximate residents would lead one to expect that communities around the lake would, as a result, enjoy improved economic welfare and suffer less poverty than communities located further away. These benefits, however, have been contested. There are, in fact, observations that the fish export is leading to poverty, manifesting itself in less nutritious food for local people (Sauper 2005), and deforestation where fishers have cut trees to make boats and smoke fish to export (Jambiya 2004). In addition, Kurien (2005) has argued that the growing export of Nile perch fisheries has posed a threat to the food security and livelihoods of the lake's riparian communities. This is because the fish exported reduces its availability to the local communities. Moreover many people have joined the lucrative Nile perch fisheries from the poorly performing agricultural sector, implying that production from agriculture has further declined. Due to the increased effort in fisheries, its production has also decreased and is, consequently, affecting fishers' livelihoods.

Banana farming is a key feature in Kasheno, having virtually covered a substantial part of the village. Nyakasenge, in contrast, is barely covered by any crop, let alone bananas. Both villages are accessible only by dirt road. Residents here depend exclusively on the Lake waters for their domestic use. They use wood and/or charcoal as their main source of energy. Both villages are void of any healthcare facility and residents therefore have to travel for about five kilometers to access such services.

Kasheno village is located in an area where residents have to climb a steep hill in order to access various services, including healthcare, schools and even the market for their fish products. They have recently improved access to some of these services by constructing a road and building a school. The road has improved the local public transport system, which is comprised of motorcycles that transport residents and fish products. Nyakasenge village, on the other hand, is located roughly 7 km away from the main highway that leads to Mwanza, the biggest city on the Tanzanian side of the Lake. The village is surrounded by flat land, typically used for rice cultivation. The Nyakasenge beach area also serves as the primary shopping area for local residents.

Nyakasenge consists mainly of fishers who have migrated from different communities around the Lake. Kasheno residents, on the other hand, are permanent settlers who claim ownership to the land where they have built their houses. The difference can even be seen in the type of housing in the respective villages. Several houses in Kasheno are of a more permanent construction; they are made of concrete bricks with iron roofs. In Nyakasenge, in contrast, the houses are made of unburned bricks with iron roofs.

Governing Poverty Alleviation

There is no single way in which the poverty problem can be addressed in communities such as those described above. The difficulty of alleviating poverty here lies in locating the poverty problem that needs to be solved. For instance, in Nyakasenge fishers understand poverty to mean the inability to use one's hands, legs and head to solve poverty problems. This perception of poverty raises the question of how many people are in the state of not being able to use these capabilities to improve their lot. Thus the poverty problem in Nyakasenge seems to be an issue of disability or perception of what that means exactly in that situation. How then can either of the two be addressed?

Moreover, there is no certainty that a chosen strategy will meet expectations. In reality, it is quite challenging for those in government and actors within the NGO community, otherwise labeled, "the poverty alleviation governors", to be certain that conducted efforts are likely to be effective in tackling poverty – in the short as well as in the long run. What has been witnessed, time and again, is that the strategies adopted for poverty alleviation have rarely succeeded. For instance, strategies to increase individual income as a means of addressing poverty have been implemented with some level of success, but have yet to yield the expected results. Similarly, the success of other efforts, such as micro finance programs, has yet to fully materialize. Ukerewe, a town near Lake Victoria where these initiatives have been implemented, is a case in point. Poverty has therefore remained a challenge.

The persistence of poverty in small-scale fisheries calls attention to the fact that this is a composite and a multi-level problem that presents an enormous challenge to governance. Perceptions of poverty in a given instance can have multi-, including non-local, loci. Poverty might not be perceived at a local level, and thus have no solution, but if the problem is extended, say regionally, then poverty exists and has a solution. Given the complexity of the poverty issues, governors struggle to uncover ways of alleviating it.

As Rittel and Webber (1973), we argue that poverty alleviation governors are presented with problems that are best understood as 'wicked'. Wicked refers to an issue that is tricky or thorny, difficult to define and delineate from other and bigger problems; its meaning is complex, elusive and problematic. Poverty is problematic to define, identify and, subsequently, resolve. In fact, preconceived solutions have a tendency to precede and define the problem.

We find this to be the case in the Lake Victoria fisheries, where it has been difficult to locate the poverty problem and even address it. This is largely because poverty is not an issue of income here; fishers on average earn about USD 4 a day (Odongkara et al. 2009) (a figure well above the World Bank USD 1 or 2 a day to qualify one to be extremely poor). Children in fishing communities at least have access to basic or primary education due to government efforts that have opened primary schools at the village level and even secondary schools at division level (a division comprises for instance two to five villages). Moreover, fishers do not believe that they are poor despite the reality that they are still void of adequate clean and safe water. They still have limited access to health facilities and poor infrastructure, among other factors. Poverty in the Lake cannot be explained either by income or expenditure. Nor can we identify one single explanation to define or trace what poverty really is in this area. It is this abstraction that challenges the efforts of poverty alleviation governors, and hence requires re-conceptualization using the governability approach.

In order to understand poverty as a wicked problem with severe governability challenges, we examine its nature in this chapter, using the interactive governance framework outlined in earlier chapters as guideline. We thus trace poverty, its causes and potential remedies within the three governance sub-systems: the systems-to-be-governed, the governing systems and the governing interactions. The first sub-system is the target of the governing system, while governing interactions refers to the ways in which the two systems actually communicate with each other.

System-to-Be-Governed

Natural System

The fishers have been affected by the changing aquatic system, which is comprised of the fish species and their habitats. Although very little is known about fish species composition at the time of the Lake's formation, it is observed that the Lake has witnessed an unusually rapid speciation since approximately 14,000 years ago. In a biologist's words, this speciation has been described as 'spectacular'

(Okeyo-Owuor 1999). Okeyo-Owuor (1999, 10) argues that "the rate of evolution and speciation of the fishes of the Great Lakes of East Africa warrant the use of the terms such as 'evolutionary avalanche, explosive evolution and explosive speciation." Numerous species have been recorded; Johnson et al. (2000) indicate that over 400 endemic species of cichlids occupied the Lake after the Pleistocene period. According to these researchers this can be explained as follows: a few species of cichlids and other fish species swam out of their fluvial refuges to colonize the new Lake, generating hundreds of new species.

Until the 1980s, the Lake was dominated by endemic fish species (cichlids), mainly Haplochromines species and two tilapiine species – a Tilapia (*Oreochromis esculentus*, Cichlidae) known in Tanzania as Ngege, Sato or Perege, and another Tilapia species (*O. variabilis,* Cichlidae) known as Mbiru (Ligtvoet et al. 1995). The Lake also played host to fish species other than cichlids (Graham 1929). In 1950s, fish species like the Nile tilapia (*O. niloticus,* Cichlidae) and Nile perch (*Lates niloticus*, Latidae) were introduced. The Lake also contained sardine (*Rastreneobola argentea*, Clupidea), known locally as Dagaa.

The high diversity implies complex trophic relations in the food web (Okeyo-Owuor 1999). Moreover, the vast number of species and rapid speciation observed have also presented a dynamic scenario that, given the amount of information required in order to fully understand the system, poses a great challenge for governing poverty. For instance, governors have faced difficulty in designing management systems for different species of fish. Fisheries management in the Lake has basically taken a more uniform character; mesh size and slot size regulations, boat size and closed areas have been formulated on the basis of the Nile perch, Tilapia and Dagaa fisheries, regardless of the presence of other species.

Social System

In Lake Victoria, the social fisheries system-to-be-governed consists mainly of fishers, fish processors and traders of Nile perch, Tilapia and Dagaa. Although fishers constitute the largest number of coastal residents in the Lake Victoria region, they have not been the face of the Lake's fisheries, especially when it comes to small-scale fishing communities to describing the livelihood impact of the Lake's lucrative fisheries. The local systems-to-be-governed are comprised of a diversity of stakeholders, including fishers, boat makers and repairers, fishing gear makers and repairers. The stories about successes in the fisheries have been dominated by the life of fish processing factory agents, their housing and material possession (www.invisiblepossibilities.com). In addition, the sudden construction of ultra-modern office and hotel buildings in major Lake cities like Mwanza, Musoma and Bukoba have been used to further elevate the success story about the Lake's fisheries. Poverty alleviation governors are faced with the difficulty of understanding why small-scale fishers in communities like Nyakasenge and Kasheno cannot address their poverty, when others who depend on the Lake's fisheries have made such lucrative steps.

Using the interactive governance perspective as a basis, in this chapter we argue that in order to understand the pervasive poverty among the fisher groups on Lake Victoria's shores, it is important to accept that these poor fisher groups are complex, diverse and dynamic, and operate at different scales. In Lake Victoria, understanding one group of poor fishers requires a wider lens. For instance, some gear owners are also crewmembers, in the same way that some gear menders are both gear owners and crewmembers. That is to say, the boundaries of each group may overlap. Nevertheless, gear owners, crewmembers and gear menders are highly dissimilar groups with differing interests.

Secondly, Lake Victoria fishers have sub-groups within each group. Among crewmembers, for example, there are fishers who target happlochromine, Nile perch, Tilapia and Dagaa species. The Nile perch fishers use a range of different gear and are subdivided accordingly. There are those who use long lines and gill-nets, as well as those who use outboard engines, wind and hand-paddled boats. The same holds true for Tilapia fishers; some use hand lines, while others opt for gillnets. Each group has its own unique set of needs and concerns. For example, happlochromine and local fishers have always complained about restrictions on using 5 in. mesh-size for fishing. This is because there are some fish species whose total size at maturity can still not be caught by those nets. This therefore means that they cannot catch such species and, consequently, their livelihood concerns are not optimally addressed.

The Nile perch fishers also have their concerns. Although this species has generated considerable earnings, due in large part to its export to, among others, European and Middle Eastern markets, the fishers have not been taken out of poverty. Their daily life is characterized by intensive labor, while their living conditions have only experienced minimal change for a sustained period of time. The proliferation of Nile perch fisheries in the Lake has brought with it incomes that has not matched fishers living conditions. In particular, local fisheries authorities have reported that fisheries generate between USD 2–4 per day per fisher (Odongkara et al. 2009), which may not equate to extreme poverty according to official definitions, but do not constitute more than a meager livelihood either. This holds true especially where fishers continue to live in conditions in which their ability to access health and school services, good roads, adequate clean and safe water, sustainable sources of energy, among others, is found wanting.

The Governing System

This system consists of various institutions that can be traced back to 1908, the year in which the Fish Protection Ordinance was enacted (Geheb 1997). In 1947, a Lake Victoria Fisheries Service (LVFS) was formed. The LVFS was focused in particular on the enforcement of fisheries laws and regulations. It was later transformed into the East Africa Fresh Water Fisheries Research Organization (EAFFRO). In the 1990s, a joint management arrangement for the Lake was proposed, and, consequently, the Lake Victoria Fisheries Organization (LVFO) (See Fig. 9.2) was formed.

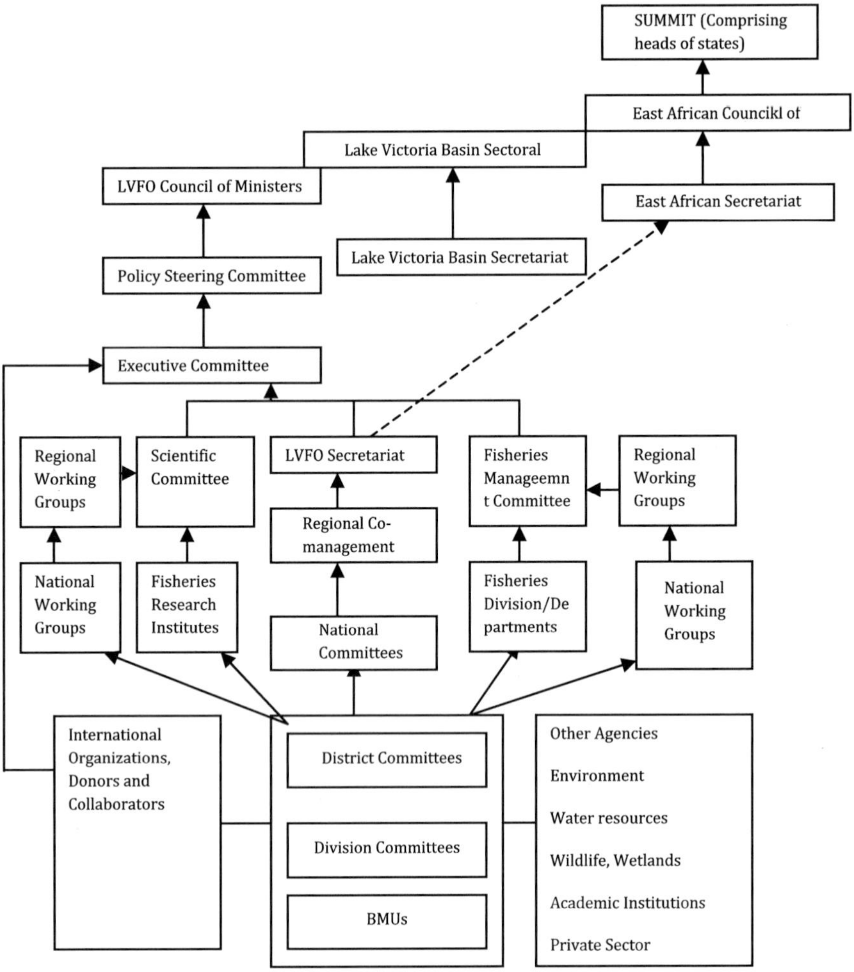

Fig. 9.2 Multiple governing actors in Lake Victoria fisheries and interactions between them (Source: Lake Victoria Fisheries Organization)

Before the formation of the LVFO, the Lake's riparian countries in Kenya, Tanzania and Uganda had formed fisheries directorates to oversee the management of the fisheries resources for both the Lake and their countries. These directorates were part of Fisheries Acts, such as the Fisheries Act No 6. of 1970 in Tanzania, The Fisheries Act Cap 378 in Kenya and the National Fisheries Policy (NFP) of 2004 in Uganda. Although they do not specifically identify how they address poverty, the Acts take the point of view that by caring for fish stocks, poverty will eventually be eradicated. The fact that, three decades later, poverty has still not been wiped out, suggests that these approaches have not worked, and that a governability problem exists that requires revisiting.

In Tanzania fisheries, the Ministry of Livestock and Fisheries Development, the district councils, non-governmental organizations like Lake Nyanza Environment and Sanitation Organization (LANESO), East African Communities Organization for Management of Lake Victoria Resources (ECOVIC), and Environment Management and Economic Development Organization (EMEDO), among others, focus on income poverty and improving infrastructure and access to services. There are some non-governmental organizations, namely Tanzania Home Economics Association (TAHEA) and Foundation for Initiative for Community Assistance (FINCA), that extend credit facilities to fishers. Others, like LANESO, are simply involved in activities that target the rehabilitation of the Lake for the benefit of the poor fishers.

The LVFO consists of the Fisheries Division, Department and Commission, as well as the Fisheries Research Institutes (Tanzania Fisheries Research Institute TAFIRI, Kenya Marine and Fisheries Research Institute KEMFRI, and National Fisheries Resources Research Institute (NAFIRRI)) (See Fig. 9.2). Other partners include: the Beach Management Units, which have been established across the entire Lake; the private sector, consisting mainly of the Fish Processing Establishments; and the riparian district councils. District Councils are government administrative structures at a district level. In Tanzania, government activities are decentralized to districts and further to villages. Each of these two levels house full-fledged district and village governments, respectively. At the district level, there is an executive council, consisting of the District Commissioner, District Administrative Secretary and other officers, and a legislative council, comprised of elected councilors. The District Council coordinates both central government activities as well as their own activities. The district is made up of several divisions, which are sub-divided into villages, the smallest administrative denomination.

The LVFO's many committees and working groups (Fig. 9.2) make the governance system increasingly complex from a poverty governance perspective. In fact, it could be argued that the governing system, in all its dimensions, poses a governability problem in itself, particularly with regard to the need for coordination and concerted action. As if these governance levels are not already enough, fishers have also been incorporated into the governing system through their Beach Management Units. At each of these governance levels poverty is understood differently. There is generally no agreement in terms of what poverty means and how it should be addressed.

Moreover, among the parties involved in poverty alleviation, there is great appeal for the use of strategies that have been used elsewhere and reports have indicated some success in addressing poverty. The extension of credit facilities to fishers, especially women, which has been practiced in Asia, is an example of one such strategy that has gained popular support (Sethuraman 1998). Credit in the form of cash or fishing gear and boats has played an important role in the Lake's fishery attempts to confront poverty. As people utilize the loan, most of the incomes are used to service the loan and not improve their livelihoods. By the time they have settled their debt, the gear can no longer be used to generate incomes and they are therefore left, more or less, in the same state as before. Where improvements in living conditions are experienced, these are largely minimal.

The Fish Processing Establishment (FPE) owners have also designed a strategy for developing infrastructural facilities such as sanitation and roads, even improving landing sites as a means of reducing poverty within fishing communities. In addition to this, some of these FPEs have embarked on digging wells for some fishing communities since 2008. These activities have been readily accepted by fishing communities and poverty alleviation governors. Although it is too early to measure the full impact of such initiatives, the rate at which poverty has been growing in these communities generates a lot of questions with regards to whether they can yield any tangible or observable changes. In other words, poverty in these communities is broader and requires a more comprehensive approach than just providing such facilities. As Krishna (2010) argues, one positive step taken to address poverty in one direction is affected by several other factors, which serve to pull the poor in the opposite direction; deeper into poverty. This presents a governability challenge to poverty alleviation governors. Poverty alleviation approaches must therefore be sensitive to the many aspects of poor fishers' livelihood conditions, an entrenched feature of broader approaches like the interactive governance perspective.

Governing Interactions

Governing interactions comprise forms of communication, participation and representation between the system-to-governed and the governing system. From our perspective, it is the nature of the interactions between the governors and the system-to-be-governed that can shed some light on the stubbornness of poverty in small-scale fishing communities. We argue that poverty persists in fishing communities, because poverty alleviation governors, with good intentions, confront the poverty problems based on each governor's image of what the problem is rather than how poverty is actually experienced within these communities. When determining policy agendas, governing interactions that target poverty alleviation are guided by interests and power. This generates inadequate and ineffective interactions, where people at the village level have little or any influence on decisions or strategies directed at the alleviation of poverty.

As we have shown above, there are several institutional poverty alleviation governors in the Lake Victoria fisheries. Each of these tends to perceive and address poverty differently. At the regional level, for example, the LVFO confronts poverty through empowering communities to manage the Lake's fisheries. They do this by undertaking activities that include, among others, capacity-building in business and management skills, empowering fishers to participate in management of the fisheries. As these governors undertake their activities, they have established working relations and a communication system that is coordinated by the LFVO (Fig. 9.2).

As Fig. 9.2 indicates, interactions between the governing system and system-to-be-governed in Lake Victoria fisheries cover various levels, beginning with fishers at the beach level where they have formed Beach Management Units (BMUs). BMUs are a form of co-management regime comprised of fishers, fish traders, gear makers and menders, women and other residents in a fishing village. They were formed in 1998 and have since been re-established across the entire Lake. The interactions here take place between and among the BMU members at the fishing village and the district councils. The communication that takes place is focused on reporting activities undertaken by the BMUs in the lower level to upper level institutions like the Ministry of Livestock and Fisheries Development, nationally, and the LVFO, regionally. Other agencies and donors also work with the BMUs, the district councils, the ministry and LVFO, to promote capacity-building, awareness creation and sustainable activities.

Interaction with higher levels has mainly been in two directions. The thematic working groups mentioned above interact with the BMUs through data collection, which is used to inform management policies. The working groups pass the same information to the higher-level governors, some of whose responsibilities we discussed in the previous section. Once decisions are made by the council of ministers, the information flows back to the lower levels for implementation.

The governance interaction in Lake Victoria described here points out to two main challenges. The first has to do with the time taken to make decisions and implement them. Whereas fishers poverty problems involve everyday life experiences, for example, the crocodile attacks in Nyakasenge village, or access to health services, markets for fish products and schools in the case of Kasheno village, the need to make daily decisions regarding how to address these poverty problems requires that the interaction process among poverty alleviation governors should be one that expedites decision-making.

Moreover, governors must address micro-problems like the frequent crocodile attacks and the difficulties with reaching a health facility at night, instead of only focusing on macro issues like improving access to the export markets for fish products (Krishna 2010). As important as the latter is, addressing the former must go hand in hand.

Secondly, the governance interaction also shows that upper level institutions, such as the Council of Ministers or the Summit (The Head of the riparian States), are generally responsible for making decisions. The difficulty or challenge with decisions made at these levels has to do with what poverty means to the decision makers and the poor, and how to address it. In other words, it depends on whose knowledge or on what grounds poverty alleviation strategies should be built. In our estimation, the governability problem would diminish if the knowledge of decision-makers or the poor fishers regarding what poverty means and what it would take to bring fishing people out of it were based on a shared understanding among both governors and stakeholder. This would require more constructive interactions than those that currently take place within this complex institutional chain of decision-making.

Governance Orders

The three systems model illustrated above for the case of Lake Victoria, i.e. the systems-to-be-governed, governing system and governing interaction, provides a means through which the poverty and poverty alleviation experiences can be analyzed. It indicates the conditions and challenges that governors face as they try to address the poverty problem through what interactive governance theory calls, the three orders of governance. As argued by Kooiman (2003), the three orders of governance provide a clearer understanding of how governing activities takes place.

At the first order, poverty alleviation governors are involved in addressing daily problems and identifying or creating opportunities. The second order regards the construction of institutions for governing in the first order. The third order, or meta-governance, is about the constitutive values, norms and principles upon which governing activities and institutions are founded. In the interactive governance perspective, understanding poverty and designing strategies to eradicate it requires initiatives at all orders. These range from the more contemplative reflections that occur about what kind of concern poverty is and what approaches would be ethically sound at the meta-order level, to the actions and institutions (organizations, rules, mechanisms) that represent instruments, right down to the actual implementation of poverty alleviation strategies at the level of the administration and community. Therefore, an assessment of poverty requires an examination of the role basic societal values, norms and principles among governing actors and fishers/stakeholders play in defining and alleviating the poverty problem (Jentoft and Midré 2011). When poverty alleviation governors and stakeholders tackle daily problems, and when institutions are formed, values, norms and principles underpin how these are formulated and addressed. Assessing governability of poverty will therefore involve examining variables such as values, norms and principles, and the ways in which they influence decision-making and the institutions within the systems-to-be-governed, governing system and governing interaction (Table 9.1).

Assessing Poverty by Governing Orders

The first order of governance regards daily decision-making in order to address the quotidian problems that fishers are confronted with as they go about doing their regular business. Here small things can make a difference, such as creating the facilities for storing fish on the beach, securing regular access to markets by providing necessary transport, keeping the school or the health clinic open and so forth. In doing so, governors are often faced with the hard choice of having to determine whether to address the immediate needs of the fishers or focus on things that matter more in the long run, such as building institutions for sustaining the fish stocks (Onyango and Jentoft 2011). This places Lake Victoria governors in a dilemma as far as poverty is concerned. Restrictions on harvesting may benefit the resource and

Table 9.1 A framework for assessing governability of poverty issues in small-scale fisheries

Governance orders	Governing system	System-to-be-governed	Governing interaction
Decision making	What governing modes are used to reach decisions? Are they top-down, interactive or bottom-up?	What are the social and ecological impacts of the decisions made to address poverty? Are these decisions considered fair and transparent from the perspective of the poor?	How do individuals, groups and governing institutions interact as they negotiate decisions?
Institutions	How are governing systems institution alized? How is power exercised, responsibilities and mandates distributed and with what outcomes?	How are social practices, interests and power institutionalized? How do they determine social relationships and the distribution of cost and benefits among stakeholders?	What are the institutional characteristics of governing interactions? What rules exist pertaining to representation, participation and communication?
Values, norms and principles	How do values, norms and principles of governing institutions relate to problem definition, agenda setting and conflict resolution?	What values, principles and norms underpin social action and local institutional formations, decision- making and power relations?	How are values, principles and norms shared among stakeholders in their interactions?

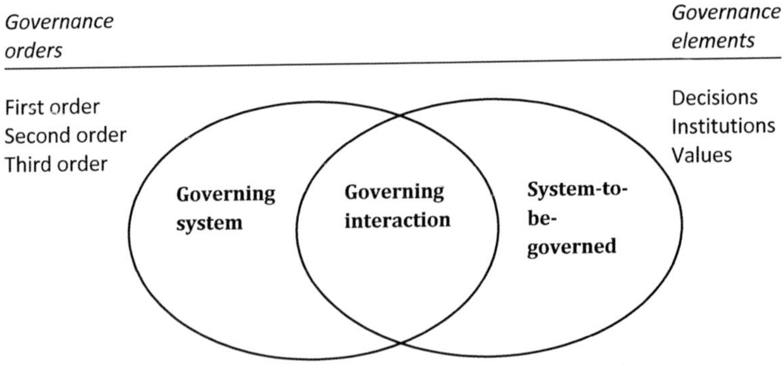

Fig. 9.3 Governability assessment framework

then help to alleviate poverty in the long run, but, in the short run, may take food out of the mouth of the poor, so to speak, if they do not have other ways of compensating for reduced fish supply or income lost.

Harris (2010) argues that making decisions (or hard choices) reflects the values and preferences of the decision maker. In this case, each poverty alleviation governor in the Lake Victoria fisheries system will be guided by the values that he or she embodies or that have been established by the organizations that they belong to and represent. In the next instance, these values are converted into which concerns should have priority over others and into operating goals for the governing system. For the Ministry of Livestock and Fisheries Development, ECOVIC and LANESO, the goal of sustainable fisheries goal takes precedence, while giving opportunity to fishers through credit and savings facilities is the primary interest of, for instance, the FINCA and Mwanza Women Development Association (MWADA).

Decision-making involves a process whereby the decisions made should generate the highest utility, while at the same time fit with values and aspirations of the people concerned. This is also to say that value judgments are at play when decisions are being made (Jentoft and Midré 2011). Our empirical information about the various governance levels discussed under the systems-to-be-governed shows that governors listed are guided by values that focus on sustainability of the fish resources, wealth generation, empowerment and fairness as they set priorities and actions to address poverty. This has crucial policy and governance implications.

In the second order governance, where institutions are established to provide mechanisms for decision-making, the assessment of governability involves examining how power is distributed within and among the poverty alleviation governors mentioned above. As indicated in Figs. 9.2 and 9.3, in the case of Lake Victoria, one would have to examine the distribution of power between the BMU and the governors at each scale: county, division, district, national or regional. Jentoft and Midré (2011) argue that social elites have both the political and economic power at the heart of definitions of poverty and the ways in which it has been addressed. Power is, therefore, a central variable in assessing the governability of poverty.

The challenge that power presents to wicked problems such as poverty is that a skewed distribution of power towards the elite makes them appear to be 'poverty experts' (see Chuenpagdee and Juntarashote 2011). Rather than the poor themselves, they are the ones who are perceived to hold the ideas for designing alleviation mechanisms and mobilizing resources for poverty alleviation. Poor fishers are therefore hardly consulted when poverty alleviation policies are developed and implemented, as they are perceived to be void of any knowledge or ability to get themselves out of poverty. Accordingly, they are people who only need to be assisted. This perception has had implications for governing poverty and could also explain why poverty has not been wiped out from the fishing communities.

Within the systems-to-be-governed, an assessment of the governability of poverty alleviation at the second order warrants an investigation of how social practices formulate their own laws, interests and power and are institutionalized. How such institutions ensure access and distribution of resources, structure relations with regards to conducting beach meetings, patrolling fishing grounds, etc. are among the things that also needs to be assessed.

Within governing interactions, the assessment looks at the characteristics of governing interactions that takes place among and between poverty alleviation governors. How do LVFO and the Ministry of Livestock and Fisheries Development interact with the non-governmental organizations previously discussed? It examines the rules of participation and representation and the modes established to facilitate communication and cooperation between the system-to-be governed and the governing system. In other words, it asks: are decisions made at the top and implemented in a top-down fashion, from bottom-up, or in a more interactive co-governance way? Do the decisions correspond to the way poverty is experienced in the fishing communities? What characterizes the processes within which poverty is defined, assessed and deliberated? Is the actual governing mode inhibiting or facilitating a constructive process? (See Table 9.1).

In the third order or 'meta' governance, the assessment of governability within the governing system must examine the ways in which LVFO, TAHEA, EMEDO and FINCA, among other organizations, define poverty in general, as well as in the specific cases of the Nyakasenge and Kasheno fishing villages. In doing so, it must ask: which values, norms and principles underpin their approach to poverty alleviation? In the systems-to-be-governed, how is poverty experienced and defined within and among these fishing communities? How do governors perceive it? Which values, norms and principles are considered crucial for poverty alleviation? Lastly, within the governing interaction, the assessment looks at how values, norms and principles are shared among governors as they interact to alleviate poverty.

Governability Challenges at the Community Level

The governability challenge is not only experienced by an external authority vis-a-vis a local community; it is also an issue for local people. In the case of the Kasheno and Nyakasenge villages, this is illustrated by issues like regulatory compliance and the banning of beach seine equipment, for which the BMUs are responsible. In both communities, fishing activities that use beach seining equipment enable people to get food on the table. They devote all their efforts towards finding food for the day. The fishers say that unless they are provided with food, they have no alternative but to catch fish in whatever way. To them, there is no logic in obeying regulations on an empty stomach. They therefore argue that they are left with little choice; it is either you obey the government's rules and starve or look for a way out and survive. Although fishery officials sometimes conduct unannounced patrols with BMUs, the fishers have been able to avoid being caught by resorting to fishing at night, fishing from areas that are inaccessible to patrolling agents, burying beach seines in the ground after fishing to hide them, and, at times, paying bribes to fisheries staff and BMU leaders.

Furthermore, locals perceive the BMUs to be an extension of the Fisheries Administration. This is grounded in the fact that, during BMU establishment, it was the division staff that conceived the idea and led the formation process. In fact, in the eyes of the fishers, the BMUs were formed to stop illegal fishing practices and ensure beach cleanliness. Fishers have therefore developed a perception that the roles of these Units run counter to their daily necessities. Nevertheless, many in the fishers' communities do not consider it a good idea to reject a government proposal, even if you are opposed to it. Instead, it is better to show approval of it. The primary explanation for this is that the government is perceived to have all the machinery and power to undertake whatever it desires to do, regardless of what people in small-scale fishing communities are thinking. There is also an economic aspect to it. When such plans involve committees, those who anticipate becoming a member of the committees always foresee a monetary reward for their membership. Moreover, elections of BMUs are at times used as springboards by people whose interests are focused on higher elective positions.

At the formation of BMUs, those who were elected were thoroughly vetted by the community members. The vetting process revolved around being able to accommodate for the community's norms, and avoid revealing the practices of the community. Given the use of an assortment of gears, including the banned ones, a BMU member from the local community carries the moral obligation to protect the local practice. This, however, runs counter to the expectation of the government, who established the BMUs as a means of deterring such practices. In both communities, some members of the BMU and women groups have been removed from office, because they were not able to stand by the manner in which the community had previously agreed.

Many of the local fishing communities conduct their activities with a great deal of independence. For example, the Tweyambe Fishing enterprise, located in Kasheno village, runs their organization through various projects, resisting any outside interference (Onyango and Jentoft 2010). Similarly, Nyakasenge residents have had a long practice of only trusting local residents rather than outsiders. In Nyakasenge, residents can protect one another if visited by fisheries authorities on patrol by concealing the names of those who own illegal fishing gear. If, for example, an unregistered boat is found at the beach by the patrol team, residents will refuse to name the owner. Instead, they suddenly claim to be visitors without any knowledge or awareness of what goes on in the village. In claiming to be visitors, they strategically adopt a stance that allows them to run their own affairs without interference.

In the event of inspection, many BMU members actively pursue a stance that demonstrates that BMUs are actually operational and effective in conducting their duties. They typically acknowledge but a few problems, those regarding a lack of proper equipment and confrontations with outsiders who do not wish to listen to the rule of law. In reality, however, these responses are merely a way of dodging inquests by the authorities. For example, if one is confronted with questions related to the existence of banned gear on beaches under an individual's jurisdiction, a person would categorically answer in the negative. For instance a beach seine owner or fisher would knowingly deny existence of beach seining in their fishing village.

Concluding Remarks

In this chapter we have discussed poverty and poverty alleviation as a 'wicked' problem to which severe governability issues are attached. Poverty is a composite concept whose understanding, boundaries and qualities are dependent on who is in charge of defining it. We believe that it makes more sense to locate the poverty problem within both the system-to-be governed and the governing system, and to examine carefully how the interaction between the two is structured institutionally and how it works as a communicative process. Somehow, the governability of poverty will hinge upon the degree to which this process helps to create consensus with regard to the values, images and norms between those who belong to and represent both systems at different scales – i.e. local, regional, national and even global, if one considers the role that international NGOs play in communities along Lake Victoria.

In this particular case, the BMUs are most likely to be part of the answer. However, as complex and dynamic governing systems operating at the community level, they also have their own governability problem that requires careful assessment if poverty alleviation strategies are going to succeed in a way that makes a positive difference in people's lives.

References

Alkire, S. (2008). Multidimensional poverty measures. Oxford Poverty & Human Development Initiative. Retrieved from http://hdr.undp.org/en/media/Alkire_Multidimensional_Poverty.ppt

Alkire, S., & Foster, J. (2008). Counting and multidimensional poverty measurement. Oxford Poverty and Human Development Initiative (OPHI), Working Article Number 7. Retrieved from http://mora.rente.nhh.no/projects/EqualityExchange/Portals/0/articles/Alkire_Foster_CountingMultidimensional_Poverty.pdf

Béné, C. (2003). When fishery rhymes with poverty: A first step beyond the old paradigm on poverty in small-scale fisheries. World Development, 31(6), 949–975.

Béné, C. (2004). Fisheries management and poverty. In A. E. Neiland, & C. Béné (Eds.), Poverty and small-scale fisheries in West Africa (pp. 61–82). Dordrecht: FAO/Kluwer Academic.

Beverton, R.J.H., & Holt, S.J. (1957). On the dynamics of exploited fish populations. London: Chapman and Hall.
Chuenpagdee, R., & Juntarashote, K. (2011). Learning from the experts: Attaining sufficiency in small-scale fishing communities in Thailand. In S. Jentoft, & A. Eide (Eds.), Poverty mosaics: Realities and prospects in small-scale fisheries (pp. 309–331). Dordrecht: Springer.
Conklin, J. (2006). Dialogue mapping: Building shared understanding of wicked Problems. West Sussex: Wiley.
Copes, P. (1989). Why are fishing incomes often low? A critical review of the conventional wisdom. Discussion Paper 21. Burnaby: Institute of Fisheries Analysis, Simon Fraser University.
Cunningham, S. (1993). Fishermen's incomes and fisheries management. Marine Resource Economics, 9(3), 241–252.
FAO. (2005). Increasing the contribution of small-scale fisheries to poverty alleviation and food security. FAO Technical Guidelines for Responsible Fisheries. No. 10. Rome: FAO.
Geheb, K. (1997). The regulators and the regulated: Fisheries management, options and dynamics in Kenya's Lake Victoria Fishery. Unpublished D.Phil. Thesis, University of Sussex, Falmer, Brighton, U.K. Reprinted as LVFRP Technical Document No. 10. LVFRP/TECH/00/10. The Socio-economic Data Working Group of the Lake Victoria Fisheries Research Project, Jinja.
Gordon, S.H. (1954). The economic theory of a common property resource: The fishery. Journal of Political Economy, 62, 124–142.
Graham, M. (1929). The Victoria Nyanza and its fisheries. A report on the fishing surveys of Lake Victoria (1927–28). London: Crown Agents Colonies.
Hanson, F.A. (1997). How poverty lost its meaning. Cato Journal, 17(2), 189–209.
Hardin, G. (1968). The tragedy of the commons. Science, 162, 1243–1248.
Harris, R. (2010,). Introduction to decision making (November 22 version). Retrieved April 6, 2011, from http://www.virtualsalt.com/crebook5.htm
Hyatt, S.B. (2001). From citizen to volunteer: Neoliberal governance and the erasure of poverty inequality. In L. Goode & J. Maskovsky (Eds.), The new poverty studies: The Ethnography of power, politics and impoverished people in the United States (p. 320). New York: New York University Press.
Jalal, K.F. (1993). Sustainable development, environment and poverty nexus. Occasional Papers No. 7. Manila: Asian Development Bank.
Jambiya, G. (2004). Study of impact of fisheries activities on resource and environmental degradation. LVEMP Technical Report. Dar es Salaam: LVEMP.
Jentoft, S., & Eide, A. (Eds.). (2011). Poverty mosaics: Realities and prospects in small-scale fisheries. Dordrecht: Springer.
Jentoft, S., & Midré, G. (2011). The meaning of poverty: Conceptual issues in small-scale fisheries research. In S. Jentoft, & A. Eide (Eds.), Poverty mosaics: Realities and prospects in small-scale fisheries (pp. 43–68). Dordrecht: Springer.
Johnson, C.T., Kelts, K., Odada, E. (2000). The Holocene history of Lake Victoria. AMBIO: Journal of Human Environment, 29(1), 2–11.
Kooiman, J. (2003). Governing as governance. New Delhi: Sage.
Kooiman, J., Bavinck, M., Jentoft, S., Pullin, R. (Eds.). (2005). Fish for life: Interactive governance for fisheries. Amsterdam: Amsterdam University Press.
Krishna, A. (2010). One illness away: Why people become poor and how they escape poverty. New York: Oxford University Press.
Kurien, J. (2005). Responsible fish trade and food security. FAO Fisheries Technical Paper, No. 456. Rome: FAO.
Ligtvoet, W., Mouse, P.J., Mkumbo, O.C., Budeba, Y.L., Goudswaard, P.C., Katunzi, E.F.B., Temu, M.M., Wanink J.H., Witte, F. (1995). The Lake Victoria fish stocks and fisheries. In F. Witte, & W.L.T. van Densen (Eds.), Fish stocks and fisheries of Lake Victoria. A handbook for field observations (pp. 11–53). Cardigan: Samara Publishing Limited.
Lewis, O. (1963). The culture of poverty. Society, 1(1), 17–19.
Lukunga, S.A. (2005). Fish quality assurance. In Y.D. Mgaya (Ed.), Synthesis report on Fisheries Research and Management, Tanzania. Final report LVEMP 2005 (pp. 212–227). Dar e Salaam: LVEMP.

Narayan, D., Chambers, R., Shah, M., Petesh, K. (2000). Voices of the poor crying out for change. New York/Washington, DC: Oxford University Press/The World Bank.

Odongkara, K., Abila, R.O., Luomba, J. (2009). The contribution of Lake Victoria fisheries to national economies. African Journal of Tropical Hydrobiology and Fisheries, 12, 47–51.

Okeyo-Owuor, J.B. (1999). A review of biodiversity and socio-economics research in relation to fisheries in Lake Victoria. Socio-economics of the Lake Victoria Fisheries. IUCN Eastern Africa Programme. Report No. 5. Nairobi: IUCN.

Onyango, P.O. (2009). Re-configuring poverty: The wickedness perspective. African Journal of Tropical Hydrobiology and Fisheries, 12, 37–46.

Onyango, P.O. (2011). Occupation of last resort? Small-scale fishing in Lake Victoria, Tanzania. In S. Jentoft, & A. Eide (Eds.), Poverty mosaics: Realities and prospects in small-scale fisheries (pp. 97–124). Dordrecht: Springer.

Onyango, P.O., & Jentoft, S. (2010). Assessing poverty in small-scale fisheries in Lake Victoria, Tanzania. Fish and Fisheries, 11, 250–263.

Onyango, P.O., & Jentoft, S. (2011). Climbing the hill: Poverty alleviation, gender relationships, and women's social entrepreneurship in Lake Victoria, Tanzania. Maritime Studies (MAST), 10(2), 117–140.

RAWG (Research and Analysis Working Group). (2007). Poverty and human development report 2007. Dar es Salaam: REPOA.

Rittel, H.W.J., & Webber, M.M. (1973). Dilemmas in a general theory of planning. Policy Science, 4, 155–169.

Sauper, H. (2005). Darwin's nightmare. (Film production). Available at http://www.coop99.at/darwins-nightmare/darwin/html/startset.htm

Sen, A. (1981). Poverty and famines: An essay on entitlement and deprivation. Oxford: Clarendon Press.

Sen, A. (1999). Development as freedom. Oxford: Oxford University Press.

Sethuraman, S.V. (1998). Gender, informality and poverty: A global review – Gender bias in female informal employment and incomes in developing countries. Draft version of a study with the World Bank (Poverty Reduction and Economic Management Department) and WIEGO. Retrieved from http://www.wiego.org/papers/sethcontents.html

Smith, I. (1979). A research framework for traditional fisheries. Manila: ICLARM.

WCED (World Commission on Environment and Development). (1987). Our common future. New York: United Nations. Retrieved from www.cmsweb1.loudoun.k12.va.us

World Bank. (1992). World development report. Oxford: Oxford University Press.

Chapter 10
South Indian Trawl Fisheries – Assessing Their Governability

Joeri Scholtens and Maarten Bavinck

Abstract As governance occurs through interaction between governors and the governed, an investigation of the conditions that facilitate 'better' interaction, and improve governability, is worthwhile. We argue on the basis of the interactive governance approach that one such condition is a proper match between the system-to-be-governed and the governing system. This match is conceptualized as having two dimensions: architectural compatibility and attunement. The possibilities of this assessment approach are studied in the context of the trawl fisheries in the Palk Bay, South India. The trawl fleet in this region developed rapidly since the 1960s, and has engaged in lively conflicts with small-scale fisheries on the Indian, and more recently on the Sri Lankan side of the Bay. The non-synchronization of fishing areas with administrative boundaries constitutes an important governance challenge, as is environmental pressure.

Keywords Match • Governability assessment • Institutional fit • India • Responsiveness • Trawl fisheries

Introduction

The global crisis in fisheries has resulted in the formulation of panaceas, the content of which tend to depend on their proponents' disciplinary backgrounds (Degnbol et al. 2006). As Jentoft et al. (2008) have pointed out, however, panaceas do not take account of contextual variation and therefore frequently fail to achieve their objectives (cf. Ostrom 2007). This chapter proceeds from the assumption that contextual

---

J. Scholtens (✉) • M. Bavinck
Department of Human Geography, Planning and International Development Studies, University of Amsterdam, Amsterdam, The Netherlands
e-mail: j.scholtens@uva.nl; j.m.bavinck@uva.nl

M. Bavinck et al. (eds.), *Governability of Fisheries and Aquaculture: Theory and Applications*, MARE Publication Series 7, DOI 10.1007/978-94-007-6107-0_10,

diversity matters, and that governance solutions should match the nature of the problems at hand.

Our conceptual entry point is interactive governance theory and the concept of governability, defined as "the overall capacity for governance of a societal entity or system" (Kooiman et al. 2008, 3). Interactive governance theory assumes that societal systems are characterized by diversity, complexity, dynamics, and scale variations, and that these features have implications for the success of a governing effort. Applied to a fisheries context, this approach suggests that there needs to be a match between the characteristics of a fishery (i.e. the system-to-e-governed) and the set of prevailing institutions (i.e. the governing system) that steer it in a required direction.

This chapter aims to contribute to the theoretical debate on the 'match' between the governing system and system-to-be-governed, thereby advancing the methodology of assessing governability. It focuses on the features of diversity, dynamics and scale,[1] and takes one specific South Indian fishery as a case-in-point. The trawler fishery of the Palk Bay provides many people in the region with economic opportunities, but has generated a range of ecological and social problems as well. Fish stocks are declining, and there are substantial conflicts between this category of fishers and others partaking in the fisheries. Rather than attempting an answer to the governability problems afflicting this fishery, we highlight the process of assessment. Our supposition is that if one can get a grip on the nature of systemic mismatches, it will be easier to make suggestions for improving governability.

The case study developed in this chapter is based on 5 months of ethnographic fieldwork on trawl fishing in the Palk Bay by the first author during two periods in 2006 and 2007. Fieldwork data were collected primarily through semi-structured interviews with trawler owners and crewmembers, who were selected by means of a stratified convenience sample. Additional in-depth interviews were held with secondary stakeholders, such as government authorities, association leaders, coastguard personnel, boat breakers, fish and shrimp traders, and NGOs. For additional details on methodology see Scholtens (2006) and Sathyapalan et al. (2007). The second author has been investigating the socio-economic character of the marine fisheries of Tamil Nadu since 1995, and currently leads a research project on fishing conflicts in the Palk Bay.

Assessing Governability: A Matter of Matching Systems

Interactive governance theory argues that fisheries systems are inherently diverse and dynamic, and that they operate at multiple scale levels (Kooiman and Bavinck 2005). Each system-to-be-governed possesses these features to varying extents and

[1] A third feature ascribed to societal systems, complexity, is relevant to the topic (see Chap. 2 by Kooiman and Bavinck, this volume), but left aside here for reasons of convenience. Not only is complexity more difficult to illustrate than diversity and dynamics – it would also have extended this chapter beyond the available word limit.

therefore places different demands on the governing system. According to Jentoft (2007), a fishery's governability, understood as the overall capacity for governance of a societal entity or system, is largely dependent on whether a governing system is capable of handling the constellation of features prevalent in the system-to-be-governed. A proper match between the natural and socio-economical characteristics of a given fishery and the institutions that are governing this fishery is therefore a condition for effective governance (Jentoft 2007; Mahon 2008).[2]

How can match be imagined? The kitchen provides a suitable metaphor in the form of a pot and a lid, which constitute a functional unit. As a badly fitting lid negatively affects the cooking process, the cook will try to improve the fit by adjusting one of the two pieces. Translated to the terminology of interactive governance, we are interested in improving the match between the system-to-be-governed (pot) and the governing system (lid). Governability enquiries ask to what extent the architecture of governing institutions is appropriate given the features and scale of the system it is to govern.

The pot-and-lid metaphor highlights the structural side of match, but ignores the fact that fisheries systems consist of people, ecosystems and institutions that not fixed, but rather in a constant state of flux. In view of the dynamic nature of the components, a proper match can therefore never be a matter of architecture alone. Without continuous fine-tuning, a governing system would soon lose its grip on the system-to-be-governed, resulting in a poor match and hence low governability.

Building upon the discussion above, we suggest defining match in terms of architectural compatibility as well as attunement. Note that match is not a black-or-white issue. The states of 'match' and 'non-match' merely represent extreme positions on the scale of the relationship between the governing system and system-to-be-governed. Many positions, in fact, exist in between them. Thus "governability is a condition fluctuating on a continuing scale without ever reaching the extremes; it is never total or complete, and [...] it is never wholly absent either" (Bavinck and Kooiman 2008).

In the following section, we consider architectural compatibility particularly from the perspective of matching the spatial scales available in a fisheries system. In addition, we consider the extent to which linkages between different governing actors, as well as between governors and those being governed, compensate for existing disjunctures. We then turn to the issue of attunement and the question of whether a governing system is able to meet the demands emanating from a system-to-be-governed.

---

[2] The notion that institutions must fit with the context in which they are to work is not unique to interactive governance. Young (2002, 54) for example, argues that: "all accounts of the problem of fit start from the same premise: the effectiveness of environmental and resource regimes ... is determined in considerable measure by the degree to which they are compatible with the biogeophysical systems with which they interact." Ostrom (2007) broadens the analyses by also including the social system. She suggests an institutional diagnostics approach that highlights critical features of a resource problem followed by an effort to specify the most suitable institutional arrangement. This corresponds with interactive governance theory on all but one count: whereas Ostrom takes the objectives of governance as externally given, scholars in the field of interactive governance consider objectives to be subjected to negotiation (Chuenpagdee and Jentoft 2009).

Finally, we explore the relationship between governability and governing modes, or styles. Modes constitute an important component in the edifice of interaction theory, and are considered to play an important role in determining governability (see Chaps. 1 and 2 of this volume).

Architectural Compatibility: Spatial Correspondence and Linkages

Primary activities like fishing take place within bounded geographical spaces. These spaces contain 'things' and 'processes' of various kinds that operate at different scale levels. We can thus distinguish ecosystems from social, economic and administrative systems, each of which functions at its own scale level. Mismatches between these scales are not only common, they can also be extremely problematic. Cumming et al. (2006, 3) therefore argue, "When the scales of social organization and environmental variation are mismatched, problems inevitably arise in either the social institutions that are responsible for management or the ecological systems that are being managed." In a similar vein, interactive governance theory argues that a match between the scale level of the governing system and that of the system-to-be-governed is crucial for governance systems to function well (Kooiman 2008).

Given the multiplicity of units and activities in each system, each with their own boundaries, a coincidence of scales is anything but a regular affair. A system-to-be-governed may function at multiple levels of scale at the same time, as ecosystem boundaries may, for example, not correspond with the boundaries of fishing in a particular locality. On the side of the governing system, a particular fishery is often subjected to multiple governing systems operating at different levels: e.g. a fishing association working at a local level and a fisheries department operating at national or regional levels. Although administrative systems are sometimes nested, situations of legal pluralism are common (Jentoft et al. 2009). The first topic for assessing the governability of fishery systems is therefore the matching, or non-matching, of scale levels within the fishery.

Given the fact that the scales of a governing system rarely coincide perfectly with a system-to-be-governed, Fanning et al. (2007) point out the importance of vertical and horizontal interactions between governors, who can act to modify discrepancies of scale. Communication channels between local, national and international governing actors constitute vertical linkages between scale levels, whereas horizontal linkages are the channels between governing bodies that co-exist at similar levels. The content of linkages varies from information sharing to full-fledged cooperation. Assessing the nature of linkages between different governing agents is important as they provide insight into the capacity of the governing system to overcome discrepancies of scale.

Q1: To what extent does the scale of the natural system-to-be-governed, social-economic system-to-be-governed and governing system coincide, and what is the nature of the linkages between the different governing system actors?

Attunement: Sensitivity, Flexibility and Responsiveness

Bavinck and Salagrama (2008) argue that rather than a mirroring of features, the important issue is the availability of a "governance induced response" to problems that occur in the system-to-be-governed (Bavinck and Salagrama 2008, 7; also see Chuenpagdee and Jentoft 2009, 112). In this section we are first interested in exploring the structural factors that facilitate such responsiveness.

Jentoft (2007, 364) argues that if diversity of a system-to-be-governed is high, it is crucial for governance to take place close to where the particular problem or opportunity is manifested. Governing a diverse system-to-be-governed requires the solicitation of expertise and experience from stakeholders; it calls for sensitive and particularistic governing approaches. Queries related to sensitivity include: are there institutionalized arrangements for collecting feedback on the effects of governance, such as regular stakeholder consultation meetings; is there a point of reference that guides the question of who is incorporated in such meetings?

On the other hand, dynamics in a system-to-be-governed, notably the systems that face a high incidence of (sometimes unpredictable) change, call for a governing system that is flexible. A flexible governing system is characterized by a continuous willingness to adapt to changing circumstances and a preparedness for sudden change and discontinuity (Jentoft 2007). Do governors consider interventions to be trial-and-error operations or fixed frames to be imposed on the system-to-be-governed? Is there a proper system of monitoring and evaluation to facilitate learning? Assessing a fishery along the criteria of sensitivity and flexibility provides insight into the availability of preconditions for attunement – the topic we turn to next.

Q2: To what extent does the available governing system possess qualities of sensitivity and flexibility required for dealing with diversity and dynamics in the system-to-be-governed?

Attunement denotes an interactive process between the governing system and the system-to-be-governed, which is intended to address the problems and opportunities that occur in a fishery. Working in the field of corporate governance, Orlitzky and Swanson (2002) argue that, in a firm, attunement is basically determined by an executive's receptivity and the extent to which he/she uses his/her discretionary powers to facilitate a response. Translated to the field of resource management, this suggests that attunement depends on particular qualities of the governing system. The first of these qualities is receptivity, while the second is the willingness to proactively deal with the information received. As Pitkin (1968) argues, "There need not to be a constant activity of responding but there must be a constant condition of responsive*ness*, or potential readiness to respond" (Pitkin 1968, quoted in Kooiman 2003, 177, italic in original).

Q3: To what extent is the governing system responsive, that means receptive and proactive, towards the demands present in the system-to-be-governed?

Attunement can also be assessed through the modes by which governance takes place. Interactive governance theory distinguishes between three ideal-typical modes or styles of governance: self-governance, co-governance and hierarchical governance (Kooiman 2003). It is argued that each mode possesses particular capacities and weaknesses, and it is therefore reasonable to assume that different systems-to-be-governed require different (mixes of) governance modes. Before investigating which governance modes are required in a particular context, we inquire as to whether some governance modes are naturally more responsive than others. For example, in instances where governors govern themselves (self-governance), governance responses to emerging problems may be faster and to the point than in the case of hierarchical governance, which is typically slow and bureaucratic. This does not, however, imply that self-governance, from the viewpoint of match, is necessarily more desirable. Self-governance bodies are typically local and their authority has a limited spatial scale – as a consequence, they are frequently unable to deal with problems that exceed this scale level. What is more, self- and co-modes of governance often need a stick in the background, "if not for other reasons then at least for the well-known free-riders who may threaten cooperative efforts" (Pascual-Fernandez et al. 2005, 221). Hence, interactive governance theory argues that many problems must be addressed by a combination of governance modes, rather than a single one.

Jentoft (2007) has proposed two ways of investigating the process of optimizing modes on the basis of the characteristics of a given fishery. The first approach inquires into the governing mode that best fits the features of the system-to-be-governed. Jentoft (2007) argues that a high measure of diversity in a system-to-be-governed is best handled in the self-governance mode, which is most sensitive to variation. Dynamics, on the other hand, requires co-governance, as this tends to be more inclusive and flexible. He adds that systems that are particularly vulnerable may well benefit from a more hierarchical style of governance.

The second approach proceeds from the principle of subsidiarity (Jentoft et al. 2011). Subsidiarity suggests that responsibility for governance should be located at the lowest possible organizational level (Bavinck and Jentoft 2011). Hierarchical governance is then the fallback alternative if neither self-governance nor co-governance are up to the tasks at hand. When assessing the suitability of the current mix of governance modes, we should thus also inquire whether there are issues that are dealt with at an unnecessarily high governance level, and whether a devolution of authority might yield benefits.

Q4: To what extent is the present mix of governance modes appropriate for dealing with diversity and dynamics in the system-to-be-governed, and to what extent are issues in the system-to-be-governed dealt with appropriate levels of governance?

The Palk Bay Trawl Fisheries

The demarcation of the boundaries of the societal system in question is a useful starting point for a governability assessment. The first boundary to be considered here is a geographical one. It is designated by the term Palk Bay, a marine space

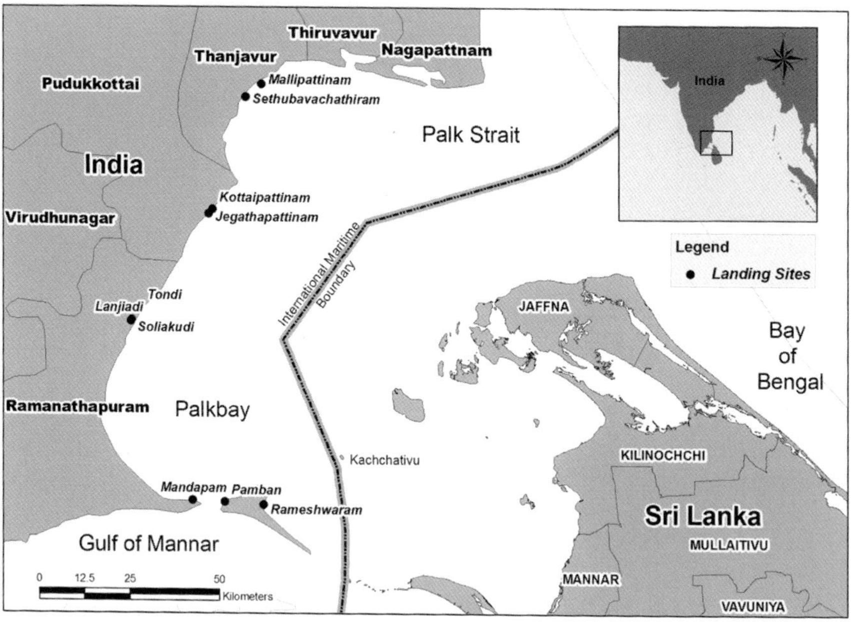

Fig. 10.1 Trawler landing centers and the International Boundary Line in the Palk Bay (Adapted from Google Earth 2005. Source: University of Jaffna, Department of Geography)

situated between southeast India and northwest Sri Lanka (Fig. 10.1). The second boundary is technological in nature: our analysis is limited to the Indian trawl fisheries, thus excluding Sri Lankan and Indian fishers who operate small-scale craft. As there are multiple interactions between these various fisheries sub-sectors in the region, however, the analysis will sometimes be more inclusive. A final demarcation follows from the fact that we focus on the capture segment of the fish chain. This implies that traders, processors, boat makers, as well as those employed in a range of other occupations directly or indirectly connected to the fishery sector, figure only in the margins of the analysis.

The Natural System-to-Be-Governed

On the Indian side, the Palk Bay covers an area of about 11,300 km^2 and a coastline that is 270 km in length (Sathyapalan et al. 2007). In the north, the Palk Bay is connected to the Bay of Bengal, while in the south it borders the Gulf of Mannar, from which it is separated by a chain of small islands and reef shoals. The Palk Bay is a relatively shallow basin with an average depth of 9 m and is known for its lack of turbulence. However, for one period each year (October to December), while the northeast monsoon rages, the waters of the Palk Bay turn rough and fishing becomes challenging.

The Palk Bay is also known in terms of its rich biodiversity. Parts of the region's coast are covered by mangroves, while just offshore, the area contains a number of coral reefs and hosts an exceptional diversity of sponges, molluscs, crustaceans, seaweeds, sea grasses and fish (Venkataraman and Wafar 2004). In her analysis of India's coastal resources, however, Bhathal (2005, 64) argues that Tamil Nadu's marine resources are in "a downward spiral with no reversal in sight". This process manifests itself in a decline of the mean trophic level. Many fishers have already signaled the alarming trend, reporting that the species they often encountered in the past now seem to have disappeared. They also point out that the average size of the species caught has reduced. More specifically, the catch per unit of effort (CPUE) appears to be diminishing, as fishers note that aggregate catches have not increased in line with the expansion of fishing effort.

In conclusion, the natural system-to-be-governed is characterized by a high diversity of species, typical of a shallow marine ecosystem in the tropics. In terms of dynamics, two aspects stand out: (a) the seasonal monsoon, which disrupts the Bay's tranquility and affects fishing patterns; and (b) the steady growth of fishing pressure that seems to exert a negative influence on the rich ecology of the area.

The Socio-economic System-to-Be-Governed

Fishing Activity

Trawlers were introduced to the Palk Bay (and India) in the late 1950s through an intensive governmental modernization program (Bavinck 2003). The size of the fleet increased gradually to about 2,800 in 2000, after which the figure reduced to the current 1,750 (during low season) and 2,170 (during high season) in 2007 (Sathyapalan et al. 2007). This trawl fleet is dispersed over ten locations (see Fig. 10.1), ranging in size from a jetty at Tondi (a mere 12 boats) to Rameswaram harbor (670 vessels).[3]

Compared to other fleets in Tamil Nadu, the trawlers operating in the Palk Bay are generally small in size and capacity (length: 28–45 ft; capacity: 50–120 hp). Most boats lack global positioning devices, echo sounders and other advanced fishing instruments (Sathyapalan et al. 2007). In addition, the majority of boats are poorly maintained.

The spatial and temporal scope of fishing activity is partly determined by a set of governmental rules and regulations. Trawl fishers have officially been allocated a fishing ground that commences at three nautical miles (nm) from the coast, and extends to the Indian-Sri Lankan International Maritime Boundary Line (IMBL) (located between 14 and 30 nm from the Indian coastline). These restrictions imply that each trawler has no more than 2.5 km^2 of fishing ground, which is considerably lower than the Indian average of 10 km^2. In practice, however, fishing grounds extend well beyond the IMBL into Sri Lanka waters and also into the inshore waters reserved for small-scale fishers.

[3] The number of trawlers per village is based on our own counts, carried out from February to April 2007], which were lower than the official figures provided by the Fisheries Department.

The temporal range of operations is determined by a governmental regulation dictating that trawl fishers are only permitted to leave port on three alternate days per week (see Box 10.1). This implies that fishing trips never exceed 24 h. As a result, the spatial range of the trawlers is practically limited to the Palk Bay.

Demography

The population operating trawl vessels in the Palk Bay is remarkably heterogeneous compared to other coastal areas in Tamil Nadu (Bavinck 2003). Trawl fishers are diverse in terms of religious persuasion, caste, occupational and geographic origin, and affluence. Muslims, Hindus and Christians are all involved, each grouping having distinct places of worship, rituals and festivals. Some have lived on the Bay for generations, while others have moved in recently.

We distinguish between two basic types of people in trawl fisheries: 'traditional fishers' who have inherited the profession and belong to a fishing caste (such as Paravar, Pattinavar or Mukkuvar), and 'non-traditional fishers' who have entered the occupation over the course of recent decades (Bavinck et al. 2008). Although members of different castes often work together and attend each other's festivals, the general trend is one of segregation. Diversity is found not only at the individual or the group level, but substantial variation can also be observed at the level of landing centers. For example, some landing centers are populated by people of single caste or religion, while others have a mixed character (Sathyapalan et al. 2007).

Three points emerge from the discussion above. In terms of scale, it first appears that even though trawlers may regularly shift to other ports, fishing activity is limited to the Palk Bay. Second, diversity is substantial at both the individual, group and landing center level. In terms of dynamics, the incidence of migration and the substantial increases in the number of trawlers over recent decades (but also their decline in recent years) stand out the most.

Key Demands from the System-to-Be-Governed

For investigating the responsiveness of the governing system to the demands of the system-to-be-governed, we now discuss four issues that threaten the sustainability of the trawl fisheries and have required continuous attention from governors over a longer time period. These issues have been selected for their severity, but also for representing different societal concerns, namely human security, social justice, ecosystem health, and economic viability.

Ecosystem Health: Overfishing and Destructive Gears

In the section above, we mentioned the evidence that fishery resources in the Palk Bay are degrading. Although systematic research on this topic is still lacking, it is clear that trawl fishing effort plays a major role.[4] Sathyapalan et al. (2007) argue that the current fleet of about 2,000 trawlers is heavily overcapitalized. First, the average fishing area available per trawler on the Indian side of the Palk Bay has steadily declined; secondly, on average one-third of the boats remain idle on days that fishing is allowed; and finally, the technical efficiency of the boats is as low as 78% (Sathyapalan et al. 2007).

Harmful fishing techniques constitute the other side of the coin. Bottom trawling is the most widely practiced trawl technique in the Palk Bay, and this technique is increasingly recognized as destructive (Pauly et al. 2002; FAO 2004).

Economic Viability: Decreasing Profitability

Diesel costs constitute 70–90% of the operating costs in trawler fishing (Sathyapalan et al. 2007). The profitability of the trawling industry is thus extremely dependent on the development of fuel prices. Reflecting international trends, the price of diesel in the Palk Bay rose by more than 50% between 2002 and 2007. At the same time, shrimp prices actually decreased. Assuming that CPUE has not gone up during these years, we can conclude that the profitability of the trawl fishing occupation has, on average, gone down severely. One of the most visible signs of this trend is that during these 5 years at least 400 trawlers were sold to boat breakers.

Social Justice: Conflicts Between Trawler and Small-Scale Fisheries

People along the Indian coastline have been involved in fisheries for many centuries, initially making use of small-scale crafts and using their catch for own consumption as well as for sale at local markets. A trawl fleet came about gradually from the 1950s onwards. Rather than replacing the traditional fleet, however, trawl fishers started operating alongside and in competition with small-scale fishers. From the very beginning the relation between these two sub-sectors has been tense, at times resulting in physical violence (Bavinck 2001, 2003; Subramanian 2009). The main reasons for conflict are:

- Small-scale fishers typically fish with passive – standing – nets, while trawl fishers make use of active (moving) gear. Especially during the night, trawl boats run the risk of cutting through standing nets, resulting in painful losses for small-scale fishers.
- Small-scale fishers are convinced that trawlers destroy the ecosystem on which they rely, thus undermining the viability of their occupation; they also contest the rights of trawl fishers operating in 'their' inshore fishing grounds.

[4] The number of small-scale vessels plying the Palk Bay too has increased dramatically in the past decades, probably contributing to the overfishing problem. Our focus in this chapter, however, is on the 'big guys' – the trawl fishermen.

Safety and Security: The 'Sri Lankan Issue'

The governments of India and Sri Lanka agreed on the demarcation of the International Maritime Boundary Line (IMBL) in 1974, with a subsequent amendment in 1976. From that time on, Indian fishers were officially prohibited from carrying out operations in Sri Lankan waters (Suryanarayan 2004). During the Sri Lankan civil war (1983–2009), the Palk Bay obtained a reputation as an important smuggling route, and the Sea Tigers (the marine division of the LTTE[5]) competed with the Sri Lankan navy for control of the area. In order to get a grip on the movement of people and goods across the Bay, the Indian Government deployed a substantial naval force. This force has been maintained despite the cessation of the civil war.

As mentioned earlier, to this day many Indian trawl fishers venture to the Sri Lankan side of the IMBL, which is known for its relative lucrative fishing grounds. In harbors located particularly close to the border (like Rameswaram), trips to Sri Lanka are the rule rather than the exception. The trawl fishers in question argue that it is both imperative and just for them to venture into the Sri Lankan waters – imperative because this is the only way they can secure a catch good enough to compensate for their operational costs, and fair because they consider the IMBL to be illegitimate.

At the same time, however, Indian fishers crossing the IMBL know that they run a significant risk of confrontations with the armed forces or with irate Sri Lankan (small-scale) fishers, who contest their right to fish in their waters. In the period between 1983 and 2003, over a hundred Indian fishers were killed in the Palk Bay area, 326 fishers were injured, and 35 fishing vessels were destroyed (Suryanarayan 2004; Scholtens et al. 2012).

The Governing System: Actors and Instruments

A governing system consists of organizations, formal and informal rules and regulations, and economic incentives. This section describes the contours of three bodies most relevant to the Indian Palk Bay trawl fisheries, i.e. the central government, the Tamil Nadu state government and the boat owner associations.

State Government: Fisheries Department

According to the Indian Constitution (1950), state governments are responsible for inshore fishing. The Tamil Nadu government's involvement in fisheries management

[5] The Liberation Tigers of Tamil Eelam (LTTE), the guerilla force that strived for an independent State, during the Sri Lankan civil war.

Box 10.1 Government-Regulations Applicable to Palk Bay Trawlers

3–4 day rule: In order to solve ongoing conflicts between artisanal and trawler fishermen, a rule that is known as the alternate night schedule dictates that trawlers are only permitted at sea during 3 days a week with a maximum of 24 h (leaving 4*24 h to the non-mechanized sector).

Three nautical mile rule: One of the main clauses of the Tamil Nadu Fisheries Act of 1983 directs that trawlers are not allowed to fish within the first three nautical miles from the shore, to allow exclusive usage by non mechanized craft.

45-day ban: In 2001 the Government of Tamil Nadu introduced an annual ban on mechanized boat fishing, lasting from April 15th until May 29th, which became generally known as the 45-day ban. During this period, any sea going activity of mechanized boats is prohibited to allow regeneration of fish stocks.

Pair trawl ban: In between 2000 and 2002 the government of Tamil Nadu banned four types of nets through Government Orders, among which the pair trawl net which due to its large catching capacity had created much uproar between the trawlers and small scale crafts.

commenced in 1983 with the introduction of the Marine Fishing Regulation Act, which began to curb the independence of the local trawl fleets, including those in the Palk Bay. The Fisheries Department's (FD) role has since become more substantial and complicated. Apart from implementing a range of regulations (see Box 10.1) and mediating conflicts, the FD is also in charge of welfare schemes. Its personnel look after the implementation of a complex diesel subsidy scheme that was launched in 2004 as a result of increasing fuel prices. Finally, the FD provides fishers in the Palk Bay with identity cards and daily tokens with which they can identify themselves as genuine fishers. This diverse package of tasks means FD officials are bookkeepers, welfare providers and policemen too. The contradictions that emerge between the various roles weigh heavily on their shoulders (Bavinck 2001). The FD faces serious constraints in carrying out its comprehensive mandate. These constraints relate to a serious lack of financial and human resources (most offices are understaffed), and to the fact that they do not have any patrol boats at their disposal.

Central Government: Coast Guard and Navy

Trawl fishers in the Palk Bay are also subjected to governance measures by Central Government agencies. Security concerns have forced the Central Government to deploy 11 navy boats and 3–4 coast guard vessels in the region. As far as fisheries in the Palk Bay are concerned, the responsibilities of the coast guard and the navy do not differ significantly. Both agencies are responsible for ensuring that fishers do not engage in the smuggling of people and goods. They therefore help to enforce the following rules:

- Trawlers are prohibited from crossing the IMBL;
- Fishers must bear an identity card issued by the FD of Tamil Nadu;
- Each trawler heading for sea requires a fishing token for that particular day, issued and signed by the FD;
- Each trawler is allowed a maximum of five members;
- Each vessel may carry only a reasonable amount of diesel (enough for a regular fishing trip).

At the height of the civil war in Sri Lanka, the Indian navy and coast guard kept a close eye on the Indian trawl fleet, inflicting severe punishments on those that violated rules. Although the civil war is now officially over, these agencies still maintain their surveillance.

Boat Owner Associations

The trawl boat owner associations constitute the third important governing actor in the Palk Bay. The 10 landing centers in the Palk Bay count a total of 30 associations, organized along the lines of caste, religion and political affiliation. Boat owner associations usually have a committee headed by a president and a secretary and impose a tax on sea-going boats (in the form of shrimp or money). They deal with matters internal and external to the industry. The latter involves lobbying with government for adequate protection and better (read more favorable) fishing regulations. Sit-down strikes are a common method.

Internal responsibilities include mediating in disputes between trawl fishers or in conflicts between trawl and small-scale fishers. Associations apply their own set of rules, norms and procedures in such instances. At the local level, boat owner associations interact substantially with the FD, in order to make complaints or negotiate the scope and applicability of a particular government rule. FD officials acknowledge that these associations are important, realizing that they require their cooperation for a variety of affairs.

The capacities of these associations, however, differ substantially from one landing center to the next. Generally speaking, the more diverse the population in a particular landing site, the weaker the association is. For example, in Rameswaram – the human diversity hotspot in terms of caste, geographical origin and religion – boat owners have established 13 associations. The competition that occurs between associations in such centers tends to hinder cooperation. On the other extreme is Jagathapatinam, where boat owners are united in a single association. In such landing centers, associations enjoy substantial weight in local affairs.

Assessing the Match Between Governing System and System-to-Be-Governed

Architectural Compatibility: Spatial Correspondence and Linkages

We have noted that the Palk Bay ecosystem, i.e. the natural system-to-be-governed, is a neatly contained spatial area with two passages to the Bay of Bengal and the Gulf of Mannar respectively. The socio-economic system-to-be-governed consists of four clusters of landing centers in three districts. Although the trawl fishers in each cluster have favorite fishing grounds, our research shows that they generally limit their activities to the Palk Bay, albeit in both the Indian and the Sri Lankan territorial waters. Hence, the socio-economic system-to-be-governed coincides largely with the natural system-to-be-governed.

The governing system on the Indian side consists of a system of government agencies that operate at the district, state and central level, and about 30 boat owner associations that have a local ambit. The fishers in the Palk Bay are thus subjected to multiple governing systems at the same time, implying a situation of legal pluralism. What stands out is that no single governing actor (or group of actors) operates at a scale level that coincides with the Palk Bay's ecosystem boundaries. What is more, due to the incidence of cross-border fishing, the authority of the Indian administration only extends over part of the areas in which fishing activity takes place; once trawlers cross the boundary line, fishers are at the mercy of Sri Lankan authorities.

Given the mismatches of scale, it is important to investigate the structural linkages between governing systems on the Indian side of the Palk Bay. We consider four instances in which linkages occur or are notably absent.

1. The ban that has been imposed on pair trawling (see Box 10.1) illustrates the nature of the linkage between the FD and the Indian navy and coast guard. Despite its prohibition by government order in 2000, pair trawling is still commonplace in six out of the ten landing centers in the Palk Bay. The FD is burdened with the responsibility of implementing this ban, but has no presence at sea and thus faces severe constraints in carrying out its responsibilities. The navy, which conducts daily patrols, could in principle spot and report transgressors to the FD. In practice, however, such cooperation is rare, as navy officers do not consider the implementation of fisheries law, which does not have security aspects, their business.
2. The relationship between the FD and the boat owner associations is one of frequent, albeit informal, interaction. In each landing center, the associations and the FD have their own patterns of communication, which vary from cooperation in one landing center to cat-and-mouse games in another.
3. Boat owner associations operate primarily at the level of the landing center. As soon as issues exceed the local level (a rise in the price of diesel or the governmental ban of a particular gear), however, cooperation with associations in neighboring

centers becomes essential. Such cooperation does take place, but primarily along the lines of caste and political affiliation. What stands out is the absence of any structural interaction at the regional level, e.g. in the form of umbrella organizations, impeding associations from linking to other local associations to discuss wider concerns.

4. A final example relates to Sri Lankan border transgressions at the time of the civil war. During periods of extreme tension between the Sri Lankan navy and the Sea Tigers, the FD made pointed efforts to keep trawl fishers informed about the risks that they were running. In these cases of life and death, the information flow between governmental and non-governmental actors proved to be remarkably smooth.

In conclusion, as different governing actors operate at different scale levels, each covering at least part of the Palk Bay area, linkages between them are crucial to achieve some level of coherence and coordination. Our research, however, demonstrates that these linkages are mostly of an ad hoc nature and fail to adequately address the discrepancies.

Attunement: Structural Conditions

We noted above that the Palk Bay trawler fisheries are diverse and dynamic in many respects. To what extent does the governing system possess the necessary qualities – i.e. sensitivity and flexibility – to deal with these properties?

What stands out is the complete absence of regular and planned meetings between the FD and the trawl fishers. Although interaction with boat owner associations is substantial, and there is always room for discussion and even negotiation, this all occurs on an *ad hoc* basis, rather than an institutionalized one. At the level of policy making, there is no system of stakeholder consultation to enrich the information base (e.g. the fisher knowledge of the state of the ecosystem dynamics) or to share opinions about policy tools. In addition, the FD lacks a mechanism to monitor and evaluate the implementation of governmental regulations.

The monthly meetings between the District Collector (the highest authority at this administrative level) and fishers in Ramnathapuram District (one of the districts adjoining the Palk Bay) to discuss prevailing conflicts and problems or register complaints is an interesting exception in this context (Jentoft et al. 2009). It is unclear whether meetings of this kind also take place in other coastal districts along the Palk Bay.

The coast guard also provides a positive example. At the time of fieldwork, this agency had launched a 'community interaction programme', which materialized in monthly visits to a landing site. These meetings were intended not only to provide information, but also to create trust among the fishing population of the Palk Bay. Although this gesture is viewed with some suspicion among members of the boat owner associations, it constitutes an interesting example of efforts to strengthen the linkages between one section of the governing system and the system to be governed.

Being membership organizations, boat owner associations exist by the grace of being sensitive and adaptive to the interests of the people they represent. Although these associations are not devoid of hierarchy, regular gatherings to discuss the important matters of the moment ensure that the leaders incorporate feedback from their members, and adjust their activities if popular opinion so requires.

Attunement to Demands from the System-to-Be-Governed

The section above presented four concerns that cause popular commotion and pose clear demands on the governing system. The various agencies that make up the governing system have reacted to each of them in various ways. In order to deal with the 'Sri Lankan issue', the coast guard and the navy keep a continuous eye on Palk Bay waters and the IMBL. Boat owner associations raise hue and cry if their members suffer violence in Sri Lankan waters, and attempt to pressurize state and central level governments to take protective action. FD officials are engaged in a variety of welfare activities, such as care for the victims of violence in Sri Lankan waters.

In order to curb the conflicts between trawler and small-scale fishers in Tamil Nadu, the government introduced the 3 nm rule, as well as the rule limiting trawling to three alternate days. The State government has also responded to evidence of overfishing by banning several types of allegedly destructive nets, as well as installing a 45-day moratorium on trawling (Bavinck et al. 2008). Finally, in order to address the problems related to the decreasing profitability of the trawling enterprise, the government launched a diesel subsidy scheme in 2004.

These examples point out that governing actors in the Palk Bay are, in a broad sense, receptive to prevailing issues, and that responses materialize in one way or another as well. Responsiveness, however, is not only about reacting, but also about thinking ahead and being proactive. Below, we examine three governmental regulations to better understand this aspect of responsiveness.

Diesel Subsidy Scheme

In response to increasing diesel prices, the Tamil Nadu Government established a subsidy scheme for fuel that reduced fishing costs by 10–20%. These subsidies have had a clear effect: the number of boat owners that went bankrupt and disposed of their boats immediately decreased, resulting in the safe-guarding of thousands of jobs in the Palk Bay trawler fishery (Scholtens 2006). Not everyone shared in the ebullient mood though. Small-scale fishers had been happy to see the trawler fleet slowly decreasing in size, a trend that would logically result in more breathing space for their fishery. Outside observers (Salagrama 2004) also questioned this scheme from the perspective of ecosystem health, pointing to the incongruity of government subsidizing a fishery sub-sector that is overcapitalized and reportedly overfishing the stock.[6] From this perspective, the government, by solving one problem, actually made others worse.

Another approach to the situation of declining profitability can be imagined. For example, in a survey one third of the trawl boat owners indicated a willingness to leave the sector if they were to receive compensation (Sathyapalan et al. 2007). Add to this the fact that many of the boats in the Palk Bay are worth less than the value of a 2-year diesel subsidy, and an opportunity to use this subsidy as a lump sum to buy out a substantial part of the fleet emerges. Although such measures are certainly easier put forward than carried out, the point here is that there has not been any occasion during which the government discussed such possibilities with boat owners. In short, this measure seems to be a perfect example of reactiveness without being genuinely responsive – in the sense of fully appreciating the different concerns at stake and attempting to provide an integrated formula to address them.

3 Nautical Miles Rule

The 3 nm rule was issued in 1983 to curb the violent conflicts between small-scale and trawler fishers throughout Tamil Nadu. Bavinck (2001, 2003) points out, however, that this rule is generally flouted. This is the case in the Palk Bay as well.

At Malipattinam, a smaller landing center in the Palk Bay that hosts 150 vessels, trawl fishers repeatedly move inshore in search of shrimp. This practice was illustrated during one of our field visits to the location when we observed about a hundred trawlers operating within 2 km from the coast. Small-scale fishers in the village were livid, but explained that they are powerless because the FD takes the side of the trawler fishers and refuses punitive action. The Fisheries Inspector in question defended himself saying that he thought the rule had lost its relevance in view of technical changes in the small-scale fisheries. What he did not say, but might have done, is that the 3 nm rule has basically proven to be unenforceable, if only because the FD lacks personnel and equipment to apprehend violators on the spot (Bavinck 2001).

Alternate Day Fishing Rule

In order to solve the conflict between small-scale and trawler fishers, district officials in the Palk Bay decided to separate the warring parties by introducing time slots: trawl fishers would fish during 3 days a week, and small-scale fishers would fish the remaining days (Bavinck 2003). Apparent here is the fit between the type of regulation and the Palk Bay's natural characteristics. Due to the limited size of the Bay, long

[6] The subsidization of unsustainable fisheries is, however, an international phenomenon (see Kurien 2006; Sumaila and Pauly 2006).

distance trips are hardly feasible and the regulation that limits a fishing trip to 24 h does not pose harsh constraints on trawl fishers. Although boat owners did take the rule to the High Court – where it was dismissed – it currently enjoys backing from the entire fishing sector.

How to explain the genesis and success of this rule? The fact that it was decided at the level of the district – and not the state or the landing center – seems to have played a major role. It can be argued that District officials are 'closer' to the scene of conflict and the parties involved than a state-level agency is, and therefore more likely to benefit from the resolution of conflict. The validation of the district regulation by the High Court in Chennai provided an umbrella to protect it against legal action. The security issue in Sri Lanka finally resulted in the establishment of a tight control system. Trawl fishers that want to go fishing on a particular day, require a daily permit issued by the Fisheries Department. Without this permit, the owner in question cannot obtain subsidized diesel and runs the risk of being apprehended by the navy or coast guard.

Attunement: Matching Through Appropriate Governance Modes

In the Palk Bay trawler fisheries we can observe elements of each of the three modes of governance. Hierarchical governance is performed by state and central governments through various rules and regulations. While informal authorities are not as conspicuous in the Palk Bay as they are in other coastal regions, boat owner associations play an important role in self-regulating the trawler fishery. Co-governance – i.e. the situation in which government and fisher representatives share governance responsibilities formally – is virtually absent in the Palk Bay fisheries.

In short, the Palk Bay trawler fishery is characterized by the relative prevalence of the hierarchical governance mode, an overall weakness of self-governance bodies and the virtual absence of institutionalized co-governance. Hence, the governance modes that are naturally more responsive are weak or absent, while hierarchical governance is relatively prevalent but not always effective. Self-governance is weak for two reasons: (a) diversity tends to inhibit effective cooperation; and (b) cooperation between landing centers to deal with larger scale issues is more of an exception than the rule. Where co-governance occurs, with the district collector as facilitator, it appears to build on the strengths of both governance modes, while modifying their respective weaknesses.

Assessing the Governability of the Palk Bay Trawl Fishery

This chapter departed from the assumption that contextual variation is to be taken seriously, and that governance solutions, contrary to panaceas, should match the nature of the problems at hand. A proper match between a governing system and system-to-be-governed is therefore a precondition for effective governance and high governability. Many scholars have emphasized the need for such a match (e.g. Ebbin 2002; Young 2002; Folke et al. 2007), but an approach to systematically assess match in a system that includes social and natural aspects has yet to be proposed. Building on interactive governance and governability assessment literature, this chapter has developed a perspective for assessing the strengths and weaknesses of match by posing four key questions. Its utility can be evaluated on the basis of its subsequent application to the Palk Bay trawl fisheries.

Q1: To what extent does the scale of the natural system-to-be-governed, social-economic system-to-be-governed and governing system coincide, and what is the nature of the linkages between the different governing system actors?

In terms of architectural compatibility, we identified mismatches at two levels. First, the natural and social boundaries of the system-to-be-governed – the entire Palk Bay – do not correspond with the operational range of the various governing actors. While trawlers operate both in Indian and Sri Lankan waters, these waters are subjected to different legal regimes. Although this chapter has focused on the situation pertaining to the Indian side of the Palk Bay, it is clear that the Palk Bay requires a transnational approach. As Vivekanandan (n.d.: p89) put forth, 'From a pure resource management perspective (ignoring the civil war and politics), only a joint management by India and Sri Lanka can ensure proper exploitation and conservation'. Such an alignment would indeed be an important step towards overcoming the discrepancy in scale.

There are, however, architectural mismatches on the Indian side of the Bay too. On the Indian side, trawl fishers are subjected to a variety of governing regimes, each with their own mandates, capacities and ranges. Although there is nothing wrong with a diversity of governors or with the fact that scales do not coincide, these discrepancies must be addressed through linkages and coordination. Our analysis demonstrates that linkages between the different governing systems, if present at all, are mostly ad hoc and non-formalized. Creating institutional convergence is certainly not easy or built overnight, especially under conditions of legal pluralism (Jentoft et al. 2009). They nevertheless remain important for the achievement of effective governance at the Palk Bay level.

Q2: To what extent do the available governing systems possess qualities of sensitivity and flexibility required for dealing with diversity and dynamics in the system-to-be-governed?

Next we considered the attunement between the governing system and system-to-be-governed. First, we asked whether the governing system possessed sensitive and flexible qualities to deal with the high levels of diversity and dynamics prevalent in the sector. We addressed this question by looking at the nature of the linkages between the system-to-be-governed and the governing system, which provided a mixed picture. While interactions are frequent and linkages are in place, especially at the level of the FD and policy, the absence of an institutionalized platform for interaction, joint problem analysis and monitoring stands out. This subsequently bears upon the capacity of the governing system to properly respond to the demands in the system-to-be-governed.

Q3: To what extent is the governing system responsive, that means receptive and proactive, towards the demands present in the system-to-be-governed?

We then looked at the responsiveness of the governing system to four demands emanating from the trawl sector. We found that the government reacted in one way or another to each of the core problems identified in the Palk Bay trawl fishery sector. This hints at the availability of receptive antennas on the governmental side, but also at the ability of the fishers to raise the amplitude of the signal. Governing system reactions, however, do not always seem inspired by an integrated appreciation of the issues at stake. Nor do they go to the roots of the problem. This analysis reveals that although the distinction between the two dimensions of match, i.e. architecture and attunement, may be sensible from a theoretical point of view, in practice they are blurred and intertwined. Architectural features prove to be directly related to the capacity for attunement.

Q4: To what extent is the present mix of governance modes appropriate for dealing with apparent diversity and dynamics in the system-to-be-governed, and to what extent are issues in the system-to-be-governed dealt with at too high levels of governance?

Finally, we considered the appropriateness of governance modes. Hierarchical governance is the most prevalent mode in Palk Bay fisheries, with government authorities – FD, navy and coast guard – playing key roles. Self-governance is generally weak, while co-governance is virtually absent. This particular mix of governance modes was argued to limit the possibilities for attunement of the governing system to the system-to-be-governed.

In the face of these mismatches and their consequences, the governing system of fisheries in the Palk Bay would seem to require restructuring. A co-governance body that brings the system-to-be-governed and the various sections of governing system together at the level of the Palk Bay seems a promising approach for attunement. More specifically, such a platform, which might have chapters in each of the districts, should include the two types of boat owners, small-scale fishers, the navy and coastguard, and the Fisheries Department. It clearly needs to be backed by government, as only government can provide such a body with a serious mandate and the requisite authority. As Fanning et al. (2007) argue, governance at the local level depends heavily on sustained inputs and technical support from the national level.

Concluding Remarks

In this chapter we have attempted to assess the match between the system-to-be-governed and governing system in a particular setting for the purpose of ascertaining governability. We commenced by exploring the concept of match as it has developed within interactive governance theory, distinguishing between two dimensions: architectural compatibility and attunement. The former highlights the structural dimensions of match, whereas the latter looks at its capacity for adaptation. We then pointed out some crucial aspects within each dimension: correspondence of scales, linkages between governing levels and actors, responsiveness to demands, and a relevant mix of governing modes.

What we have thus tried to portray is (a) the way in which issues of scale, diversity and dynamics of the system-to-be-governed, in combination with its specific demands, provide a unique landscape to be governed, and (b) how an assessment can reveal the extent to which the governing system possesses the characteristics to adequately deal with that particular landscape, which is done along the lines of four guiding questions.

The subsequent application to Palk Bay fisheries demonstrated the utility of the approach, identifying weaknesses in a particular societal system instead of quantifying governability. These can be interpreted as areas for governing attention. Our analysis did not emphasize direct suggestions for policy, but rather provided ideas that inspire the enlargement of governance potential.

These guiding questions, provided by interactive governance theory, in their present state clearly leave ample discretionary power to the assessor. There is certainly scope for focusing the questions further. The next step would therefore be to develop more focused evaluative instruments, providing better guidance on 'where to look' and 'what to look for' (see Chap. 18 by Chuenpagdee and Jentoft, this volume).

References

Bavinck, M. (2001). Marine resource management: Conflict and regulation in the fisheries of the Coromandel Coast. New Delhi: Sage.

Bavinck, M. (2003). The spatially splintered state: Myths and realities in the regulation of marine fisheries in Tamil Nadu. Development and Change, 34(4), 633–657.

Bavinck, M., & Jentoft, S. (2011). Subsidiarity as a guiding principle for small-scale fisher-ies. In R. Chuenpagdee (Ed.), Contemporary visions for world small-scale fisheries (pp. 311–320). Delft: Eburon

Bavinck, M., & Karunaharan, K. (2007). Managerial pluralism in the marine fisheries of Ramnad District, Tamil Nadu, India. Working paper. New Delhi: IDPAD.

Bavinck, M., & Kooiman, J. (2008). Slippery as fish – How variations in capture fisheries affect opportunities for governance. (Paper presented at the American Fisheries Society Congress, Ottawa, Canada).

Bavinck, M., & Salagrama, V. (2008). Assessing the governability of capture fisheries in the Bay of Bengal. The Journal of Transdisciplinary Environmental Studies, 7(1), 1–13.

Bavinck, M., de Klerk, L., van Dijk, D., Rothuizen, J.V., Blok, A.N., Bokhorst, J.R., van Haastrecht, E.K., van de Loo, T.J.C., Quaedvlieg, J.G.J., Scholtens, J. (2008). Time-zoning for the safe-guarding of capture fisheries: A closed season in Tamil Nadu, India. Marine Policy, 32(3), 69–378.

Bhathal, B. (2005). Historical reconstruction of Indian marine fisheries catches, 1950–2000, as a basis for testing the 'Marine Trophic Index'. Fisheries Centre Research Reports, 13, 4. Vancouver: Fisheries Centre, University of British Columbia.

Chuenpagdee, R., & Jentoft, S. (2009). Governability assessment for fisheries and coastal systems: A reality check. Human Ecology, 37, 109–120.

Chuenpagdee, R., Kooiman J., Pullin, R. (2008). Assessing governability in capture fisheries, aquaculture and coastal zones. The Journal of Transdisciplinary Environmental Studies, 7(1), 1–20.

Cumming, G.S., Cumming, D.H.M., Redman, C.L. (2006). Scale mismatches in social-ecological systems: Causes, consequences, and solutions. Ecology and Society, 11(1), 14.
Degnbol, P., Gislason, H., Hanna, S., Jentoft, S., Raakjær Nielsen, J., Sverdrup-Jensen, S., Wilson, D.C. (2006). Painting the floor with a hammer: Technical fixes in fisheries management. Marine Policy, 30, 534–543.
Ebbin, S.A. (2002). Enhanced fit through institutional interplay in the Pacific northwest salmon co-management regime. Marine Policy, 26, 253–259.
Fanning, L., Mahon, R., McConney, P. (2007). A large marine ecosystem governance framework. Marine Policy, 31, 434–443.
FAO. (2004). The state of world fisheries and aquaculture. Rome: FAO.
Folke, C., Pritchard, L., Berkes, F., Colding, J., Svedin, U. (2007). The problem of fit between ecosystems and institutions: Ten years later. Ecology and Society, 12(1), 30.
Jentoft, S. (2007). Limits of governability: Institutional implications for fisheries and coastal governance. Marine Policy, 31, 360–370.
Jentoft, S., Chuenpagdee, R., Kooiman, J. (2008). Fisheries and coastal governance as a wicked problem. (Paper presented at the conference, "Coping with global change in marine social-ecological systems," Rome, Italy).
Jentoft, S., Bavinck, J.M., Johnson, D.S., Thomson, K.T. (2009). Fisheries co-management and legal pluralism: How an analytical problem becomes an institutional one. Human Organization, 68, 1.
Jentoft, S., Eide, A., Bavinck, M., Chuenpagdee, R., Raakjær, J. (2011). A better future – Prospects for small-scale fishing people, In S. Jentoft, & A. Eide (Eds.), Poverty mosaics: Realities and prospects in small-scale fisheries (pp. 451–470). Dordrecht: Springer.
Kooiman, J. (2003). Governing as governance. London: Sage.
Kooiman, J. (2008). Exploring the concept of governability. Journal of Comparative Policy Analysis, 10(2), 171–190.
Kooiman, J., & Bavinck, J.M. (2005). The governance perspective. In J. Kooiman, M. Bavinck, S. Jentoft, R. Pullin (Eds.), Fish for life: Interactive fisheries for fisheries (pp. 11–24). Amsterdam: Amsterdam University Press.
Kooiman, J., Bavinck, M., Chuenpagdee, R., Pullin, R., Mahon, R. (2008). Interactive governance and governability: An introduction. The Journal of Transdisciplinary Environmental Studies, 7(1), 1–11.
Kurien, J. (2006). Untangling subsidies, supporting fisheries – The WTO fisheries subsidies debate and developing country priorities. OCSF Occasional paper. Chennai: ICSF.
Mahon, R. (2008). Assessing governability of fisheries using the interactive governance approach: Preliminary examples from the Caribbean. The Journal of Transdisciplinary Environmental Studies, 7(1), 1–12.
Orlitzky, M., & Swanson, D.L. (2002). Value attunement: Toward a theory of socially responsible executive decision-making. Australian Journal of Management, 27, 119–128.
Ostrom, E. (2007). A diagnostic approach for going beyond panaceas. Proceedings of the National Academy of Sciences, 104(39), 15181–15187.
Pascual-Fernandez, J.J., Jentoft, S., Kooiman, J., Trinidad, A. (2005). Institutional linkages. In J. Kooiman, M. Bavinck, S. Jentoft, R. Pullin (Eds.), Fish for life: Interactive fisheries for fisheries (pp. 217–238). Amsterdam: Amsterdam University Press.
Pauly, D., Christensen, V., Buénette, S., Pitcher, T. J., Sumaila, C. J., Walteres, R. et al. (2002). Towards sustainability in world fisheries. Nature, 418, 689–695.
Pitkin, H.F. (1968). The concept of representation. Berkeley: University of California Press.
Salagrama, V. (2004). Policy research: Implications of liberalisation of fish trade for developing countries – A case study for India. Unpublished report. Kakinada: ICM.
Sathyapalan, J., Srinivasan, J.T., Scholtens, J. (2007). Fishing fleet reduction and its livelihood implications: A case study of Palk Bay resource users in the east coast of Tamil Nadu, India. Chennai: FAO/UNTRS Publication.
Scholtens, J. (2006). Fishermen in a dire straight: A study on the pressures faced by Rameswaram trawler fishermen and the diverse scope of implications for the continuation of their livelihood. Dissertation, BSc thesis, University of Amsterdam.

Scholtens, J., Bavinck, M., Soosai, A.S. (2012). Fishing in dire straits: A Sri Lankan perspective on trans-boundary incursions of Indian trawlers in the Palk Bay. Economic and Political Weekly, XLVII(25), 87–96.

Subramanian, S. (2009). Shorelines: Space and rights in South India. Stanford: Stanford University Press.

Sumaila, U.R. & Pauly, D., (Eds.). (2006). Catching more bait: A bottom-up re-estimation of global fisheries subsidies. Fisheries Centre Research Reports 14, 6. A Research Report from the Sea Around Us Project. Vancouver: Fisheries Centre, University of British Columbia.

Suryanarayan, V. (2004). Conflict over fisheries in the Palk Bay region. New Delhi: Lancer Publishers & Distributors.

Venkataraman, K., & Wafar, M. (2004). Coastal and marine biodiversity of India. Indian Journal of Marine Sciences, 43(1), 57–75.

Vivekanandan, V. (2001). Crossing maritime borders: The problem and solution in the Indo-Sri Lankan context. Proceedings from the ICSF conference 'Forging Unity: Coastal Communities and the Indian Ocean's Future' (pp. 76–89), Chennai.

Young, O.R. (2002). The institutional dimensions of environmental change: Fit, interplay, and scale. Cambridge: MIT Press.

Chapter 11
Salmon Aquaculture in Canada and Norway – Appraising Governability

Yajie Liu, Ratana Chuenpagdee, and U. Rashid Sumaila

Abstract Salmon aquaculture is one of the world's fastest growing food producing industries that provides many benefits for producers and society. Its rapid development has raised some concerns related to environmental and economic impacts. Is sustainable development an achievable governance goal for salmon aquaculture? We explore this question by examining key features that may hinder or foster its governability. We base the discussion on a review of the evolution of the industry, current practices and trends, emerging concerns, management challenges, and existing policies and regulations. As a system-to-be-governed, salmon aquaculture is a highly capital-intensive and industrialized commercial activity, with well established technology and markets. The governing system is rather complex, involving multi-layered authorities, international, regional, national, provincial, state and local departments and agencies, and numerous regulations, policies and standards. Environmental issues and concerns related to salmon aquaculture production add to the difficulty in governing this industry to achieve sustainability. All of these have effects on the governability of salmon aquaculture industry, which can be improved through technological improvement, economic-based instruments, and governance solutions.

Keywords Salmon aquaculture • Evolution • Challenges • Policies • Regulations • Governability • Sustainability

Y. Liu (✉)
SINTEF Fisheries and Aquaculture (Fiskeri og havbruk),
Postboks 4762, Sluppen, 7465 Trondheim, Norway
e-mail: yajie.liu@sintef.no

R. Chuenpagdee
Department of Geography, Memorial University of Newfoundland, St. John's, NL, Canada
e-mail: ratanac@mun.ca

U.R. Sumaila
Fisheries Economics Research Unit, Fisheries Centre,
University of British Columbia, Vancouver, BC, Canada
e-mail: r.sumaila@fisheries.ubc.ca

M. Bavinck et al. (eds.), *Governability of Fisheries and Aquaculture: Theory and Applications*, MARE Publication Series 7, DOI 10.1007/978-94-007-6107-0_11,

Introduction

Salmon aquaculture is one of the fastest growing food production sectors in the world. In just over a few decades, farmed salmon production has increased from around 500 metric tons in 1970 to over 1.32 million metric tons in 2009. Global farmed salmon production has exceeded wild salmon production since 1998 (FAO 2007). Salmon aquaculture has provided a number of benefits for society, including employment and income for coastal communities, foreign exchange earnings, and affordable seafood for consumers. While the supply of fish from capture fisheries is levelling off, aquaculture is viewed by many as a source that can bridge the gap between wild fish supply and increasing market demand (Bailey et al. 1996; Tidwell and Allan 2001; Garcia and Grainger 2005; FAO 2007; Asche 2008).

As a resource-intensive and industrialized activity, salmon aquaculture generates concerns related to environmental and socio-economic impacts. Potential environmental problems include disease and parasite transfer, escapees, waste discharge, introduction of exotic species (specifically, Atlantic salmon, *Salmo salar,* into the Pacific Ocean), residues from chemical and drug uses, and the utilization of fish as raw materials for fishmeal and fish oil (Naylor et al. 2000; Kautsky et al. 2001; Milewski 2001; Pauly et al. 2002; Morton and Routledge 2005; Naylor and Burke 2005; Krkošek et al. 2006). In addition to environmental issues, salmon aquaculture creates potential market conflicts by competing with products from the wild and driving down prices (e.g., Asche et al. 1999a ; Naylor et al. 2003; Knapp et al. 2007). Increasing the supply from salmon aquaculture has led to a decline in market prices, consequently resulting in profit losses for both wild salmon fisheries and salmon aquaculture (Naylor et al. 2003; Knapp 2005; Knapp et al. 2007). Other sectors potentially affected by the aquaculture industry include recreational fishing and boating, navigation, upland property development, conservation (of marine mammals such as seals) and archaeological studies (SAR 1997).

There is little doubt that salmon aquaculture will continue to grow, although this will likely occur at a slower rate than anticipated (Liu and Sumaila 2008). Concerns over negative environmental and socio-economic impacts associated with the industry suggest, however, that a balance between the growth of this industry and environmental sustainability is required. Finding such an equilibrium has been the biggest challenge for both fish farmers and policy-makers. Different levels of governing authorities ranging from international organizations such as the United Nations to local communities have attempted to set regulations and regulatory frameworks to address these concerns. While improvement has gradually been achieved, the outcomes have been far from satisfactory. As a contribution to the ongoing debate over the concerns about its sustainability and following the suggestion by Pullin and Sumaila (2005), we examine salmon aquaculture from a governance perspective. Specifically, we employ the concept of 'governability' to systematically examine the characteristics of salmon aquaculture industry. Defined simply as the 'overall quality for governance' (Kooiman 2003, 2008), the operationalization of the concept involves asking questions about the properties of a system that are conducive to governing, the ability of the governing system to perform its tasks, and the existence and quality of governing interactions that may foster or limit governance (Jentoft 2007; Chuenpagdee et al. 2008; Chuenpagdee 2011). Applying it to salmon aquaculture, we ask what aspects of the industry pose limitations to governance, and what aspects offer opportunities to the industry and policy makers in attaining their sustainability goals.

The rest of the chapter is organized as follows. In the next section, we present an overview of the evolution of salmon aquaculture development and its associated problems and challenges, with some emphasis on the Canadian and Norwegian industry. This is followed by a description of current policies and regulations governing the industry. Based on this, we discuss the salmon aquaculture industry in the context of governability. In the final section, we suggest ways to improve governability and promote sustainability in salmon aquaculture development.

Evolution of Salmon Aquaculture

Salmon aquaculture first began as part of a stock enhancement program, which involved keeping the fish in rearing facilities for a certain period before releasing them as smolt or adults into rivers or coastal areas. This means of enhancing and restoring declining or depleted wild salmon stocks was initially conducted in Japan, Canada and the US (Thorpe 1980). In the 1960s, aquaculturists in Norway and Scotland started to rear salmon in open floating cages in areas close to seashores (Willoughby 1999). Breakthroughs with respect to biological and technological bottlenecks, such as smolt rearing and the formulation of dry feed, came in the early 1970s with a fully developed technology for commercial-scale salmon aquaculture. Although a small amount of farmed salmon still comes from land-based facilities, the majority is nowadays produced in open net-cage or net-pen systems.

Over 95% of the total global farmed salmon production is Atlantic salmon, while the remaining share is chum (*Oncorhynchus keta*), chinook (*O. tshawytscha*) and coho (*O. kisutch*). Norway, Chile, the UK and Canada are the four largest producers of farmed salmon in the world, contributing over 85% of the total production and values. Norway has long been the leading farmed salmon producer, as well as the pioneer in technological innovation and development of new markets (Aarset 1998). In the 1980s the Norwegian government launched a licensing system limiting farm size and fish density in order to avoid over-production and domination by a few big companies (Willoughby 1999; Sønvisen 2003). This rigorous system, however, drove Norwegian investments overseas to places like Chile and Canada, where there were fewer regulations and more potential for profit-making. Salmon aquaculture in Chile, for instance, has continued to expand since its introduction in the 1980s.

It benefits from government support, a suitable coastal environment, a cheap labor force and large feed availability. Chile is the world's fastest growing salmon aquaculture producer. Given its current rate of growth, Chilean production will likely surpass that of Norway (Liu and Sumaila 2008). In the last few years, however, this growth has plunged due to the outbreak of deadly infectious salmon anemia (ISA) (Asche et al. 2010).

In Canada, salmon aquaculture technology was introduced in the 1970s to help the economies of coastal communities recover from the decline of the fishing and forestry sectors. Salmon aquaculture initially began along the Sunshine Coast in British Columbia, western Canada, with native salmon species such as chinook and coho. Atlantic salmon was later introduced to this area because of its biological resilience and fast growth rate. Farmed salmon production from British Columbia contributes about two-thirds of the total farmed salmon production in Canada. The rest comes from the Bay of Fundy on the east coast. In recent years, salmon producers in British Columbia have been challenged by many environmental activists due to growing concerns over related environmental issues such as sea lice.

Like other aquaculture industries, salmon aquaculture has experienced remarkable growth over the years as a result of the expansion of farming areas in new locations, improved productivity, enhanced husbandry practices and management, economies of scale, and growing global markets (Bjørndal et al. 2002, 2003; Asche and Khatun 2006; Asche 2008). In the beginning, salmon farms were small family businesses, scattered along sheltered inlets and fjords, and targeting local markets (Holm and Jentoft 1996; Willoughby 1999; Hjelt 2000). Some of these small family farms eventually went bankrupt and many were replaced by farms owned by a few multinational companies and joint cooperatives (Willoughby 1999). Today, salmon aquaculture has become a vertically-integrated and export-oriented commercial, for-profit enterprise (Bjørndal et al. 2003; Asche and Khatun 2006). It is estimated that 70–80% of farmed salmon production comes from a dozen multinational companies (Naylor et al. 2003). Farmed salmon products used to be luxury seafood available mainly in restaurants. Nowadays, these products have become more affordable and are readily available in many food markets worldwide (Forster 2002). As previously mentioned, the controversy over environmental impacts and low profitability may have slowed down salmon aquaculture development in most producing countries. However, the drastic decline in farmed production in Chile has driven up the market prices for the last two years, resulting in a steady increase in production and profitability outside Chile, in places like Norway.

Salmon Aquaculture Industry as a System-to-Be-Governed

Salmon aquaculture is an industrialized, commercial industry with established technologies and markets. An open net-cage production system is normally used, although size, shape and associated structures differ slightly between one location and the next. For instance, the production systems in Norway tend to be more capital intensive, while those in Chile are more labor-intensive (due to lower wages). Larger enterprises normally hold multiple licenses and are vertically integrated. That is, they have full control of the whole fish chain, ranging from broodstock, nursery, growout, harvesting to marketing. Most products are fresh or frozen, and are traded in three international markets: the European Union, Asia (e.g., Japan) and the United States.In terms of economics, production and market, salmon aquaculture is considered one of the most successful aquaculture practices in the world.

While salmon producers are concerned with maximizing profits and maintaining competitiveness in international markets, the industry suffers from an increasingly negative public opinion. This poses major challenges to the salmon aquaculture industry that, unless properly addressed, will most likely affect its future growth and development. Key issues related to salmon aquaculture industry occur throughout the 'fish chain' and as part of the natural and social systems-to-be-governed, as described below.

Farming Environment

The public, especially seafood consumers, are becoming increasingly aware and suspicious of the environmental issues associated with salmon aquaculture (Whitmarsh and Wattage 2006). For instance, news about sea lice from salmon farms, associated with the declines in wild salmon stocks in Scotland (Gargan et al. 2002) and the west coast of Canada (Morton and Routledge 2005; Krkošek et al. 2006), has led to poor public perception of the salmon aquaculture industry. This perception may have a negative effect on the market, as some people may no longer want to purchase farmed salmon products.

Production Factors

The productivity of salmon aquaculture has improved dramatically over the years due to rapid technological innovations. Production costs have also declined (Asche 2008). The feed conversion ratio, a measure of fish's efficiency in converting feed into increased body weight, has greatly improved, and can now be as low as 1.2 – i.e., salmon can gain 1 kg body weight with 1.2 kg of feed. Operations are capital intensive, however, especially in Norway and Scotland. Furthermore, many salmon producers suffer economic losses from the high mortality rate caused by diseases and parasites, as well as prevention and control costs (Costello 2009; Asche et al. 2010; Aunsmo et al. 2010). Although the development of vaccines has significantly reduced the use of antibiotics (Asche et al. 1999b), disease and parasite problems continue to be the major risk factor facing salmon aquaculture. Escape is also a problem resulting in economic losses for salmon farms (Jensen et al. 2010).

Resource Competition and User Conflict

Space limitation and increased use of coastal waters mean higher levels of conflicts among local resource users (Sønvisen 2003; Maurstad et al. 2007). In competing for the same passages or watercourses, for instance, salmon farms may have conflicts with wild salmon fisheries and recreational activities (Liu et al. 2011). Recreational anglers were willing to pay substantially less for the high share of escaped farmed salmon in rivers (Olaussen and Liu 2011). At an international level, global competition and demand for fishmeal and fish oil have intensified with the rapid expansion of salmon farming and intensive aquaculture of other carnivorous species, such as turbot, halibut and cod. This competition has led to increasing prices of fishmeal and fish oil, consequently resulting in increasing production costs for salmon aquaculture producers. This competition will likely intensify, because almost 90% of the global fish oil production is currently used by industrial aquaculture (Tacon and Metian 2008), unless alternative sources to replace fish oil in aquafeed are discovered.

Market

Existing international markets such as the United States, the European Union and Japan may be approaching saturation, and the development of new markets is still at an early stage. Hence, the marketing of farmed salmon products has become very competitive. The falling market prices of farmed salmon, due largely to increasing global production, have resulted in low profitability for producers, driving some of them out of business. Some countries (like Norway and Chile) have been accused of dumping their farmed products in European Union and United States markets (Asche 2001). This may result in higher trade barriers being placed on them as other producing countries try to protect their domestic market shares. Since producers in Chile have the lowest production costs, they generally have a trade advantage with respect to competitors like Norway and Canada.

Food Safety

Consumers have been increasingly concerned about seafood safety and the quality of farmed fish. Since salmon aquaculture farms use chemicals and drugs to treat diseases and parasites, and require fishmeal and fish oil as protein sources, there are concerns about the potential accumulation of contaminants in the salmon. Health risk is another issue affecting salmon farming. Studies suggest, for instance, that persistent organic pollutants (e.g., PCB, DDT) are higher in farmed salmon than in wild salmon (e.g., Hites et al. 2004).

Current Governing Systems and Policies

Aquaculture in general, and salmon aquaculture in particular, is managed and governed by multi-layered authorities, including international, regional, national, provincial (or state) and local departments and agencies. At the international level, management and governance is led by the Food and Agriculture Organization of the United Nations (FAO). Article 9 of the FAO's Code of Conduct for Responsible Fisheries presents a guideline for the development and management of aquaculture (FAO 1995). Regional organizations such as the North Atlantic Salmon Conservation Organization (NASCO) and the European Union have also given specific guidance to salmon farming. For instance, NASCO signed the 'Oslo Resolution' agreement in 1998 calling for "*the Parties to the Convention for the Conservation of Salmon in the North Atlantic Ocean to minimize impacts from salmon aquaculture on the wild salmon stocks*" (Porter 2005, 6). Through the involvement of the salmon aquaculture industry, the resolution has led to guidelines on the physical containment of farmed operation and husbandry practices. The Oslo Resolution was later supplanted by the "Williamsburg Resolution" in 2003 (NASCO 2007).

Canada is a good example of governing complexity at the national and provincial levels. Salmon aquaculture is administered by a combination of federal, provincial and local authorities. At the federal level, 17 departments and agencies share responsibilities for the regulation of the salmon aquaculture industry. The Department of Fisheries and Oceans is the principal agency. Other important agencies include Environment Canada, Canadian Food Inspection Agency, Canadian Environmental Assessment Agency and Health Canada. These departments and agencies oversee the conservation and protection of wild fish stocks and fish habitats, international trade, as well as general research and development, and address any issues beyond provincial boundaries. Responsible governing institutions vary at the provincial level. In British Columbia, for instance, three governmental agencies – namely, the Ministry of Agriculture and Lands, the Ministry of Environment, and the Integrated Land Management Bureau - are directly involved in salmon aquaculture management. They are responsible for issuing licenses and permits, managing and regulating operation facilities, training and education, and collecting production and financial data. The primary legislative instruments for aquaculture at a national level include Fisheries Act Regulations (1973), Fisheries Act (1985, 1996), Aquaculture Regulation (2002) and Environmental Management Act (2003). Based on these instruments, specific management strategies and policies are formulated and implemented.

Norway is another good example of how regulations and policies for marine aquaculture have evolved over the years due to a variety of concerns. Salmon aquaculture is governed by the Ministry of Fisheries and Coastal Affairs, and primarily administered by the Directorate of Fisheries. The Norwegian Food Safety Authority, among others, is responsible for animal health, food safety and quality (Maroni 2000). The Aquaculture Act is the major legislation, and has been modified with emerging issues over time (Maroni 2000; Sønvisen 2003). At the beginning, the policy was to ensure regional development and secure local ownership through, for instance, ‘owner-operator’ licensing system, together with limitations on farm size and number. Increasing competition in the industry resulted in making the ‘owner-operator’ license system less stringent, which helps the local industry compete in international markets. It implies that one company can hold multiple licenses at various locations. Later on, the policies shifted towards environmental protection due to emerging problems such as escapees and diseases (Liu et al. 2011).

Although different countries and regions have their own governing systems, governing actors tend to concentrate on similar issues and problems, namely siting/licensing criteria and regulating operation facilities (including waste management, escape prevention, fish health and uses of chemicals and drugs). The following highlights some key policies.

Siting/Licensing Criteria

Most countries have developed guidelines describing a number of environmental and social criteria to determine whether to issue a license or allow for aquaculture operations. In the case of British Columbia, such environmental criteria include minimum distance from a First Nation reserve, salmon stream, herring spawning ground, shellfish beds, habitats for marine mammals, ecological reserve, protected parks and areas, and existing aquaculture farms. Furthermore, a farm cannot be situated in an important commercial or recreational fishing ground, and in or near heritage sites (MAFF 2006). In Norway, existing salmon farms have to be relocated from, and new licenses will not be issued near, the National salmon watercourses and fjords, which have been established to protect wild salmon stocks (Sønvisen 2003; Porter 2005). There are also limits on the number of fish farms that can be situated in a specific area/region.

Aquaculture Operations

Rules may focus on daily operation, best husbandry practice, compliance with disease outbreak or parasite infection, and escapes and containment management. Other management strategies have been developed and implemented to target various environmental problems. For instance, Norway has special regulations on stocking density, farm size, production level, and year-class separation. Most producing countries also impose a period of fallowing for farms, especially those located in sensitive areas such as wild salmon migration pathways.

Waste Management

The waste control regulations may include waste discharge standards, best waste management practices, monitoring and reporting, site remediation, fees and penalties.

There are requirements for reporting quantities and types of feed, chemicals and drugs used. Cameras are normally installed to monitor feeding and waste. Regular monitoring (or inspection) and reporting are mandatory for all producers. In many cases, such as in Norway and Canada, operators are required to install nets to prevent birds from diving for salmon, as well as set up surrounding nets to avoid marine mammal attacks.

Escape Prevention

The management and regulation for escapes include installation and maintenance of facilities and staff training, prevention plans (e.g., site inspection), and response plans (MAFF 2006). Fines, and in some cases criminal charges, may be imposed on salmon farm operators who violate the regulations or standards. If the case is serious, a license may be suspended.

Disease and Parasite Prevention and Control

This includes special guidelines or standards to deal with disease prevention, dead fish and risk factors, as well as monitoring and recording of disease, hygiene and the use of drugs and chemicals. In British Columbia, a Fish Health Management Plan has been developed based on the federal and provincial guidelines, providing key principles for individual farms to follow. In Norway, a Fish Disease Act has been created to manage and prevent disease and parasite problems. In order to reduce or minimize risks, Norway also controls the maximum numbers of sea lice per fish in different development stages. Drugs for treating sea lice are mainly administered through bathing, while antibiotics are administered through feed. Both amount and type of drugs and chemicals are determined by veterinary prescriptions. The sale of medicated feeds and drugs are also monitored.

In sum, these management strategies and policies largely consist of guidelines, standards and principles. Some are mandatory, while others are voluntary. Over time, these regulatory frameworks have become more comprehensive. Consequently, some problems have reduced, but others, like disease and parasites, persist. Strategies and policies do not necessarily guarantee desirable outcomes regarding social and environmental benefits. Likewise, they may not be as economically efficient as expected.

Improving the Governability of the Salmon Aquaculture Industry

Based on the description above, Table 11.1 summarizes several characteristics of salmon aquaculture that affect the governability of this system. At an individual farm level, the salmon aquaculture industry is rather homogenous, not very complex

Table 11.1 Characteristics of salmon aquaculture throughout the fish chain that affect governability

Features	Aquatic environment	Production system	Marketing system	Governing system
Diversity	Limited types of aquatic environment suitable for aquaculture	Mostly using similar technology and operation systems aimed at improving productivity and reducing environmental problems	Export-oriented with emphases on a few relatively saturated international markets: United States, European Union and Japan	Varying management strategies and policies to address different aspects of salmon aquaculture operations
Complexity	Highly complex with salmon spawning and growing in separate aquatic environments, such as freshwater and ocean environments	Moderately complex among individual farms due to competition for raw materials (e.g., aquafeed) and markets	Difficulty in the labelling of the products due to consumers' concerns over food safety issues and environmental problems	Highly complex institutional arrangements at all levels with involvement from various agencies
Dynamics	Slowly incorporating changes in natural environments such as temperature, salinity, dam etc.	Gradual modification in the netpen production systems, but a tendency toward adopting enclosed systems	Slow development for new products and/or markets	Imperceptibly modifying rules and regulations to keep up with emerging environmental problems
Scale	Especially sensitive due to uncontained and/or controlled environment and systems	High competition for space with other farmers and ocean users	Global markets	Overlapping responsibilities and jurisdictions

and has low dynamics. Thus, on its own, it does not pose major governance challenges. Complications arise, however, in the interaction between the farms and the aquatic environment, as well as in the marketing of products (as a result of consumer awareness and concerns). The governing system is, however, very diverse and complex, and it struggles to keep up with the emerging environmental concerns at local, national and international levels. Taken together, these features contribute to lowering the governability of the industry.

Nevertheless, numerous opportunities to improve the governance and governability of salmon aquaculture exist. Aquaculture is a sector with a mixed identity; it can fit in both fisheries (ocean) and agriculture (inland), but belongs to neither. Hence, the regulations and policies for salmon aquaculture are generally adapted or modified from, or in some cases combinations of, capture fisheries and agriculture. Generally speaking, the majority of current regulations and policies are command-and-control types. Such governing systems may not be appropriate for a number of reasons, including but not limited to: (1) limited knowledge and uncertainties of the effects of the farming system resulting from diversity, complexity and dynamics; (2) inability of regulations to catch up with the problems caused by the industry's rapid expansion; (3) lack of economic incentives for producers to improve their performance on their own initiatives; and (4) cost ineffectiveness of uniform regulations such as pollution standards (Naylor et al. 2003).

In the next section, we propose three ways that may be used to mitigate some of these problems and improve governability: technological improvement, economic-based instruments and governance solutions.

Technological Improvement

A number of technological innovations in salmon aquaculture development have been made over the years. The most groundbreaking ones are feed formulation and the development of vaccines. Advances in the feed conversion ratio and feeding in general have greatly improved feed efficiency (Asche 2008). The development of vaccines has significantly reduced the use of antibiotics and other chemotherapeutics (Tveterås 2002). Another important development is the closed containment system (either land-based or sea-based such as sea-bag system) to replace the open netcage system. Closed systems can mitigate environmental problems and associated impacts by reducing waste discharges, avoiding marine mammal interactions, protecting wild fish from disease transmission and parasites, and preventing escapes. It is worth noting, however, that the closed containment systems are not financially feasible at the current market price of farmed salmon. This is largely due to the high capital investment and operating costs necessary for aquaculture ventures (Liu and Sumaila 2007). Without a price premium for environmentally-friendly farmed salmon, adoption of the closed containment system is unlikely, especially after land-based salmon farms went bankrupt some years back in Norway, Iceland and Scotland. Unless that the newly developed technologies can reduce capital costs of such systems.

Another possible solution to the problems is the application of an integrated production system, i.e., farming of species with different trophic levels in one production system so that each species uses distinct niches and resources (Troell et al. 2009). The waste generated by one organism can be converted into inputs (e.g., nutrients) for other organism(s), resulting in the optimal use of resources and reduced environmental impacts, such as waste discharges (Troell et al. 2009). For instance, seaweed and/or mussel can be reared in the same production facility as salmon, acting as a bio-filter that absorbs the dissolved wastes (e.g., nitrogen and phosphorus) released by salmon (e.g., Buschmann et al. 1996; Buschmann et al. 2001; Chopin et al. 2001; Neori et al. 2004). Studies indicate that the nutrients around salmon netcages are greatly reduced, and seaweed grows much faster near salmon netcages than at a far distance from them (e.g., Troell et al. 1997). Seaweed can also be harvested to generate extra income. Studies show that the net present value in a salmon-mussel integrated farm is 35% higher than a salmon monoculture, and 11% higher than the sum of salmon monoculture and mussel monoculture (Whitmarsh et al. 2006). An integrated production system is thus biologically and economically promising, socially acceptable and risk reducing (Ridler et al. 2007), and is gradually being adopted globally (Barrington et al. 2009). Further research on this topic needs to be conducted on a larger scale, however (Troell et al. 2003).

There are some important caveats related to technological development. First, technologies can reduce or eliminate some environmental problems, but not all of them at once. This is mainly because such technologies take time to develop and apply in order to get desired results. Second, the development of technologies requires comprehensive information regarding problems and their associated impacts, which are neither completely understood nor well studied. Finally, although some technologies are environmentally sound and technically feasible, their application is often hampered by the need for substantial investments and a lack of organizational support.

Market-Based Instruments

Market failure is a common problem in salmon farming since the environmental costs associated with production are often not accounted for in producers' decision-making. Incorrect pricing can be remedied by incorporating the environmental costs to reflect the full cost of salmon production. This can be done in the form of taxes, tradable permits, performance bond (deposit-refund) and eco-labelling. Market-based instruments force producers and consumers to take environmental concerns into account.

Environmental tax is a fee levied on a producer. Ideally, it should be equal to the environmental damages caused by the activity in question, e.g., salmon aquaculture. Environmental taxes can induce producers to reduce their pollution to a point at which the marginal abatement cost is equal to the tax. While such a tax has the potential to address some environmental concerns (Sylvia et al. 1996), setting an appropriate tax level is challenging because it requires a full understanding of the source of an environmental problem and its associated impacts and costs. Pollution abatement costs may be used as a reference point to establish an environmental tax level (Liu and Sumaila 2008). For instance, lawmakers in Chile have filed a bill to tax salmon producers roughly 5% of the monthly profit to cover the environmental costs caused by salmon farming (Carvajal 2007).

When environmental damages are highly uncertain and complex, a performance bond/deposit-refund system may be employed (Mathis and Baker 2006). As with penalty and reward programs, a deposit is collected from producers before production begins. This deposit can be fully, partially or non-refundable depending on environmental performance. This instrument has not been widely applied except in shrimp farming in Texas (Mathis and Baker 2006). For salmon aquaculture, fines or prosecution are more commonly used to address violations such as delayed or unreported escapees. Although these regulations are not very effective and the process of fine payment is costly and time consuming, they may give producers incentives to mitigate environmental damages.

Eco-certification and labelling have recently received much attention from environmental organizations and consumers. They aim to harness consumers' purchasing behavior directly by creating market-based incentives for environmentally friendly seafood, taking into account product attributes other than price (Wessells et al. 1999; Cochrane and Willmann 2000). Eco-labelling is an effective way to provide consumers with awareness about the seafood they buy (Naylor et al. 2003). In theory, consumers are willing to pay a higher price for sustainable seafood to compensate for the increase in production costs that a product entails. Thus, seafood with an eco-label can command a premium price. This is not always the case, however, as the recent debate about Marine Stewardship Council's certification program demonstrates (Jacquet and Pauly 2008). For farmed seafood, the Aquaculture Certification Council has been widely recognized as a default eco-label. To the best of our knowledge, there are currently no eco-labelled farmed salmon products on the market. However, farmed salmon produced in land-based systems in British Columbia have been self-labelled as "eco-salmon," labels that are accepted by some retailers and consumers.

Governance Solutions

Salmon aquaculture has gained considerable institutional support and political recognition due to its contribution to the national economy (Liu et al. 2011). With the development of the industry, management strategies and environmental policies have become more complex and rigorous in order to reduce and minimize the environmental and economic effects associated with salmon aquaculture. Complying with these regulations and policies represents costs to salmon producers. Authorities have the obligation to set regulatory frameworks that balance the social, economic and environmental concerns of producers and the society at large. While a weak legal framework is problematic, excessive regulations or inappropriate policies may suffocate producers, especially the smaller ones. For instance, obtaining a new license in British Columbia can take up to two years and cost up to CAD 500,000 (Liu and Sumaila 2007). The standardization and streamlining of policies and regulations at the local and national levels can help improve sustainability of salmon aquaculture development. Importantly, an integrated approach to coastal ecosystem management is required to address the existing and emerging environmental concerns related to salmon aquaculture.

International governance of salmon aquaculture is necessary given that it is a global industry. A first concern is the fact that the fishmeal and fish oil used for fish feed are traded internationally (even though they are primarily produced in few countries, such as Peru, Chile, Denmark and Norway). The battle for fishmeal and fish oil will probably intensify, potentially resulting in the overexploitation and depletion of small pelagic fish stocks if appropriate management regimes are not implemented (Naylor et al. 2000; Pauly et al. 2001, 2002). Secondly, trade disputes between exporting and importing countries will continue to take place on account of the protection of domestic markets. Thirdly, some environmental problems associated with salmon aquaculture are of a transboundary nature. Escapes from British Columbia may, for example, cross the United States – Canada border and swim into Alaskan waters (Naylor et al. 2003). Hence, salmon aquaculture requires international standards or guidelines. Such international governance necessitates, however, collaboration from multiple stakeholder groups at civil society, market and state levels.

Conclusions

Despite its short history, salmon aquaculture has been rapidly and successfully developed (Asche 2008). There is little doubt that salmon aquaculture will continue its growth, albeit not at the current rate (Liu and Sumaila 2008). The industry is facing a number of challenges concerning social, economic and environmental aspects. Various governing systems have been established to address different issues and problems along the way. The overall governability of the salmon aquaculture industry can be improved through technological innovations, economic-based instruments and governance solutions.

Technological innovations have been considered as the top option for improving productivity and reducing environmental impacts. At present, closed containment production and integrated aquaculture system have been promoted as "green" technologies that minimize environmental problems in salmon aquaculture. They are environmentally friendly and technically promising. However, closed systems require extensive capital investment and have high operating costs. Market instruments have attracted increasing attention and demonstrated a promising potential to policy makers, because they can cost-effectively regulate salmon aquaculture. Environmental costs are internalized through economic instruments such as taxes, performance bonds and eco-labelling. Further, they provide incentives for producers to develop and adopt cleaner technologies. So far, market-based instruments have not been used widely in salmon aquaculture. To do so is very challenging, because the sources of environmental problems and their associated impacts and costs have not been determined and estimated correctly.

From a governance perspective, the multi-layered governing authorities and the numerous regulations, policies and standards provide the stringency required to achieve the sustainable development of the salmon aquaculture industry. Different stakeholder concerns – including those of relevant government agencies, aquaculture industry and allied associations, commercial and recreational fishing sectors, non-governmental organizations, local residents, First Nations, secondary supporting sectors and the general public – need to be taken into consideration. International governance and integrated coastal management may offer solutions leading to improvement of governability and sustainability.

Acknowledgement The authors would like to thank the participants of the governance symposium at the annual conference of the American Fishery Society held in August 2008, Ottawa, Canada for their helpful comments. Canada's Research Network in Aquaculture (AquaNet) and Norwegian Research Council are acknowledged for financial support to the first author. R. Chuenpagdee acknowledges support from the Canada Research Chairs program and Social Sciences and Humanities Research Council of Canada. R. Sumaila thanks the *Sea Around Us and Global Ocean Economic Projects* for the support. Any errors, opinions or conclusions expressed are those of the authors alone and should not reflect the funding agencies.

References

Aarset, B. (1998). Norwegian salmon-farming industry in transition: Dislocation of decision control. Ocean and Coastal Management, 38, 187–206.

Asche, F. (2001). Testing the effect of an anti-dumping duty: the US salmon market. Empirical Economics, 26, 343–355.

Asche, F. (2008). Farming the sea. Marine Resource Economics, 23(4), 527–547.

Asche, F., & Khatun, F. (2006). Aquaculture: Issues and opportunities for sustainable production and trade. Issue Paper No. 5. Geneva: International Centre for Trade and Sustainable Development.

Asche, F., Bremnes, H., Wessells, C.R. (1999a). Product aggregation, market integration and relationships between prices: An application to world salmon markets. American Journal of Agricultural Economics, 81, 568–581.

Asche, F., Guttormsen, A., Tveterås, R. (1999b). Environmental problems, productivity and innovations in Norwegian salmon aquaculture. Aquaculture Economics and Management, 3(1), 19–29.

Asche, F., Hansen, H., Tveteras, R., Tveteras, S. (2010). The Salmon disease crisis in Chile. Marine Resource Economics, 24, 405–411.

Aunsmo, A., Valle, P.S., Sandberg, M., Midtlyng, P. J., Bruheim, T. (2010). Stochastic modelling of direct costs of pancreas disease (PD) in Norwegian farmed Atlantic salmon (salmo salar L.). Preventive Veterinary Medicine, 93, 233–241.

Bailey, C., Jentoft, S., Sinclair, P. (1996). Social science contributions to aquacultural development. In C. Bailey, S. Jentoft, P. Sinclair (Eds.), Social dimensions of an emerging industry (pp. 3–20). Boulder: Westview Press.

Barrington, K., Chopin, T., Robinson, S. (2009). Integrated multi-trophic aquaculture (IMTA) in marine temperate waters. In D. Soto (Ed.), Integrated mariculture: A global review. FAO Fisheries and Aquaculture Technical Paper. No. 529 (pp. 7–46). Rome: FAO.

Bjørndal, T., Tveterås, R., Asche, F. (2002). The development of salmon and trout aquaculture. In P. Paquotte, C. Mariojouls, J. Young (Eds.), Seafood market studies for the introduction of new aquaculture products (pp. 101–115). Cahiers Options Méditerranéennes 59. Paris: CIHEAM.

Bjørndal, T., Knapp, G.A., Lem, A. (2003). Salmon – A study of global supply and demand. Report No. 92. Bergen: Centre for Fisheries Economics, Institute for Research in Economics and Business Administration.

Buschmann, A.H., Troell, M., Kautsky, N., Kautsky, L. (1996). Integrated tank cultivation of salmonids and Gracilaria Chilensis (Gracilariales, Rhodophyta). Hydrobiologia, 32, 775–782.

Buschmann, A.H., Troell, M., Kautsky, N. (2001). Integrated algal farming: A review. Cahiers De Biologie Marine, 42(1–2), 83–90.

Carvajal, P. (2007). Chile seeks tax on salmon profits. Bergen: Intrafish.

Chopin, T., Buschmann, A.H., Halling, C., Troell, M., Kautsky, N., Neori, A., Kraemer, G.P., Zertuche-Gonzalez, J.A., Yarish, C., Neefus, C. (2001). Integrating seaweeds into marine aquaculture systems: A key toward sustainability. Journal of Physiology, 37(6), 975–986.

Chuenpagdee, R. (2011). Interactive governance for marine conservation: An illustration. Bulletin of Marine Science 87(2), 197–211.

Chuenpagdee, R., Kooiman, J., Pullin, R.S. (2008). Assessing governability in capture fisheries, aquaculture and coastal zones. The Journal of Transdisciplinary Environmental Studies, 7(1), 1–20.

Cochrane, K., & Willmann, R. (2000). Eco-labelling in fisheries management. Proceedings of the 2000 Conference by the Centre of Ocean Law and Policy, University of Virginia, and FAO on Current Fisheries Issues and the Food and Agriculture Organization of the United Nations. Rome: FAO.

Costello, M.J. (2009). The global economic cost of sea lice to the salmonid farming industry. Journal of Fish Diseases, 32, 115–118.

FAO. (1995). Code of conduct for responsible fisheries. Retrieved August 6, 2007, from http://www.fao.org/docrep/005/v9878e/v9878e00.HTM

FAO. (2007). The state of world fisheries and aquaculture 2006. Rome: FAO.

Forster, J. (2002). Farming salmon: An example of aquaculture for the mass market. Reviews in Fisheries Science, 10(3&4), 557–591.

Garcia, M.S., & Grainger, R.J.R. (2005). Gloom and doom? The future of marine capture fisheries. Philosophical Transactions of the Royal Society B, 360, 21–46.

Gargan, P.G., Tully, O., Poole, W.R. (2002). The Relationship between sea lice infestation, sea lice production and sea trout survival in Ireland, 1992–2001. In D. Mills (Ed.), Salmon on the edge (pp. 119–135). Oxford: Blackwell Science Ltd.

Hites, R.A, Foran, J.A., Carpenter, D.O., Hamilton, M.C., Knuth, B.A., Schwager, S.J. (2004). Global assessment of organic contaminants in farmed salmon. Science, 303(5655), 226–229.

Hjelt, K.A. (2000). The Norwegian regulation system and the history of the Norwegian salmon farming industry. In C.I. Liao, & C. Kwei (Eds.), Cage aquaculture in Asia (pp. 1–17). Proceedings of the First International Symposium on Cage Aquaculture in Asia. Quezon City, Philippines: Asian Fisheries Society.

Holm, P., & Jentoft, S. (1996). The sky is the limit? The rise and fall of Norwegian salmon aquaculture, 1970–1990. In C. Bailey, S. Jentoft, P. Sinclair (Eds.), Aquaculture development (pp. 23–42). Boulder: Westview Press.

Jacquet, J.L., & Pauly, D. (2008). Trade secrets: Renaming and mislabeling of seafood. Marine Policy, 32, 309–318.

Jensen, Ø., Dempster, T., Thorstad, E.B., Uglem, I., Fredheim, A. (2010). Escapes of fishes from Norwegian sea-cage aquaculture: Causes, consequences and prevention. Aquaculture Environment Interaction, 1, 71–83.

Jentoft, S. (2007). Limits of governability: Institutional implications for fisheries and coastal governance. Marine Policy, 31, 360–370.

Kautsky, N., C. Folke, Ronnback, P., Troell, M., Beveridge, M., Primavera, J. (2001). Aquaculture and biodiversity. In S. Levin, G. Daily, R. Colwell, J. Lubchenco, H. Mooney, E-D. Schultze, D. Tilman (Eds.), Encyclopedia of biodiversity (pp. 1185–1198). San Diego: Academic Press.

Knapp, G. (2005). Implications of aquaculture for wild fisheries: The case of Alaska wild salmon. (Presentation for the Bevan Sustainable Fisheries Lecture Series, University of Washington, Seattle, Washington). Retrieved August 6, 2007, from http://www.iser.uaa.alaska.edu/iser/people/knapp/Knapp_UW_Bevan_Series_Salmon_Lecture_050210.pdf

Knapp, G., Roheim, C., Anderson, J.L. (2007). The great salmon run: Competition between wild and farmed salmon. TRAFFIC North America. Washington, DC: World Wildlife Fund.

Kooiman, J. (2003). Governing as governance. London: Sage.

Kooiman, J. (2008). Exploring the concept of governability. Journal of Comparative Policy Analysis, 10(2), 171–190.

Krkošek, M., Lewis, M.A., Morton, A., Frazer, L.N., Volpe, J.P. (2006). Epizootics of wild fish induced by farm fish. Proceedings of the National Academy of Sciences of the USA, 103(42), 15506–15510.

Liu, Y., & Sumaila, U.R. (2007). Economic analysis of netcage versus sea-bag production systems for salmon aquaculture in British Columbia. Aquaculture Economics and Management, 11, 413–437.

Liu, Y., & Sumaila, U.R. (2008). Can farmed salmon production keep growing? Marine Policy, 32, 497–501.

Liu, Y., Olaussen, J.O., Skonhoft, A. (2011). Wild and farmed salmon in Norway – A review. Marine Policy, 35, 413–418.

MAFF. (2006). Template for development of facility – Specific fish health management plans. Vancouver: British Columbia Ministry of Agriculture, Food and Fisheries.

Maroni, K. (2000). Monitoring and regulation of marine aquaculture in Norway. Journal Applied Ichthyology, 16, 192–195.

Mathis, M.L., & Baker, P. (2006). Assurance bonds: A tool for managing environmental costs in aquaculture. Aquaculture Economics and Management, 6(½), 1–18.

Maurstad, A., Dale, T., Bjorn, P.A. (2007). You wouldn't spawn in a septic tank, would you? Human Ecology, 35, 601–610.

Milewski, I. (2001). Impacts of salmon aquaculture on the coastal environmental: a review. In M.F. Tlusty, D.A. Bengston, H.O. Halvorson, S.D. Oktay, J.B. Pearce, J.R.B. Rheault (Eds.), Marine aquaculture and the environment: A meeting for stakeholders in the Northeast. Falmouth: Cape Cod Press.

Morton, A., & Routledge, R. (2005). Mortality rates for juvenile pink Oncorhynchus gorbuscha and chum O. keta salmon infested with sea lice Lepeophtheirus salmonis in the Broughton Archipelago. Alaska Fisheries Research Bulletin, 11, 146–152.

NASCO. (2007). The williamsburg resolution. The North Atlantic Salmon Conservation Organization. Retrieved May 5, 2008, from http://www.nasco.int/pdf/cnl_06_48.pdf

Naylor, R.L., & Burke, M. (2005). Aquaculture and ocean resources: Raising tigers of the sea. Annual Review of Environmental Resources, 30, 185–218.

Naylor, R.L., Goldburg, R.J., Primavera, J.H., Kautsky, N., Beveridge, M.C.M., Clay, J., Folke, C., Lubchenco, J., Mooney, H., Troell, M. (2000). Effect of aquaculture on world fish supplies. Nature, 405(6790), 1017–1024.

Naylor, R.L., Eagle, J., Smith, W.L. (2003). Salmon aquaculture in the Pacific Northwest – A global industry. Environment, 45(8), 18–39.

Neori, A., Chopin, T., Troell, M., Buschmann, A.H., Kraemer, G.P., Halling, C., Shpigel, M., Yarish, C. (2004). Integrated aquaculture: Rationale, evolution and state of the art emphasizing seaweed biofiltration in modem mariculture. Aquaculture, 231(1–4), 361–391.

Olaussen, J.O., & Liu, Y. (2011). On the willingness-to-pay for recreational fishing-escaped farmed versus wild Atlantic salmon. Aquaculture Economics and Management, 15, 245–261.

Pauly, D., Tyedmers, P., Froese, R., Liu, Y. (2001). Fishing down and farming up the food web. Conservation Biology in Practice, 2, 25.

Pauly, D., Christensen, V., Guenette, S., Pitcher, T.J., Sumaila, U.R., Walters, C.J., Watson, R., Zeller, D. (2002). Towards sustainability in world fisheries. Nature, 418(6898), 689–695.

Porter, G. (2005). Protecting wild Atlantic salmon from impacts of salmon aquaculture: A country-by-country progress report. 2nd Edn. St. Andrews: World Wildlife Fund and Atlantic Salmon Federation.

Pullin, R., & Sumaila, U.R. (2005). Aquaculture. In J. Kooiman, M. Bavinck, S. Jentoft, R. Pullin (Eds.), Fish for life: Interactive fisheries for fisheries (pp. 93–107). Amsterdam: Amsterdam University Press.

Ridler, N., Wowchuk, M., Robinson, B., Barrington, K., Chopin, T., Robinson, S., Page, F., Reid, G., Szemerda, M., Sewuster, J., Boyne-Travis, S. (2007). Integrated multi – trophic aquaculture (IMTA): A potential strategic choice for farmers. Aquaculture Economics and Management, 11(1), 99–110.

SAR. (1997). Salmon aquaculture review (Vol. 3). Victoria: Environmental Assessment Office.

Sønvisen, S.A. (2003). Integrated coastal zone management (ICZM): The allocation of space in Norwegian aquaculture – From local lottery to central planning? Tromsø: Norwegian College of Fishery Science, University of Tromsø.

Sylvia, G., Anderson, J.L., Cai, D. (1996). A multilevel, multiobjective policy model: The case of marine aquaculture development. American Journal of Agricultural Economics, 78(1), 79–88.

Tacon, A.G.J., & Metian, M. (2008). Global overview on the use of fish meal and fish oil in industrially compounded aquafeeds: trends and future prospects. Aquaculture, 285(1–4), 146–158.

Thorpe, J.E. (1980). The development of salmon culture towards ranching. In J.E. Thorpe (Eds.), Salmon ranching (pp. 1–12). London: Academic Press.

Tidwell, J.H., & Allan, G.L. (2001). Fish as food: Aquaculture's contribution – Ecological and economic impacts and contributions of fish farming and capture fisheries. EMBO Report, 2(11), 958–963.

Troell, M., Halling, C., Nilsson, A., Buschmann, A.H., Kautsky, N., Kautsky, L. (1997). Integrated marine cultivation of gracilaria Chilensis (Gracilariales, Rhodophyta) and salmon cages for reduced environmental impact and increased economic output. Aquaculture, 156(1–2), 45–61.

Troell, M., Halling, C., Neori, A., Chopin, T., Buschmann, A.H., Kautsky, N., Yarish, C. (2003). Integrated mariculture: Asking the right questions. Aquaculture, 226(1–4), 69–90.

Troell, M., Joyce, A., Chopin, T., Neori, A., Buschmann, A.H., Fang, J.G. (2009). Ecological engineering in aquaculture – Potential for integrated multi-trophic aquaculture (IMTA) in marine offshore systems. Aquaculture, 297(1–4), 1–9.

Tveterås, S. (2002). Norwegian salmon aquaculture and sustainability: The relationship between environmental quality and industry growth. Marine Resource Economics, 17(1), 121–132.

Wessells, C.R., Johnston, R., Donath, H. (1999). Assessing consumer preferences for ecolabeled seafood: The influence of species, certifier and household attributes. American Journal of Agricultural Economics, 81(5), 1084–1089.

Whitmarsh, D., & Wattage, P. (2006). Public attitudes towards the environmental impact of salmon aquaculture in Scotland. European Environment, 16, 108–121.

Whitmarsh, D.J., Cook, E.J., Black, K.D. (2006). Searching for sustainability in aquaculture: an investigation into the economic prospects for an integrated salmon-mussel production system. Marine Policy, 30(3), 293–298.

Willoughby, S. (1999). Manual of Salmonid farming. Oxford, MA: Fishing News Books.

Chapter 12
Marine Protected Areas in the Canary Islands – Improving Their Governability

Raquel De la Cruz Modino and José J. Pascual-Fernández

Abstract MPAs are complex institutional arrangements that should be analyzed from a governance perspective taking into account the serious challenges posed about their capacity to cope with the problems of implementation or effectiveness. In this paper we emphasize the huge and diverse advantages of MPAs initiated by local communities. This trend is increasing lately with the involvement and demands of traditional users, such as artisanal fishers, requesting the implementation of marine reserves. Frequently, they want to ensure the sustainability of fishing activities and avoid the pitfalls of rising numbers of other users. In Spain, many of thc latest proposals for Marine Reserves (MRs) were designed for this purpose by local fishers' organizations in partnership with biologists and social scientists, and some of these initiatives learned precisely from the inception process of La Restinga MPA, the case we are analyzing in detail here.

Keywords Marine protected areas • Marine reserves • Artisanal fisheries • Scuba diving • *Cofradías* • Governability • Step zero

Introduction[1]

Marine protected areas (MPAs) are institutional arrangements that are being promoted worldwide as solutions to the marine resource crisis and, to a certain extent, as a consequence of applying the ecosystem perspective to the preservation of sea

[1] The research summarized in this chapter has been funded with the project *Governability analysis applied to the creation process of Marine Protected Areas* (GOBAMP, CSO2009-09802), supported by the Spanish Ministry of Economy and Competitiveness. The authors acknowledge the collaboration of the Network of Marine Reserves of the General Secretary of Fisheries, Ministry of Agriculture, Food and Environment of Spain.

R. De la Cruz Modino (✉) • J.J. Pascual-Fernández
Institute of Social and Political Sciences, University of La Laguna, Tenerife, Spain
e-mail: rmodino@ull.es; jpascual@ull.es

M. Bavinck et al. (eds.), *Governability of Fisheries and Aquaculture: Theory and Applications*, MARE Publication Series 7, DOI 10.1007/978-94-007-6107-0_12,

resources. However, MPAs are complex systems that should be analyzed from a governance perspective that takes into account the serious questions posed about their capacity to cope with the problems of implementation or effectiveness (Jentoft et al. 2007). MPAs include a system-to-be-governed that basically consists of the ecosystem and its resources on the one hand, and human populations and stakeholders groups that depend on these areas, usually capable of building institutions and political organizations, on the other. We can also analyze an MPA by examining its governing system, which, in terms of its social nature, is formed by institutions and mechanisms of control, and nested into larger institutional and political settings (Thorpe et al. 2011). Both systems interact dynamically, and both the systems and their interactions should be given equal consideration in MPA research. Moreover, the research must take into account elements such as their diversity, complexity, dynamics and scale (Jentoft et al. 2007). Any of these systems may introduce limitations into the governability of MPAs, and consequently their implementation has proven more difficult than expected. In most cases, an MPA cannot be declared and implemented in a short period of time. Its establishment can take considerable time and energy; all too often more than 5 or even 10 years (Jentoft et al. 2011). Additionally, MPA establishment must always overcome governability challenges, such as those exemplified by the case of the Marine Reserve of La Restinga-Mar de las Calmas in El Hierro (Canary Islands, Spain), the focus of this chapter.

Marine conservation has a long history around the world and takes many different forms. The conservation practices in Oceania described by Johannes (1978, 1982, 2002), for example, included protected areas where fishing was considered taboo for various reasons. For centuries, these practices, together with closed seasons and many other examples of customary-based marine resource management, preceded some of the most sophisticated modern management measures, including the existing variety of MPAs. Such measures had been developed without input from the modern sciences or support from states or international donors. Instead, these were mainly conceived by taking into account traditional knowledge. Unfortunately, many were compromised by contact with western management models that attempted to impose new styles of relationships on human societies and resources (Johannes 1978). In recent decades, however, the rediscovery of these deeply rooted measures in Oceania has encouraged the allocation of territorial rights to local populations in many of these states. This has, in turn, resulted in the recovery of traditional models of fisheries management and the promotion of 'organic' MPAs, in addition to other measures focused on preserving fishing resources (Johannes 2002).

Examples of territorial use rights in fishing are present in many continents and coasts. Most of these include controls over outsiders usage by fishing communities. As suggested by Charles, examples of territorial use rights and customary usage are widespread around the world, and are identifiable in both modern and traditional fisheries. They generally have considerable potential to provide a relatively stable, socially-supported fishery management system (Charles 2002). In Spain, the location of our case study, the fishing sector is organized into *cofradías*, long-standing institutions that have survived since medieval times in some areas of the country.

The range and type of possible co-management systems may vary substantially in Spain and around the world, but many *cofradías* can, in practice, be regarded as horizontal co-management models (Symes et al. 2003). The *cofradías* (or "guilds") of fishers are local, non-profit corporations with public rights and exclusive territories (Alegret 1996). They represent the interests of the whole fishing sector by acting "as consultative and cooperative bodies for the administration, undertaking economic, administrative and commercial management tasks", and with the ability to "cooperate in matters of regulating access to the resources and informing over infractions occurring in their territory" (Pascual Fernández 1999, 71). In nineteenth-century Spain, and probably throughout Europe, these local arrangements were eroded systematically by the State, as they constituted a hindrance to the capitalist development of fisheries in the context of rising liberalism. For decades, these processes were driven by large state subsidies to industrial fisheries and, to a certain extent, by confidence in the inexhaustible condition of the oceans. Another important factor was the belief in the capacity of science to manage and predict the future states of marine species, as exemplified by the many models of single species recruitment used extensively in traditional fisheries management. All these elements had one thing in common: the disregard for local institutions, traditions and knowledge. As a consequence, local, community-based institutional arrangements were marginalized by the State. Instead, the top-down management of natural and marine resources, supported by the scientific models of fisheries biology, acquired an increasingly important role. Accordingly, the increased capacity of industrial fleets in Europe and in other areas of the world has driven a number of stocks to extinction, as well as deeply modifying coastal and marine ecosystems to the point where some predictions anticipate a jellyfish future for the world's oceans (Pauly et al. 1998; Pauly and Watson 2003). MPAs are one of the leading measures devised to prevent this scenario.

The literature contains a number of different definitions of MPAs. Perhaps the most cited is the one proposed by the 4th World Wilderness Congress in 1987. It refers to the MPA as, "*an area of intertidal or subtidal terrain, together with its overlying water and associated flora, fauna, historical and cultural features, which has been reserved by law or other effective means to protect part or all of the enclosed environment*" (Kelleher and Kenchington 1992, 44). In the United States of America, MPAs are legislated with some emphasis on the relevance of cultural issues and are defined as, "*any area of the marine environment that has been reserved by Federal, State, territorial, tribal, or local laws or regulations to provide lasting protection for part or all of the natural and cultural resources therein*".[2] Following the recommendation of the Committee on the Evaluation, Design and Monitoring of Marine Reserves and Protected Areas in the United States (National Research Council), protected areas can be classified into four categories with increasing levels of protection: Marine Protected Area, Marine Reserve, Fishery Reserve and

[2] Presidential Documents, Executive Order 13158 of May 26, 2000. Retrieved January 27, 2012 from http://ceq.hss.doe.gov/nepa/regs/eos/eo13158.html

Ecological Reserve (2001). The six categories of protected areas proposed by the International Union for Conservation of Nature (IUCN) are also widely cited (IUCN Commission on National Parks and Protected Areas 1994). All these definitions have some common general traits. To a certain extent, an MPA may be regarded as a territorial measure, characterized by the exclusion of some uses and/or users of specific resources in delimited areas. The key factor is the exclusionary capacity of these areas or resources, which needs to be enforced in some way so as to avoid being labeled a "paper park" with scarce practical relevance. In 2008, protected areas covered approximately 0.65% of the world's oceans and about 1.6% of the total marine area within Exclusive Economic Zones (EEZ) (Wood et al. 2008). Notwithstanding the progress made in the previous decades, these figures fall far short of the targets set by international organizations, such as the Convention on Biological Diversity (CBD), which suggests the protection of 10% of all eco-regions in the world (including marine and coastal areas) before 2010 (CBD-UNEP 2006). At the end of 2010, the data suggested that only 1.17% of the world's oceans was protected, and probably for this reason the countries that signed the CBD extended the deadline until 2020 (Cressey 2011).

On the whole, MPAs comprise a territorial model that has been propelled, to a certain degree, by resource crisis and conservation paradigms. MPA goals can be multiple and diverse, and need to be researched empirically (Jentoft et al. 2011). In the 1970s, the primary driver for MPA creation was related to conservation (Noël and Weigel 2007; Thorpe et al. 2011). All too frequently, this led to the marginalization of traditional users linked to those areas. The human side of the ecosystem was often considered irrelevant despite the considerable evidence suggesting that humans play an important role in many contemporary ecosystems (Vitousek et al. 1997; Stepp et al. 2003). As of the 1980s, the perception of MPAs began to change towards multiple use and sustainability. The implementation of MPAs, or protected areas in general, means that certain capacities for control of the space are assigned to the State or to a variety of institutional arrangements. Specific stakeholders may play a leading role in such arrangements. In some cases, there may be conservation-related non-governmental organizations (NGOs), while in others fishers' organizations may take the lead, since traditional extractive uses of artisanal fisheries may feasibly be considered central elements of MPA goals. Analyzing the creation processes of any MPA from step zero may help to understand how different stakeholders, but also local communities, negotiate their own future or the future of resources considered essential for their social reproduction and continuity (Chuenpagdee and Jentoft 2007).

MPAs have an impact on local community resource governance mechanisms, transforming the conventional fisheries management systems (Thorpe et al. 2011). It is not unusual to find conflicts arising between traditional users of marine resources and conservationists in relation to MPAs, especially during creation processes. Perspectives on marine environment conservation about what should be achieved with these MPAs and how are likely to vary between conservationists, scientists, governments and fishers. In some cases, MPAs may be conceived solely to preserve the ecosystem, excluding the human side and traditional extractive activities. In other cases, the proposal may be to substitute artisanal fishing activities with non-consumptive uses, such as scuba diving, or to inhibit any human activity in the area. It is, however, also possible to identify examples of MPAs designed and implemented in support of artisanal fisheries and to ensure their sustainable development (Kalikoski and Vasconcellos 2008; Pascual Fernández and De la Cruz Modino 2008, 2011). MPAs may reinforce or create new territorial use rights. For that reason, we argue that MPAs may improve local fishing management conditions when they are locally-driven and when the sea and natural resource protection policies generate some kind of community-based response rather than a simple, imposed tool. The examples of the Eastport MPA in Canada (Charles and Wilson 2009) and the Actam Chuleb Marine Reserve in Mexico (Chuenpagdee et al. 2002) illustrate the huge and diverse advantages of MPAs initiated by local communities in comparison to others that are introduced and imposed externally. This trend has been increasing of late, with the involvement and demands of traditional users, such as artisanal fishers, requesting the implementation of marine reserves. Frequently, they want to ensure the sustainability of fishing activities and avoid the pitfalls of rising numbers of other users, such as recreational fishers and intensive trawlers, taking control of the area and jeopardizing the viability of traditional activities. In Spain, many of the latest proposals for Marine Reserves (MRs) have been designed for this purpose by local fishers' organizations in partnership with biologists and social scientists. Some of these initiatives have learned precisely from the inception process of La Restinga MPA.

What is remarkable about the case study of La Restinga (Canary Islands, Spain) is that MPAs may represent a way of retaining control of resources in the hands of local fishers and their institutions by excluding any new entrant considered a threat to the health of the ecosystem or to the fishers' livelihood (Pascual-Fernandez and De la Cruz Modino 2011). This chapter begins by reflecting on the historical and legal context of MPAs in Spain, and the specific conditions that favor the involvement of fishers' organizations in their governance. It is followed by a general description of the Canary Islands and the analysis of the fishing community of La Restinga, in the south of El Hierro, where the MR was established in 1996. Taking into account the diversity, complexity, dynamics and scale framework provided by the governability theory, we provide a detailed description of some of the historical, economic and general traits of this area, as well as the specific situation and challenges faced by the local community at the moment of MR inception (Jentoft 2007; Jentoft et al. 2007). We develop a summarized analysis of the creation process and comment on how this MR has increased governability in the area in the discussion and conclusion. We demonstrate how the MR has enabled the community to clearly influence the control of local resources and economic activities, contributing to local empowerment and the slow and controlled pace of local development that favors local people. In this sense, it has increased the capacity of the community to cope with their most urgent concerns, thereby increasing governability (Jentoft 2007).

Spanish Marine Protected Areas

Biogeographically speaking, Spain is an extremely diverse coastal country. It is comprised of most of the Iberian Peninsula, some archipelagos and other islands and islets. Here, MPAs have taken on many different forms, including No-take zones, Maritime & Terrestrial National Parks, Marine Reserves, and Fishing Reserves. One of the first Spanish protected areas was created around the archipelago of the Chafarinas Islands, off the Mediterranean coast of Africa. The archipelago was declared an "Area for National Defence" in 1920, and then converted into a "National hunting refuge" from 1979 to 1982. In 1982, the area was declared a "No-take zone". In Spain, MPAs can be regarded as multiple-use areas for purposes other than ecological conservation. Spanish MPAs have already been analyzed in the literature, and their goals and images are as diverse and dynamic as the socio-ecological context in which they are implemented (Jentoft et al. 2011). Consequently, denominations or forms may change or overlap within the same area. The Medes Islands (Catalonia, Spain) are an interesting example, because the MPA has undergone several transformations. It originated as a "No-take zone" in 1983, was made a "Protected Area" in the 1990s, after which the protected area was extended and declared a marine and terrestrial "Natural Park" in 2010. Another emblematic and early Spanish MPA is Doñana (Andalusia), which encompasses various terrestrial and aquatic habitats and ecosystems. The area has been declared a Natural Park, National park, Biosphere Reserve, World Heritage Site and Special Area for the Protection of Wild Birds (Pain et al. 1998, 46). It is also reserved for several human, recreational and non-recreational (including apiculture, shell-fishing, pilgrimage route, raising livestock, etc.) uses and receives thousands of visitors per year.[3]

Article 148 of the 1978 Spanish Constitution specifies that regional governments have the capacity to legislate and manage maritime areas (Suárez de Vivero and Frieyro de Lara 1994); hence, fisheries responsibilities are shared in Spain. Both national and regional governments legislate on the protection of the marine environment under their jurisdiction, creating a need for coordination that has not always been successful. In addition, after Spain's inclusion in the European Union (EU) in 1986, several protectionist policies were implemented as a result of the European environmental agenda and the maritime and coastal directories. Several aspects affect the implementation and rising numbers of Spanish MPAs: different designations, diverse goals, and legal frameworks not always connected with any fishing interest or issues. MPA responsibilities can also be shared by the different administrations (national or regional). This may be a result of location, for example, within or

[3] In 1987 J.M. Granados Corona presented an extensive study in his doctoral thesis about the historical transformations of the Ecosystem of Doñana National Park; available at http://fondos-digitales.us.es/tesis/tesis/1555/transformaciones-historicas-de-los-ecosistemas-del-parque-nacional-de-donana/#description

outside the waters where the regional governments have competences.[4] In some cases, the characteristics of the ecological systems involved must be considered for specific conservation goals, such as the Special Protection Areas (SPAs) for birds or the Special Areas of Conservation (SACs), arising from the EU Habitats and Birds Directives. There is no clear distribution of responsibilities between national and regional governments; recently, the management of Spanish National Parks has been transferred from the State to the regional governments even though the main responsibility remains with the State (Law 5/2007 of 3 April, of the National Parks Network, Official State Gazette, number 81, of Wednesday 4 April 2007, 14639–49). In short, Spanish MPAs are the result of negotiations in different decision-making environments and contexts. For the general purposes of this chapter, we will refer to one type of MPA in particular, "Marine Reserve with fishing interest", whereby the "main goal is the sustainability of artisanal fisheries" (Revenga 2003, 101) and which allows some types of small scale fishing activities.

The legalization of the Marine Reserves (MRs) in Spain appeared for the first time in a "Ministerial Order of Maritime Restocking", published as a fishing restoration tool in 1982 (Order 11, of May 1982, Official State Gazette, number 125, 13824–5). The State would be required to consult with the National Federation of *Cofradías* and the Spanish Institute of Oceanography prior to the establishment of the MRs. The first MR created under this Ministerial Order was the Marine Reserve of Tabarca[5] in Alicante, off the Spanish Mediterranean Coast in 1986. The fishing identity of the Spanish MRs was written into Spanish Marine Fisheries Law 3/2001, which explicitly stated that "those areas[,] because of their particular characteristics[,] deemed appropriate for the regeneration of fish stocks", would be declared marine reserves (Law 3/2001, Official State Gazette, *BOE* number 75, Wednesday 28 of March 2001, 11516). Finally, the Marine Protected Area definition was drawn into Spanish Law 42/2007 on Natural Heritage and Biodiversity, of 13 December (Official State Gazette, *BOE*, number 299, Friday 14 December 2007, 51275–327). This was a legal definition proposed by the Ministry of the Environment. The Marine Reserve definition, on the other hand, was proposed by the Ministry of Agriculture and Fisheries, and had a clearer emphasis on the sustainability of fishing activities as one of the goals.

---

[4] This is due to Spanish decentralization process that provides regional governments with some competences over internal waters. As Suárez de Vivero et al. affirms: "This division of competences also affects territorial distribution: the Central Administration have exclusive competences over the Territorial Seas (TS) and the Exclusive Economic Zone (EEZ) – where most national fishing areas are located – whereas the regional governments restrict their action to Internal Waters (IW)" (1997, 199)

[5] Two years earlier the Spanish Government had published a "Royal Decree for Fisheries Management" (R.D. 681/1980, 28 May) whose main objective was to restock marine areas and resources of commercial and ecological interest. Under this decree the first no-take zones were established in Spain: the Chafarinas Islands (Melilla, 1982) and Medes Islands (Catalonia, 1983). Listing the first Spanish Marine Reserves can give rise to some confusion between the first no-take zones and the MRs created according to the 1982 Order.

At present, there are three MPAs with the label "Marine Reserves with Fishing Interest" and one area designated as "Fishing reserve". All were created under the full responsibility of regional governments. However, national and regional governments have recognized the "fishing interests" of a number of MRs created by the State or under a regime of shared responsibility. The selection of protected areas specified in Table 12.1 comprises all those that are explicitly linked with artisanal fisheries, either in their label or in public discourse.

The label "Marine Reserves with Fishing Interest" is not included in the aforementioned national legal definition, but this is a special condition assumed by the State in the public discourse. The fish-restocking goal is expressed in the national legal definitions that prompted the establishment of MRs in Spain. It is therefore highly likely that the fisheries administration, as being responsible for promoting the initial MRs, was aiming for the involvement of the artisanal fishing sector in its creation and functioning. Marine conservation in Spain has strong ties with fisheries administration for a long time, and not with environmental administration—until recently.

La Restinga and the Sea of Calms Case Study

The village of La Restinga is located on the southwest coast of the island of El Hierro in the archipelago of the Canaries. This is the main fishing community on the island, and the location of the island's only *cofradía.*

The Canary Islands are a region of Spain, located around 100 km west of the Saharan Coast of Northwest Africa and 1,500 km south of the Spanish mainland. There are seven islands and four islets, covering a total surface area of 7,446.95 km^2. Tenerife is the largest island (2,034.38 km^2) and El Hierro is the smallest (268.71 km^2) (Fig. 12.1).[6] The economy of the Canary Islands depends significantly on tourism, especially since the 1960s, when a combination of specific policy decisions at regional and national levels during the Franco dictatorship, in conjunction with the changing nature of tourism on a wider scale, precipitated the massive build-up of tourism infrastructure along the arid coastal plains in the south of each island (Bianchi 2004). At present, tourism income represents roughly 30% of the Gross Domestic Product (GDP) of the Archipelago, while the service sector as a whole reaches 77% of the GDP.[7] More than ten million tourists visit the Canaries every year, while the permanent population only slightly exceeds two million inhabitants.[8]

---

[6] Source: Institute of Statistics of the Canary Islands, see http://www2.gobiernodecanarias.org/istac/estadisticas.html

[7] Source: Economic and Social Council of the Canary Islands: Annual rapport: http://www.cescanarias.org/?q=informes_anuales

[8] Data from 2009 retrieved December 12, 2000 from http://www2.gobiernodecanarias.org/istac/estadisticas.html

Table 12.1 Number and characteristics of MRs associated with fishing activities in Spain

Name	Description	Year	Area (has)	Location	Responsibilities
Isla de Tabarca	MR	1986	1,400	Mediterranean Sea	SHARED between the State and Regional Governments
Islas Columbretes	MR	1990	4,400	Mediterranean Sea	STATE
La Graciosa e islotes del Norte de Lanzarote	MR	1995	70,700	Atlantic Ocean	SHARED
Cabo de Palos-Islas Hormigas	MR	1995	1,898	Mediterranean Sea	SHARED
Cabo de Gata-Níjar	Natural Park	1987	12,200	Mediterranean Sea	STATE
	MR	1995			
Punta de La Restinga-Mar de las Calmas	MR	1996	750	Atlantic Ocean	SHARED
Isla de Alborán	Fishing protected area	1997	429	Mediterranean Sea	STATE
	MR – Fishing Reserve	1998	425,645		
Masía Blanca	MR	1999	280	Mediterranean Sea	STATE
La Palma	MR	2001	3,719,1	Atlantic Ocean	STATE
Irta	Natural-Marine Reserve	2002–2003	No data	Mediterranean Sea	Regional Government (RG)
	Marine Reserve with Fishing Interest (MRFI)				
Desembocadura del Guadalquivir	Fishing Reserve	2004	22,200	Atlantic Ocean	RG
Cala Ratjada	MR	2007	5,900	Mediterranean Sea	STATE
Os Miñarzos	MRFI	2007	2,200	Atlantic Ocean	RG
Cedeira	MRFI	2009	720	Atlantic Ocean	RG

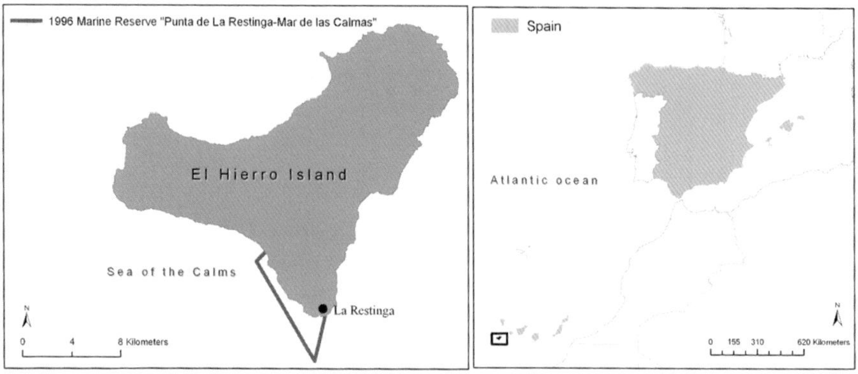

Fig. 12.1 Map of Canary Islands and El Hierro (Credit: A.J. Rodríguez-Darias)

The distribution of this human pressure around the territory is however, not balanced: Gran Canaria has 537 inhabitants per km^2 versus El Hierro's[9] 41 per km^2, the least populated island of the archipelago with only 10,892 total inhabitants.[10] Tourist and services-related development is largely concentrated on only a few of islands. Each island government, or *Cabildo*, has a role in this process. In the case of El Hierro the island government rejected the idea of mass tourism development, keeping tourist infrastructure to a minimum. The airport, for example, only allows propeller airplanes arriving from other regional islands, and there are no plans for expanding to international flights. The *Cabildo* does not want to follow the patterns of rapid growth and the models of mass tourism development of the other islands, such as Fuerteventura, which has almost tripled in population in less than 20 years.

The story of La Restinga is marked by its recent creation. Located in a peripheral and uninhabited area surrounded by volcanic lava flows but with excellent year-round climatic and environmental conditions, the fishing families who founded the village in 1940 came from Valle Gran Rey, in La Gomera. Before their arrival, the area was largely uninhabited,[11] and was used as a place of temporary settlement by farmers from the neighboring village of El Pinar (who spent several weeks a year farming and fishing on the coast) and for fishing trips from La Gomera. In the late 1970s, the total population of La Restinga counted 124 inhabitants. Since its foundation, the main economic activities in the village have involved fisheries.

[9] Data from 2009 retrieved December 12, 2010 from http://www2.gobiernodecanarias.org/istac/estadisticas

[10] Data from 2009 retrieved December 12, 2010 from http://www2.gobiernodecanarias.org/istac/estadisticas.html

[11] Without electricity or fresh water supply, the first families who came to La Restinga lived in caves on the coast.

La Restinga was founded as a fishing village despite the fact that its peripheral location made selling fish very difficult. For a long time, the fishing community depended on factories or intermediaries who practically monopolized the catches. In 1989, the fishers rejected this traditional monopoly when intermediaries refused to buy some catches because of market issues. In 1991, with the support of the *Cabildo*, local fishers set up a Fishermen's Cooperative in the village. This initiative also reflected their desire to obtain more control over tuna fishery development (Galván Tudela 1990).

The growth of scuba diving for tourists in La Restinga has changed some of these aspects. In the absence of foreign investments in tourist infrastructures, certain fishing families have taken advantage of the presence of tourists by setting up various business initiatives (Pascual Fernández et al. 2001; Pascual 2004). Female employment has also risen as a result of increasing tourism. A case in point is a commission-based system of accommodation available for tourism, which is managed by some fishermen's wives and entails building maintenance and housekeeping, client reception and direct attention, and accommodation booking. This activity takes place through informal channels and provides an important source of income for families, enabling them to improve their standard of living (De la Cruz Modino and Pascual-Fernández 2005a; Pascual-Fernández and De la Cruz Modino 2005).

There is a strong territorial identity within the fishing community, which is based primarily on a common origin. After all, the founders of the village all came from the island of La Gomera. There is also common socio-economical background that links local inhabitants to fishing activities, with shared concerns, troubles and development strategies. The community is isolated from the rest of the island and, to a certain extent, from the rest of the Canary Islands. Considering the role of the *Cofradía* of La Restinga in local fisheries management, the local fishing identity is further fed by different experiences of self-governance or co-governance. This feeling is also bound up with the main fishing area, known as the Sea of Calms, where fishers traditionally worked and learned to fish. The name of this section of the coast near La Restinga reflects the continuously calm state of the ocean, which is evident from the shore. The towering land mass offers protection against the prevailing northeasterly winds (Pascual 2004). The absence of winds and currents allows for ongoing fishing and tourism activities in the area at most times of the year. There is considerable diversity in the tropical and subtropical characteristics of the sea, but a remarkably low density of species. In general, the marine ecosystems that surround the archipelago are characterized by biodiversity and fragility; the subtropical location results in a surface water temperature of around 21 °C for El Hierro. Along the coast of La Restinga, it is possible to find pelagic and subtropical species that are less frequent in the rest of the Canary Islands. These include the whale shark (*Rhincodon typusk*), trumpet fish or atlantic cornet fish (*Aulostomus strigosus*), and the ocean triggerfish (*Canthidermis sufflamen*), all of which are a great attraction for scuba divers. The Sea of Calms is extremely important for fishing, because it is especially rich in coastal pelagic, semi-pelagic and benthonic species, with tuna stocks such as yellowfin tuna (*Thunnus* spp.) and *bonito* (*Katsuwonus pelamis*) arriving on a cyclical basis. The natural conditions, however, also facilitate poaching or spear fishing in the sea. These conditions—absence of wind, diverse but limited numbers of sea species and rising human uses—were emphasized in the proposal for the protection of the Sea of Calms from a preventive point of view.

At present, the population of La Restinga stands at around 600 inhabitants, and the community is mainly composed of close families. There are between 37–43 artisanal fishers, organized into 33 productive units, and aged between 30–40 years on average—the youngest group of fishers in the Canary Islands. There are approximately 28 families (including fishers' and fish sellers' families, and other personnel involved in commercial activities) whose economy is directly linked to the fishing sector. On the whole, fishers own their own boats and many (between 30–50%) have more than one—different fishing techniques require smaller or larger boats. The tuna fishery is tremendously important for the village and affects the rest of the fisheries present in the Sea of Calms. Depending on the fishery, tourist demand and institutional support, fishers either sell their catch themselves, through the cooperative or through other sellers. Roughly 50% of the fishers are currently involved in the cooperative and those who are not have moved to other economic sectors (construction in particular) (De la Cruz Modino 2008).

In 2007, we estimated a total of 223 tourist apartments offering accommodation for up to 829 persons (there are no hotels in La Restinga). For many, however, occupation is not year-round and mainly on summer and public holidays (De la Cruz Modino and Hernández Barbuzano 2007). Most of the apartments available for rent are owned by local people. In addition, there are four restaurants (three of them serve fresh fish) and seven bars (De la Cruz Modino 2008). Around ten scuba diving businesses cater to Spanish and European scuba-diving tourists all year-round; most are family-run businesses and there are no tour operators on the island. All of the island's scuba-diving businesses are owned by people not born in El Hierro.

The Marine Reserve of La Restinga

Proposals for Marine Reserves in the Canary Islands increased considerably in the 1980s. These were prompted by a group of marine biology researchers based at the University of La Laguna (Tenerife). One of the group leaders was born and raised in a fishing community and therefore had a deep understanding of the constraints involved in establishing protection measures for fishers. In 1987, the MR proposal was first presented to local fishers at the *Cofradía* in La Restinga. Their initial reaction was anything but positive, nevertheless the discussion remained on the table for some time and the early proposal will have an impact later on. The idea was considered interesting from the outset by some local fishers, but perhaps needed a while to mature. The intervention of a local fisher's son, who studied Marine Biology with the leader of the research group, was also important.

In 1994, the *Cofradía* of La Restinga rescue the MR proposal and discussions began again. This time, the proposal included the possibility of protecting the Sea of Calms. In the 1990s, the MR was presented as a tool to address the problems and demands previously identified by local fishers who had banned the use of certain gear in the Sea of Calms between 1980 and 1990. Fishers were aware of the area's ecological characteristics and fragility, and reached local agreements to ban gear they considered unsustainable for the ecosystem: fishing pots, long-lines and trammels. Developing countermeasures against illegal fishing activities was also one of the arguments proposed in support of the MR. In the 1990s, then, the MR appeared as an extension of actions and decisions already initiated by the fishers.

The project in the 1990s was led by the vice-president (*Vice-Patron Mayor*[12]) of the *Cofradía*. After a discussion period, fishers agreed on the MR design and voted for it at the *Cofradía*; in 1996, the *Reserva Marina Punta de La Restinga-Mar de Las Calmas* was created. It is important to bear in mind that the MR proposal was discussed extensively (always within the *Cofradía*) for almost 2 years. Time is a highly relevant variable in the governance of social systems.

Local fishers decided and voted on a range of key aspects involved in the MR design, such as boundaries, characteristics, surveillance services, gear and users allowed, and not merely on its acceptance. Various administrations and scientists participated in the decision-making process, but fishers always played the most important role. For example, in 1995, fishers rejected the first official proposal for the MR sent to the *Cofradía*, and the national Ministry of Agriculture, Fisheries and Food had to correct it. Fishers complained about the composition of the *Commissions* designed to manage the MR locally, because they were not recognized as members. Throughout the entire process, local administrations supported fishers' decisions. Responsibility for this MR is shared between the national and regional governments, based on two norms.[13] The decree issued by the regional government clearly specifies the status of "Marine Reserve of Fishing Interest". The public discourse of the Ministry of Agriculture, Fisheries and Food recognizes this specific aim: "its main goal is the sustainability of the artisanal fisheries" (Revenga 2003, 101).

After the MR was declared in 1996, fishers continued discussing objections or doubts inside the *Cofradía*, and also participated in implementing and managing the MR. In 1999, for example, surveillance activities were introduced at the sea and on land, and several fishers were employed as inspectors. In 2001, coordination activities began with the creation of Commissions for MRFI monitoring, an activity in which fishermen actively participated. At the same time, other stakeholders of the Sea of Calms, such as scuba diving entrepreneurs, were neither invited nor considered at any point in the entire process. Despite voicing their concerns and opinions about the MR to the government, they were often ignored. The MR was considered

[12] *Patron Mayor* (president) and *Vice-Patron Mayor* (vice-president) are positions of responsibility and representation within Spanish *cofradías*; both are elected positions.

[13] Order of 24 February 1996 that establishes the *Reserva Marina Punta de La Restinga-Mar de las Calmas*. Official State Gazette, *BOE,* number 30, 3 February 1996, pp. 3765–6. Decree 30/1996 of February 16 that creates a Marine Reserve of Fishing Interest in the area of "Punta de La Restinga-Mar de las Calmas". Official Bulletin of the Canary Islands, published Monday 11 March 1996, pp. 1472–4.

a "fishermen's issue" by all administrations with decision-making power during the process, generally reflecting the thinking of the villagers and local residents (De la Cruz Modino 2003, 2008).

Inside the *Marine Reserve of Punta de La Restinga-Mar de Las Calmas,* all traditional uses by artisanal fishers from La Restinga have been maintained at varying levels of regulation. Recreational fishing by boat is forbidden throughout the MR, and angling from the shore is only allowed in some areas (see Table 12.2). Scuba diving has also been restricted. Small scale fishing boats wishing to access the MR must be registered in a census. Two years of fishing experience in the area must be demonstrated in order to access the census. Consequently, fishers from other areas of El Hierro or from other islands are severely limited and only allowed fish for tuna under special permits and conditions.

Discussion: MPAs Increasing Governability

Since the MR was created, it has often been labeled the best example of a well-functioning MPA in the Canary Islands and used as an exemplar for later initiatives in mainland Spain. The natural environment, which is subject to annual scientific evaluation, has been improved since then. After some years, researchers from the University of La Laguna have recognized that, despite its size, "La Restinga MR is the best, maybe due to fishers' participation". The fishers believe that they proved decisive in the creation of the MR and consider it "their own". In 2004, a survey revealed that fishermen considered the surveillance service and the *Cofradía* as responsible for governing the MR (De la Cruz Modino and Pascual-Fernández 2005b).

The challenges to governance faced by fishers in La Restinga before the establishment of the MR were closely related to natural conditions. The multi-specific ecosystem, on a very small continental shelf, made the area extremely sensitive to depletion, requiring a diversity of fishing strategies as means of adaptation. The ecosystem, is extremely complex and dynamic, with important relations between individuals and populations. The MR may have helped to sustain high catches of key species whilst ensuring sustainability, even in the case that the MR is relatively small and isolated.

Although many new activities and enterprises in the services sector are carried out by fishing families, including fresh fish restaurants or tourist accommodation management, fisheries have remained the main economic activity in La Restinga for decades. Local people are certainly interested in some degree of tourism development, but only in low numbers and in agreement with certain parameters that enable them to continue being the main suppliers of tourism services. In the Sea of Calms, fishers have reduced and limited scuba-diving activity in the MR and created new rules in the area affected by the MPA. Moreover, it must be said that almost all scuba-diving center owners, managers and instructors in the area were born outside the island, and that the ownership and staff of some of these centers change frequently.

Table 12.2 Characteristics of the MR *Punta de La Restinga-Mar de las Calmas*

	TOTAL MR
Responsibility	SHARED: National & Regional Government
Year of declaration	1995–1996
Effective implementation	1998–1999
Depth range	0–400 m
Habitats	Rocky reefs, caves, sandy substrates
Protection Objectives	Fisheries enhancement & conservation
Management body	*Comisión de Gestión y Seguimiento* (Advisory body)/*Comisión de Control*
General Services	Surveillance service/Visitors center
Scientist & management activities	Signposting of diving points/Monitoring of underwater activities/Monitoring of angling from the shore
Forbidden in all MPA	Anchoring/Recreational fishing by boat/Spear fishing/Scuba Diving with propulsion systems/extractive uses other than those described & allowed uses

Zone classification	Total MR	Maximum restricted area	Buffer zone	Multiple uses areas
Size (ha)	750	180	90	480
Uses (General uses allowed)	Small scale fishing uses/ recreational uses	Small scale tuna fishery	Small scale fisheries/ Scuba Diving	Small scale fisheries/Scuba diving/ Angling from the shore/other recreational uses
Gears allowed	Traditional fishing gears	Hook & line and cane for tuna fishery	Hook & line gears, tuna fishery gears, harpoon	Hook & line gears, tuna fishery gears, harpoon, traps for shrimps, traps for moray (mainly), nets for bait and for shoals of *Salma salpa* that cannot be caught by hook & line

As a result, their capacity to act collectively is not comparable, and despite having joined some associations, their recognition is not particularly significant.

Analyzing the governing system in La Restinga reveals a remarkable number of new stakeholders in recent decades. These include entrepreneurs, neighborhood associations and administrations (local, insular, provincial, regional and national) involved in managing the area, users and resources. The role of the *Cofradía* and its leaders in facilitating collective action must certainly be emphasized, and fishing remains the main identity-marker of the local community. Being the only legally recognized public rights institution based in the community, the *Cofradía* has long been the channel for local demands to insular, regional or national administrations. With that said, there have been conflicts among fishers, some of considerable importance. However, the *cofradía* has generally served as a reference point or mediator during such conflicts.

In La Restinga, the MR is responsible for ensuring that fishers' decisions prevail in the Sea of Calms. In this case, fishers have successfully managed all parts of the process, including decision-making. This was exemplified in 1995 when the fishermen blocked the first official proposal for an MR and requested its revision. Bottom-up processes have produced successful results. In some cases, governing initiatives may certainly come from outside the community, but processes can still be managed or influenced from within. The MR could be interpreted as an institutional arrangement devised to prevent changes in the area, such as the growth of scuba-diving tourism, from escaping local control. As a governing tool, the MR helps confront changes, such as the extension of recreational fisheries or other dynamics and developments linked with tourism, by providing a framework within which the local community and administrations can negotiate solutions and opportunities. The MR is currently being affected by a volcanic eruption, active since July 2011, whose consequences for the ecological system-to-be-governed are becoming extremely serious for local fisheries. All fishing and scuba-diving activities have been halted in La Restinga and in the Sea of Calms, with far-reaching effects on the socio-economic system. This process is still ongoing at the end of 2011, and evaluations about the consequences for the natural and the socio-economic systems have yet to be concluded. However, the MR does still exist, the governing system is maintained, and all governing interactions are focusing on the new situation. The fishing community of La Restinga is facing new challenges that we will continue to follow in the near future (Table 12.3).

Conclusions and Recommendations

Successful co-governance in specific scenarios may be dependent on many factors. Not all scenarios can be equally governable, because real systems differ in key characteristics that, from the perspective of interactive governance, may be summarized as diversity, complexity, dynamics and scale. Furthermore, governability arrangements—in this case institutions related to the governing of MPAs—should be analyzed

Table 12.3 Governance analysis framework on La Restinga case of study

Challenges related to La Restinga & the Sea of Calms	System to be governed			
	Natural system	Socio-economic system	Governing System	Governing interactions
Diversity	There is considerable ecosystem diversity in a multi-specific ecosystem; diverse fishing strategies as a means of adaptation. The Sea of Calms presents a typical situation of tropical and subtropical islands, whose coastal ecosystems are noted for their diversity (although the density of species is low in numbers).	Economic activities in the area related to fisheries have prevailed since the 1940s. There are approximately 600 inhabitants in La Restinga and 28 families associated with small scale fishing activities (including families involved in fishing trade chain). Strong local close bonds. Scuba diving and tourism businesses have appeared more recently but they are important to local economy.	In recent decades, there has been a rise in the number of stakeholders and agencies involved in managing the Sea of Calms. For example, the Sea of Calms has been included in a range of programs linked with the MR creation and the designation of El Hierro as a Biosphere Reserve. Also, numerous administrations (local, insular, provincial, regional and national) are involved in managing the area, users and resources.	Conflicts between different fishers' groups in connection with collective action and commercialization prior to the inception of the MR and the creation of the cooperative. Also, one scuba-diving entrepreneur legally challenged the MR model because of not being consulted during its design. Ministry of Agriculture and Fisheries disregarded this legal challenge, as the courts did too.
Complexity	There is a complex ecosystem, caused by a small coastal platform, the location in the middle of the ocean, the existence of multiple habitats generated by rocky environments. These factors explain the existence of important relations between individuals and populations in the Sea of Calms. The MR may have added stability to the system.	There is average socio-economic complexity. Fishing activities have dominated the area, and tourism is a recent development. Activities involving the services sector and fishing have been developed by fishing families (fresh fish restaurants; tourist accommodation management).	Despite the growth of stakeholders and uses in the area, as fishing organization the *Cofradía* has a strong presence in the community, influencing many important issues concerning the Sea of Calms. Channels many local demands to other administrations. The *Cofradía* has presence in all governing bodies created or related to managing Sea of Calms and resources.	There have been some conflicts between *Cofradía*-fishers and scuba-diving centers, during set up and the process of limiting the number of annual dives in some of the most sought-after sites in external waters of the MR.

(continued)

Table 12.3 (continued)

Challenges related to La Restinga & the Sea of Calms	System to be governed			
	Natural system	Socio-economic system	Governing System	Governing interactions
Dynamics	There are complex and intense interactions between coastal pelagic, semi-pelagic and benthonic species in this area. Climate change influences the appearance of new species relevant for fishing and tourism. The MR may have helped sustain a high level of catches of key species, assuring sustainability at the same time.	Fishing families maintain control of some key tourist resources and activities in the area. In the context of the MR, fishers have reduced and limited scuba-diving activity. Locals are highly interested in some degree of tourism development, but with low numbers and with them maintaining control.	Fishers' decisions prevail in all management contexts. Administrations appear on scene supporting fishers' decisions. Other initiatives developed in the area, such as the declaration of the Biosphere Reserve, have not affected MR management of the area with fishers' participation.	The MR could be interpreted as an institutional arrangement devised to prevent changes in the area escaping local control. Also, the MR has reinforced some communal agreements related to banning fishing gears (long-lines, traps, trammels) made before its implementation. No decisions about the Sea of Calms and their resources are made without fishers or outside the MR.
Scale	The Sea of Calms is relatively small and isolated. Despite this, each year it is visited by oceanic pelagic species such as whale sharks, turtles or tunas during their migration, and is connected to tropical ecosystems due to climate change.	There are some factors limiting tourist growth in La Restinga and in El Hierro, due to its isolation. Nevertheless, the village has become one of the most popular destinations for scuba-diving tourism in Spain today. Fishers have also reduced and limited scuba-diving activity in the Sea of Calms, by using the MR and creating new rules in the area affected by the MR.	The Sea of Calms management model with fishers' participation has been exported as an example for other areas where new MPAs have been proposed. Also, the cooperative system implemented by fishers to commercialize their products in 1990s with island government support has been extended to other sectors (local farmers, ranchers, carriers) on the island.	The *Cofradía* has been demanding MR expansion to the rest of the island, but lack of funding and conflicts with other sectors have hindered the process.

by taking into account the 'step zero' suggested by Chuenpagdee and Jentoft (2007), which asks the following questions: who wants to establish the MPA; who constitutes the driving force; and how and to whom can the idea be communicated?

Often, agents external to the local areas where the MPA is discussed, such as national or international conservation organizations, academics and state or provincial government institutions, bring the idea forward and push it through certain agendas or mandates. But the case of La Restinga exemplifies the relevance of the capacity of civil society, the social side of the system-to-be-governed, for evaluating the possibility of building a successful MPA. All too frequently feasibility studies in this area are centered on the non-human side of the ecosystem, disregarding the relevance of governability conditions. Of course, the existence of previous institutional arrangements with legal recognition and the intervention of the *Cofradía* cannot be underestimated; besides, strong leadership reinforced their role in the process.

By examining how the MR of La Restinga was prompted, established and implemented, we can affirm that it enhance "the governing system's ability to address the most urgent concerns" (Jentoft 2007, 362). In some way, the MR acts as a territorial measure that provides institutional support for the preservation of the will of local stakeholders against new entrants or free riders who can endanger the key resources that support the local way of living. The governing system devised to cope with this arena has favored a slow pace of development, which permits locals to stay in their village to work as fishers and develop a livelihood they enjoy. The example shows how diverse, complex and dynamic local contexts are for artisanal fisheries, but also how some populations are capable of using global tendencies to assure their control of local scenarios. In this case we have described a well-organized fishers' group that exerts a clear leadership in governing local fisheries and consequently obtains government support. It is possible to observe that the general system has improved its governability if we realize that, despite a degree of social conflict, it did not disturb fishing management or the agreements made around the Sea of Calms.

An MPA is not simply a technical fix. Although many scientists may only focus on its capacity to protect ecosystems, an MPA is also a social institution that has been devised to allocate rights, preserve uses and/or exclude users (Degnbol et al. 2006; Pascual-Fernandez and De la Cruz Modino 2011). Relatively small and coastal MPAs, such as La Restinga MR, constitute a good opportunity for co-governance, where societal parties (state, local communities and institutions, stakeholders) join hands to build institutional arrangements and propose specific goals for the protected area (Kooiman and Bavinck 2005). One of these goals may be the conservation of marine resources, but other goals, such as preventing new users from taking control of an area or developing new activities like tourism, are usually also intermingled. This makes goal formation an empirical research issue that is especially relevant for MPA governability analysis.

The Spanish legal framework that provides the *cofradías* with a consultative role for fisheries administration, and which links marine reserves with small-scale fishers as traditional users in the protected areas, has made this entire process possible.

Once again, involving local communities or supporting their will when they clearly propose a conservation measure, constitute the best foundations for protected areas. Pure conservation of the natural environment is not the goal pursued by local inhabitants; other goals are always intermingled (Jentoft et al. 2011). In this process, they can, of course, use the globalization patterns that generalize protected areas in the sea for their own benefit, all the while assuring ecosystem conservation and preserving a way of life with a practical perspective on their own problems. This agenda should not be regarded as illegitimate; it constitutes an effort to secure a livelihood, reducing present or future risks. Planners need to take into account the broader, highly contextual situation that influences people's lives (Gonzalez and Jentoft 2011); lives that depend on natural environment factors, but also on many other circumstances at the same time. This broader perspective is compelling when planning a protected area and evaluating its governability.

References

Alegret, J.L. (1996). Co-management and legitimacy in corporate fishermen's organizations. The confraries de pescadors de Catalunya, Spain. In R.M. Meyer, C. Zhang, M. Windsor, B. McCay, L. Hushak, R. Muth (Eds.), Fisheries utilization and policy. Proceedings of the World Fisheries Congress, Theme 2 (pp. 342–348). Oxford: Lib. Pub. Co., Pvt, Ltd.

Bianchi, R.V. (2004). Tourism restructuring and the politics of sustainability: A critical view from the European periphery (The Canary Islands). Journal of Sustainable Tourism, 12(6), 495–529.

CBD-UNEP (2006). Report of the eighth meeting of the parties to the convention on biological diversity (Reissued for technical reasons), Curitiba, Brazil. Curitiba: UNEP. Retrieved November 15, 2009, from http://www.cbd.int/doc/meetings/cop/cop-08/official/cop-08-31-en.pdf

Charles, A.T. (2002). Use rights and responsible fisheries: Limiting access and harvesting through rights-based management. In K.L. Cochrane (Ed.), A fishery manager's guidebook. Management measures and their application (pp. 131–157). Fisheries Technical Paper 424. Rome: FAO.

Charles, A., & Wilson, L. (2009). Human dimensions of marine protected areas. ICES Journal of Marine Science, 66(1), 6–15.

Chuenpagdee, R., & Jentoft, S. (2007). Step zero for fisheries co-management: What precedes implementation. Marine Policy, 31(6), 657–668.

Chuenpagdee, R., Fraga, J., Euan-Avila, J.I. (2002). Community perspectives toward a marine reserve: A case study of San Felipe, Yucatan, Mexico. Coastal Management, 30(2) 183–191.

Committee on the Evaluation Design and Monitoring of Marine Reserves and Protected Areas in the United States, N. R. C. (2001). Marine protected areas: Tools for sustaining oceans ecosystems. Washington, DC: National Academy Press.

Cressey, D. (2011). Ocean conservation: Uncertain sanctuary. Nature, 480(7376) 166–167.

De la Cruz Modino, R. (2003). Gestión de los recursos: Turismo, usos y apropiación del patrimonio natural. Tesis de Licenciatura, Universidad de La Laguna, Tenerife, Spain.

De la Cruz Modino, R. (2008). Turismo, pesca y gestión de recursos en la Reserva Marina Punta de La Restinga- Mar de Las Calmas (El Hierro- Islas Canarias) y el Área Natural Protegida de las Islas Medas (Girona, Cataluña). Tesis Doctoral, Universidad de La Laguna, Tenerife, Spain.

De la Cruz Modino, R., & Hernández Barbuzano, I. (2007). Comercialización y distribución de productos turísticos marinos en El Hierro a través de Internet. La Laguna, Tenerife: Trabajo Fin de Máster. Universidad de La Laguna, Facultad de Ciencias Económicas y Empresariales. Máster en Gestión de Empresas Turísticas (MGET) [Sin publicar].

De la Cruz Modino, R., & Pascual-Fernández, J.J. (2005a). Mujeres, diversificación económica y desarrollo del turismo marino. En torno a la Reserva Marina Punta de la Restinga-Mar de las Calmas (El Hierro – Islas Canarias). In K. Frangoudes & J.J. Pascual-Fernández (Eds.), AKTEA Conference: Women in fisheries and aquaculture: Lessons from the past, current actions and ambitions for the future (pp. 263–275). La Laguna: Asociación Canaria de Antropología.

De la Cruz Modino, R., & Pascual Fernández, J.J. (2005b). Reservas marinas, ¿herramientas de gestión pesquera? In J. Pascual Fernández & D. Florido del Corral (Eds.), ¿Protegiendo los recursos? Áreas protegidas, poblaciones locales y sostenibilidad (Vol. VIII, pp. 83–101). Sevilla: Fundación El Monte, FAAEE, Asociación Andaluza de Antropología.

Degnbol, P., Gislason, H., Hanna, S., Jentoft, S., Raakjaer Nielsen, J., Sverdrup-Jensen, S., Clyde Wilson, D. (2006). Painting the floor with a hammer: Technical fixes in fisheries management. Marine Policy, 30(5), 534–543.

Galván Tudela, A. (1990). 'Pescar en grupo': De los azares ambientales a los factores institucionales (La Restinga, El Hierro). Eres (Serie de Antropología), 2, 9–60.

Gonzalez, C., & Jentoft, S. (2011). MPA in labor: Securing the Pearl Cays of Nicaragua. Environmental Management, 47(4), 617–629.

IUCN Commission on National Parks and Protected Areas. (1994). Guidelines for protected area management categories. Cambridge: IUCN/UICN.

Jentoft, S. (2007). Limits of governability: Institutional implications for fisheries and coastal governance. Marine Policy, 31(4), 360–370.

Jentoft, S., van Son, T. C., Bjorkan, M. (2007). Marine protected areas: A governance system analysis. Human Ecology, 35(5), 611–622.

Jentoft, S., Chuenpagdee, R., Pascual-Fernandez, J.J. (2011). What are MPAs for: On goal formation and displacement. Ocean & Coastal Management, 54, 75–83.

Johannes, R.E. (1978). Traditional marine conservation methods in Oceania and their demise. Annual Review of Ecology and Systematics, 9(1), 349–364.

Johannes, R.E. (1982). Traditional conservation methods and protected marine areas in Oceania. AMBIO – A Journal of the Human Environment, 11(5), 258–261.

Johannes, R.E. (2002). The renaissance of community based marine resource management in Oceania. Annual Review of Ecology and Systematics, 33(1), 317–340.

Kalikoski, D., & Vasconcellos, M. (2008). Marine protected areas and reconciling fisheries with conservation: Insights from the common property theory. Reconciling Fisheries with Conservation, I and II, 49, 1211–1219.

Kelleher, G., Kenchington, R.A., Great Barrier Reef Marine Park Authority. (1992). Guidelines for establishing marine protected areas (A marine conservation and development report). Gland: IUCN in collaboration with Great Barrier Reef Marine Park Authority.

Kooiman, J., & Bavinck, M. (2005). The governance perspective. In J. Kooiman, M. Bavinck, S. Jentoft & R. Pullin (Eds.), Fish for life: Interactive governance for fisheries (pp. 11–24). Amsterdam: Amsterdam University Press.

Noël, J.F., & Weigel, J.Y. (2007). Marine protected areas: From conservation to sustainable development. International Journal of Sustainable Development, 10(3), 233–250.

Pain, D.J., Sánchez, A., & Meharg, A.A. (1998). The Doñana ecological disaster: Contamination of a world heritage estuarine marsh ecosystem with acidified pyrite mine waste. Science of the Total Environment, 222(1–2), 45–54.

Pascual, J.J. (2004). Littoral fishermen, aquaculture and tourism in the Canary Islands: Attitudes and economic strategies. In J. Boissevain & T. Selwyn (Eds.), Contesting the foreshore: Tourism, society and politics on the coast (pp. 61–82). Amsterdam: Amsterdam University Press.

Pascual Fernández, J. (1999). Participative management of artisanal fisheries in the Canary Islands. In D. Symes (Ed.), Southern waters: Issues of management and practice (pp. 66–77). London: Blackwell's Science/Fishing New Books.

Pascual-Fernández, J.J., & De la Cruz Modino, R. (2005). Mujeres, reservas marinas y estrategias de diversificación en las poblaciones litorales: El caso de los restaurantes de pescado. In K. Frangoudes & J.J. Pascual-Fernández (Eds.), AKTEA conference: Women in fisheries and aquaculture: Lessons from the past, current actions and ambitions for the future (pp. 247–262). La Laguna: Asociación Canaria de Antropología.

Pascual Fernández, J.J., & De la Cruz Modino, R. (2008). Los espacios marinos protegidos en España: ¿nuevas formas institucionales para las estrategias de apropiación? In O. Beltrán Costa, J. Pascual Fernández, I. Vaccaro (Eds.), Patrimonialización de la naturaleza: El marco social de las políticas ambientales (pp. 199–221). Donosti: Ankulegi Antropologia Elkartea.

Pascual-Fernandez, J.J., & De la Cruz Modino, R. (2011). Conflicting gears, contested territories: MPAs as a solution? In R. Chuenpagdee (Ed.), World small-scale fisheries contemporary visions (pp. 205–220). Delft: Eburon.

Pascual Fernández, J., Santana Talavera, A., Batista Medina, J.A., Dorta Morales, C., Hernández Armas, R., Díaz de la Paz, A., Martín de la Rosa, B., Macías González, J. (2001). Pescatur: Un modelo de desarrollo integral de poblaciones litorales. La Laguna: Instituto U. de Ciencias Políticas y Sociales, Viceconsejería de Pesca del Gobierno de Canarias (Unpublished).

Pauly, D., & Watson, R. (2003). Counting the last fish. Scientific American, 289(1), 42.

Pauly, D., Christensen, V., Dalsgaard, J., Froese, R., Torres, F., Jr. (1998). Fishing down marine food webs. Science, 279(5352), 860–863.

Revenga, S. (2003). Las Reservas Marinas Canarias (España). In D. Moreno & A. Frías (Eds.), Actas de las I Jornadas sobre Reservas Marinas y I Reunión de la Red Iberoamericana de Reservas Marinas (RIRM). Cabo de Gata, Almería 17–23 de Septiembre de 2001 (pp. 101–111). Madrid: Publicaciones del MAPA, Secretaría Técnica, Madrid.

Stepp, J.R., Jones, E.C., Pavao-Zuckerman, M., Casagrande, D., Zarger, R.K. (2003). Remarkable properties of human ecosystems. Conservation Ecology, 7(3), 11.

Suárez de Vivero, J.L., & Frieyro de Lara, M. (1994). Spanish marine policy – Role of marine protected areas. Marine Policy, 18(4), 345–352.

Suárez de Vivero, J.L., Frieyro de Lara, M., Jurado Estevez, J. (1997). Decentralization, regionalization and co-management: A critical view on the viability of the alternative management models for fisheries in Spain. Marine Policy, 21(3), 197–206.

Symes, D., Steins, N., & Alegret, J.L. (2003). Experiences with fisheries co-management in Europe. In D.C. Wilson, J.R. Nielsen, P. Degnbol (Eds.), The fisheries co-management experience: Accomplishments, challenges, and prospects (pp. 119–132). Dordrecht: Kluwer Academic Publishers.

Thorpe, A., Bavinck, M., Coulthard, S. (2011). Tracking the debate around marine protected areas: Key issues and the BEG framework. Environmental Management, 47(4), 546–563.

Vitousek, P.M., Mooney, H.A., Lubchenco, J., Melillo, J.M. (1997). Human domination of Earth's ecosystems. Science, 277(5325), 494–499.

Wood, L.J., Fish, L., Laughren, J., Pauly, D. (2008). Assessing progress towards global marine protection targets: Shortfalls in information and action. Oryx, 42(3), 340–351.

Chapter 13
Gender in Galician Shell-Fisheries: Transforming for Governability

Katia Frangoudes, Begoña Marugán-Pintos, and José J. Pascual-Fernandez

Abstract This chapter analyzes the transformation of on-foot shellfish gathering in Galicia, a traditional activity developed primarily by women under a free access regime. Nowadays, many of the areas where this activity flourishes are in a situation of active co-governance that employs a type of licensing system. Shellfisher organizations plan and control exploitation with the support of the government, even using seeding techniques to regenerate areas that were previously degraded. This process has not been easy, as it has resulted in a strong reduction of persons linked to the activity. Moreover, in the current economic crisis it is menaced by a surge in poaching. Despite the advantages of including women in co-management processes, in terms of equity, legitimacy and inclusiveness, enhancing the capacity of user groups and civil society to cope with the tasks involved in co-governance situations constitutes a serious challenge.

Keywords Governance and governability • Gender • Women empowerment • Resource management • User participation

K. Frangoudes (✉)
Center for the Law and Economics of the Sea, University of Brest, Brest, France
e-mail: katia.frangoudes@univ-brest.fr

B. Marugán-Pintos
Department of Sociology, University Carlos III of Madrid, Getafe, Madrid, Spain
e-mail: bmarugan@polsoc.uc3m.es

J.J. Pascual-Fernandez
Institute of Social and Political Sciences, University of La Laguna, Tenerife, Spain
e-mail: jpascual@ull.es

M. Bavinck et al. (eds.), *Governability of Fisheries and Aquaculture: Theory and Applications*, MARE Publication Series 7, DOI 10.1007/978-94-007-6107-0_13,

Introduction[1]

This chapter introduces a gender lens to the governability perspective, illustrated by the case of on foot women shellfish gatherers (*mariscadoras)* in Galicia (Spain). The role of women in fisheries, especially in resource management and governability, is not sufficiently documented, as few social scientists have studied this topic. The case of the *mariscadoras* in Galicia is an opportunity to examine female participation in shellfishing and the process by which women's organizations are established for resource governance.

A few comments on the concept of gender will help distinguish the difference between 'women', a sexual connotation, and gender, a social construct. Simone de Beauvoir's statement that a woman is not born but made[2] characterizes the field of feminist research that evolved in the 1970s. The first anthropologist to theorize about gender as a social construct was Gayle Rubin (1975). In, "The Traffic in Women: Notes on the Political Economy of Sex," Rubin (1975) explained that males and females are biologically similar and that there is less difference between humans than in any other species. The difference is, therefore, not biological but cultural.

For feminists, using the word 'woman' in development programs reflected a desire to integrate women's issues into the larger discourse about development without taking into account the existing inequalities in power. As a concept, gender has made it easier to denounce occasions when the obvious sexual difference between women and men becomes discrimination against women. The concept of gender helps us understand that we must differentiate the fact of being born as one sex from that of having a markedly different destiny. Ushering the gender concept into the discussion had important advantages: it permitted academics to speak about the socially constructed relations between males and females and about the issue of power. Gender is considered a social construct that refers to the inequality between women and men, the difference in sex being the most important feature. Bennett et al. observe, "The perception of the roles, duties and obligations of male and female has been reinforced through centuries by institutions: the household, the market, the community and the state amongst other things" (2004, 7).

The gender perspective makes it easier to understand that male and female identities may be transformed if they are neither natural or innate, but instead cultural or social. This capacity to transform gender identities has analytical potential. Yet despite this potential, few fishery researchers have fully developed such an

---

[1] The research summarized in this chapter is related to the project *Governability analysis applied to the creation process of Marine Protected Areas* (GOBAMP, CSO2009-09802), supported by the Spanish Ministry of Economy and Competitiveness.

[2] In the French original: "On ne naît pas femme: on le devient" (de Beauvoir 1990 [1949], 13).

approach. In Europe, only a handful of studies have dealt with women in fisheries and aquaculture.[3]

A new dimension is opened up when the gender concept is extrapolated to the governance of fisheries: the observer is able to distinguish between the different roles of women and men in fisheries, and realizes the benefits of integrating women into decision-making processes (Bennett 2005; Pascual-Fernández et al. 2005). A gender lens is therefore useful when analyzing governability, or any of its elements. The manner in which women and men are involved or marginalized in governing processes related to solving society's concerns is an important issue. Too frequently women are simply obviated in these processes, and their participation in problem-solving tasks is regarded as irrelevant. The interactive governance perspective (Kooiman et al. 2005a) is adopted in this paper in order to illustrate the ways in which civil society can be employed to solve societal problems in fisheries. This extends beyond stakeholder participation in the form of collaborations between the state, civil society and market (Kooiman et al. 2005a; Jentoft 2007; Mikalsen et al. 2007) to include women as key actors in the scenario (Frangoudes et al. 2008b). Taking into account the wicked nature of some of the problems confronted, it is questionable whether civil society—the social sphere located between the state and individuals, where autonomous groups are organized and activities take place—is always up to this task (Mikalsen et al. 2007; Jentoft and Chuenpagdee 2009). This is of central concern, and we contend that, while civil society may not be up to the task of co-managing the resources in many cases, as we show in this paper, steps may be taken to improve civil society's capacities. Shellfish gathering in Galicia exemplifies how the capacity of civil society can improve governability, and how the implementation of an elaborate co-governance scheme has made shellfish gathering more sustainable within a decade (Kooiman and Chuenpagdee 2005; Jentoft 2007). The scheme was spread out across the entire region, with local populations assuming important responsibilities for the planning and marketing of the sustainable use of resources, as well as the management of the activity (Marugán Pintos 2004; Pascual-Fernández 2005; Frangoudes et al. 2008b).

[3] At the beginning of the twenty-first century, European funding helped overcome some of these shortcomings. "The role of women in the fisheries sector," was drawn up by Elliott MacAlister for the European Commission (MacAlister 2002). In January 2003, the Commission sponsored a conference on the role of women in the fisheries sector. The FEMMES project (contract no. Q5Tn.2002-01560-CGE, D.G Fisheries), coordinated by the University of Brest (France), was particularly significant in this area, both analytically and politically. The project staged three workshops for women in fisheries from different parts of Europe and also served as the basis for the conference, "Women in fisheries and aquaculture: lessons from the past, current actions and ambitions for the future", held in Santiago de Compostela (Spain) and funded jointly with the Xunta de Galicia (www.fishwomen.org). Data gathered during the project form the basis of this chapter. A more recent publication, "The role of women in the sustainable development of European fisheries areas" (Frangoudes et al. 2008a), is also related to the FEMMES project and to the findings of this chapter. Special mention must be made of Begoña Marugán Pintos's (2004) work, one of the key empirical sources for this chapter.

The process of building *mariscadoras* associations and empowering women, previously disorganized and hostile, is the central theme of this chapter. It focuses on how governability can even be improved in situations shaped by free access and a strong discrimination against certain users — in this case women. It also aim to show that gender constitutes a key factor for understanding the institutional dynamics surrounding shellfish harvesting in Galicia and its historical development. Furthermore, this chapter highlight the problems faced by women in the governance of this practice, how they have overcome them, and their contribution to resource governance. A comprehensive understanding of the process, however, first warrants a discussion of the contextual socioeconomic background.

Background

Spanish fisheries, especially those in Galicia, excel with respect to other regions in Europe. The total annual catch landed by Spanish fisheries amounts to 795,461 tons. Additionally, there are 11,394 registered vessels, and the fleets (off-shore, industrial, artisanal, etc.) and activities related to shellfishing and aquaculture are highly diversified (Ministry of Agriculture, Fisheries and Food, data 2008). Nearly 40% of all affiliates to the special system of Social Security of the Sea in Spain are found in Galicia[4] (official data, December 2010). The Galician fisheries play a vital role in the local economy and account for 10.7% of the regional gross domestic product (GDP) (Galician Institute of Statistics, data for 2003–2004).

In Spain, shellfishing is part of the fisheries sector, and Galicia is home to the majority of on foot shellfishers. In fact, 91% of the 6,177 shellfishers affiliated to the special system of Social Security of the Sea in Spain are from this region.[5] In other areas of Spain, shellfishing has almost disappeared. In Galicia, however, it continues thanks in large part to the natural conditions of the *rias* (drowned river valleys) and their high coastal productivity. The main shellfish species consumed by the local population are grooved carpet shell (*Venerupis decussatus*), carpet shell (*Venerupis pullastra*), and cockle (*Cardium edule*). The importance of this activity has long been disparaged, despite a market value of 21,035.42 million euro and catches of 2,700 t of clams in 1995 (Marugán Pintos 2004). In 2011 the market value of clams landings was 67.486.733 €, with a total weight of 8.722 tons (Galician Government, official data 2011).

Shellfishing comprises two different activities: resources are harvested along the beaches or rocky coastlines either onboard vessels or on foot. In June 2008, 89% of on foot shellfish gatherers were women. Shellfishing conducted from boats,

---

[4] This Social Security of the Sea constitutes a special branch of the social security in Spain devoted only to workers of the sea, including many other ancillary activities. It is managed by the Social Institute of the Sea (*Instituto Social de la Marina*), see http://www.seg-social.es/prdi00/idcplg?IdcService=GET_FILE&dID=72917&Rendition=Web, accesed 28 November 2011.

[5] Data provided by the Social Institute of the Sea (SIS) in 2003.

however, was primarily a male activity. The knowledge and skills have been passed down from generation to generation, from mother to daughter, a cultural heritage that has disappeared in other areas of Spain. Until recently, however, the status of harvesting shellfish on foot was low due to conflicts and resource depletion. The fact that women undertook the activity only served to worsen its public image and may have influenced the authorities' lack of interest or capacity to regulate the activity.

The transfer of (fisheries) responsibilities to the regional government of Galicia in the mid 1980s created new opportunities for the governability of shellfishing. The activity was considered important in the region and administrative attention increased significantly with the aid of European funding.

The entire process has been marked by the co-management of resources between the administration and local shellfisheries associations (which are part of local fisher organizations). The chapter will also discuss the transformation of an activity from informal to regulated, resulting in full recognition and social protection for *mariscadoras*. Professionalizing the activity boosted the confidence of on foot women shellfishers or *mariscadoras*, so much so that they created their own organization, managed the resource, developed extensive shellfish farming, and joined established male organizations, assuming various positions of responsibility within the governing bodies of these organizations.

Development of Shellfish Activity in Galicia

In Spain, shellfish has historically been consumed by people with a low income. Coastal populations practiced shellfishing, because the intertidal zone was open to all, and people sought sources of protein and sustenance. Shellfish was regarded as a marginal food and was mainly harvested on foot by women and children for home consumption. In the early 1960s, shellfish gathering became a commercial activity when the local cannery industry, which processed sardines, could not secure sufficient supplies. Stocks were critical and landings of sardines had decreased considerably. The industry needed to replace sardines with other species for several months of the year in order to maintain a certain production level (October to February), and shellfish provided a solution. The onset of the sardine crisis forced canneries to create the first big commercial market for shellfish in Spain (Meltzoff 1995). The shift of the local cannery industry towards shellfish and the opening of new markets for fresh seafood raised the value of this product. As a consequence, the number of gatherers increased, and the constant pattern of harvesting quickly provoked an overexploitation of shellfish beds. Men, women, and children flocked to the beaches to gather the resource. When the tides were favorable, women would leave home and children would forego school. Even factory workers (e.g., the Citroën factory in Vigo, Galicia) would take their holidays in October, the beginning of harvesting period. At the time, shellfish gathering was an informal activity and there were no limits on the number of gatherers or the quantity and size of shellfish that could be harvested. Any person facing financial difficulties or seeking an income could become a shellfish gatherer. In the words of one female shellfisher: "It was a way to survive (…), by selling clams and cockles we could buy six sacks of wheat flour" (Marugán Pintos 2004, 45).

Until the mid 1980s, the central government was responsible for regulating shellfish practices. In 1963, the first ministerial order concerning the governance of this activity awarded management of the first sale of seafood to local fisher organizations (*cofradías*), thereby strengthening their position. The role of the *cofradías* was further reinforced by Law 59/1969, which granted them the right to establish and exploit shellfish farms. *Cofradías* are backed by the state as local nonprofit corporations with public rights. They represent the interests of the entire fishing sector by acting as consultative and cooperative bodies for the government and undertaking of economic, administrative and commercial management tasks. They also have the ability to "cooperate in matters of regulating access to resources and reporting infractions occurring in their territory" (Pascual Fernández 1999, 71). As consultative bodies, *cofradías* may propose specific regulations to the administration (e.g., regulations about gears), collaborate in the surveillance of shellfish areas, and manage the first sale of catches.[6]

Despite the 1963 regulation requiring a license for shellfishing, resources were overexploited because there was no limit to the number of licenses issued. On top of that, they were extremely easy to obtain.[7] As a result, the number of harvesters, and consequently the level of overexploitation, increased year after year. During the first days of gathering following the end of the closed season, all manner of persons would come to the shore to profit from the high catches. However, once the amounts gathered declined and the activity became less profitable, women were the main harvesters. In this sense, shellfishing by women shared similar characteristics to those often attributed to preindustrial women's work: inconstant and irregular (Vara Miranda and Maquieira D'Angelo 1996; Marugán Pintos 2004), and regarded as an extension to their domestic tasks. In fact, a high percentage (42.4%) of women

[6] Until 1978, *cofradías* were the sole organizations representing ship-owners and crewmembers. Royal Decree 670/78 ended the compulsory membership of *cofradías* for industrial fleet owners and crews. Subsequently, owners of industrial fleets created their own organizations and crewmembers joined trade unions. Today, members of *cofradías* are boats owners and crewmembers of artisanal fisheries. They are equally represented in their governing bodies. In Galicia, shellfish gatherers are included in artisanal fisheries and are therefore members of *cofradías*. Owing to their importance and specificities, shellfishers can create their own section (*agrupación de mariscadoras*) inside *cofradías*. The *cofradía* is managed by a chair (*patrón mayor*) and by two other governing bodies: the general committee and the *cabildo* (Pascual-Fernández et al. 2005: 161). In the past, the position of *patrón mayor* was usually held by experienced fishermen but today this is no longer the case and some time women involved in shellfishing have assumed this role in some *cofradías*. In Galicia, *cofradías* manage fisheries activity, first-sell auctions, and provide accounting services for their members. See also Alegret (1996, 1998, 1999), Pascual-Fernandez (1999), and Erkoreka Gervasio (1991).

[7] Harvesters only needed to be Spanish, aged over 16, live in the maritime province where licenses were issued, and be included in the shellfisher census of the local *cofradía*.

who gathered shellfish before the 1990s described themselves as housewives when asked about their occupation (Sequeiros 1995). The introduction of a shellfish license had no clear effect on shellfish activity and did not prevent the overexploitation of stocks. Women in possession of licenses, as well as those without any documentation, competed for the same resources, often resulting in conflicts that were sometimes "resolved" with physical aggression.

Law 59/69 was the first attempt to organize shellfishing. It was, however, met with strong opposition. The coastal population perceived the sea as a free access space and shellfish as a common resource that could be gathered by all. In 1970, a shellfishing plan was devised by the State. Resource management, however, still constituted a significant problem, a factor reflected in the diminished production of clams and cockles in the 1980s (Consellería de Pesca Marisqueo e Acuicultura 1993). Neither the *cofradías* nor the national government were interested in a non-professional fishery practiced by housewives and, in some sense, the government ignored their responsibilities in shellfish resource management.

The political change in Spain following the end of the dictatorship in the late 1970s led to the creation of powerful regional governments with the capacity to legislate coastal shellfishing areas, and was therefore a key factor in improving the situation. The democratization process modified "the administrative structure of the country and distributed new competencies to the regions (*autonomías*)" (Frangoudes et al. 2008b, 226). Regions were autonomous and had the competency to intervene in many public policies within their territory, fishing activity being one of them. Since 1982, regional fisheries authorities have regulated fishing activities that take place within their inland waters (*rías*, bays, etc.) (Suárez de Vivero and Frieyro de Lara 1997; Suárez de Vivero et al. 1997). This case provides a good example of the subsidiarity principle in that the proximity of the regional government to the conflicts rendered it, to a certain extent, more capable of improving the governability of the sector. Moreover, the regional government created a specific ministry to manage fisheries, aquaculture, and shellfishing.

Main Regional Legal Framework on Shellfish Gathering

From 1985 onwards, regional fisheries authorities made several attempts to reinforce existing regulations by introducing a number of new laws: Law 2/1985 on February 2nd, concerning the Regulation of Maritime Fishing in the Waters of the Regional Government of Galicia; Law 15/1985 on August 23rd, concerning the Regulation of Shellfishing and Aquaculture; Decree 116/1987,[8] regulating the requirements for shellfishing; Law 6/93 on May 11th, concerning fisheries in

---

[8] Decree 116/1987, on May 14th, of the regional government of Galicia established new requirements for obtaining a shellfishing license. It was now necessary to follow a training course; this marked a turning-point for the professionalization of shellfishing in Galicia.

Galicia; and several decrees in 1993 that established permits for exploitation (127/93) and streamlined legislation in the sector (Mahou Lago 2008, 99; Frangoudes et al. 2008b). Though perhaps not as efficient as the circumstances required, these regulations paved the way for managing the activity. Without these foundations, the radical transformation of the sector after 1995 would have simply been impossible.

Before 1995, conflicts in the sector were ongoing. Overexploitation of stocks meant that women seeking to increase their production would move to the territory of other *cofradías*. Articles regarding the conflicts between *mariscadoras* from different *cofradías* or between women with licenses and those without appeared in regional newspapers on a daily basis. Such publications painted a negative picture of shellfish gatherers in regional society and linked them to conflicts related to resource access. This negative picture not only permeated their public image, but also the image that *mariscadoras* held of themselves.

The attitudes of *cofradías* towards the activity were diverse, and some decided to react by establishing daily quotas for each gatherer and imposing a minimum size for harvested shellfish. The first *cofradía* to establish stricter rules than the regional authorities was Illa de Arousa, studied by Meltzoff in 1990 (Meltzoff 1995; Meltzoff and Broad 1996). In an attempt to recover control of the activity, the regional government created a surveillance service against poachers and issued a new decree that made the closed seasons more flexible, avoiding the concentration of catches in a short period and reducing the power of the canning industry. Commercialization improved, because production was no longer concentrated in a few months, and incomes were more evenly distributed throughout the year. An annual market reduced pressure on stocks, provided the women with a year-round income, and created an enhanced pricing structure. The successful development of the shellfish market, however, put new demands on management.

Law 6/1993 (on May 11th), concerning fisheries in Galicia, established a licensing system, which led to new decrees being issued with stipulations for obtaining shellfishing licenses (called *permex*) that were far more specific. The new regulations allowed year-round shellfish extraction, except on Saturdays, Sundays and public holidays, and established guidelines for *cofradías* to submit a management plan for shellfish resources. The regional administration and *cofradías* formally agreed to regulate the shellfish sector, but much work was required to turn these intentions into reality.

Factors Contributing to Mariscadoras' Empowerment

In the period before 1994, public opinion deemed shellfish gathering an economically marginal activity. This explains the scant importance afforded by the fisheries administration to *mariscadoras* and the little interest shown by *cofradías* to women employed in shellfish gathering. Female shellfish gatherers faced discrimination within fisheries organizations and by the fisheries administration. Their small economic contribution to the management of the *cofradías* (a percentage of the catches) in comparison to fishers on boats served to justify situations of gender discrimination. This attitude can be explained by the fact that *cofradías* were established by the guilds to protect fishermen, and many did not consider shellfish gathering as fishing at all. Disregard for women in fishing was widespread, as was resistance to women entering organizations traditionally reserved for men. Many *cofradías* barred shellfish gatherers and, in the event of admittance, they refused them the right to vote (Marugán Pintos 2004).

The situation began to change after 1995, when the regional fisheries administration modified their position towards *mariscadoras*. The arrival of new administrative staff, many of who were women, played a crucial role in *mariscadoras'* empowerment and in shellfish management. "When I arrived at the fisheries administration I listened to them (mariscadoras) and felt concerned. My general director was a woman" (Pencha Santamarinas, quoted in Frangoudes et al. 2005, 12). The presence of women with an awareness of the need for equality in the management (or at the head) of the fisheries administration constituted a key factor for the improvements to the situation of the *mariscadoras*. The fisheries administration strove for a better understanding of the situation by spending time with shellfish gatherers and discussing issues that concerned them. In 1994, there were an estimated number of 9,263 persons with exploitation permits, 90% women or *mariscadoras*.

These female civil servants were supported by politicians in the regional fisheries administration and collaborated with other influential public servants well versed in the problems in question. Their combined efforts aimed to put an end to the conflicts between rival *mariscadoras* within the community. The first priority was to professionalize the sector, regulate resource use and terminate conflicts between women. First and foremost, resource depletion needed to be addressed and, secondly, the *mariscadoras* needed to be organized and protected. With these goals in mind, a regional meeting of Galician *mariscadoras* was organized in 1995. The regional fisheries administration (*Consellería)* was, however, attempting to disrupt the prevailing power structure within the fisheries sector and, instead of consulting representatives from the male-dominated *cofradías*, they sought advice directly from women shellfishers. The public servants and politicians leading the change in regional fisheries administration were convinced that the provision of vocational training for *mariscadoras* was vital for improving their situation. Training would enable women to view their activity as employment and not as a marginal activity. Moreover, it would allow them to collaborate actively in resource management. In summary, the key actors in the regional fisheries administration opened a dialogue with *mariscadoras* based on mutual trust that would prove to be productive in the short term.

The administrators believed the *mariscadoras* should take the initiative and formulate their own training needs. The main challenge, however, was to find a way to convince shellfish gatherers to meet and discuss the future of their activity and to identify their needs without being entangled in the conflicts between women within—or between—communities. Experience showed that when meetings were organized during a crisis period, the discussion would always focus on the immediate problem and not on ideas about the future.

Regional Meetings: Problems, Constraints, and Solutions

The regional meeting organized in 1995 brought together *mariscadoras* who overcame cultural barriers that would have prevented them from attending such meetings in the past. "When I went to the first meeting of mariscadoras organized by the Consellería de la pesca, I asked my husband's permission. I prepared the food for all the family and … for future meetings I just told him" (Frangoudes et al. 2008b, 228). During the meeting, women struggled to understand why they should unify their efforts, because, up until that point, they had been competing for the same resource. When the women met at that first meeting, they became aware of the discrimination they had been subject to (Marugán Pintos 2004).

Finally, at the end of the meeting consensus was reached on the importance of the following issues:

- Shellfish gathering on foot was considered a marginal activity by *cofradías*;
- Gatherers themselves failed to consider the occupation as a profession and lacked social security protection;
- Income was poor;
- Women wished to obtain a brand name for their catches;
- Poaching presented a significant problem, as many people considered the intertidal zone as open access;
- Need for training courses.

Following these agreements, the fisheries administration dedicated European funds to first prepare trainers and then train the women. The trainers were equipped to instruct women in the building of organizations, management of resources, development of shellfish farming activities and, more importantly, the benefits of being organized. The initial training program was held at Cambados, the main shellfish production area. The event was not overly publicized, because participants did not want the wider community to know of their efforts at self-improvement (Santasmarinas 2003; Frangoudes et al. 2005).

The women who participated in the first regional meeting and ensuing training program occasionally needed to convince their husbands, family, and even themselves of the importance of these events. With that said, 2,888 women overcame these cultural obstacles and participated in 152 training courses organized by the fisheries administration.

Regional meetings were organized biennially and were always productive. At the second meeting, in 1997, the women opted for a more professional approach to their work, establishing their own organizations (*agrupación de mariscadoras*) at the local scale and as part of the *cofradías*. The third meeting focused on the legal recognition of their work in order to ensure access to the social security system, ways to act against poaching, and means of obtaining brand names for their catches. The meeting in 1999 examined their achievements to date and identified their future objectives.

Training Contributes to *Mariscadoras'* Empowerment

The massive participation of women in the various training courses considerably enhanced the organization of shellfishing. Women who did not participate either in training courses or regional meetings benefited from the knowledge of those who had. Training enabled the women to improve their skills and establish their own organizations. In fact, the enhancement of women's private and public roles is invariably linked.

Re-organization of Shellfish Activity: Resource Management, Marketing, and Restocking of Clams

The training courses provided women with basic knowledge for the sustainable development of resources. They followed courses on the biological and ecological aspects of bivalves, bivalve productivity, and harvesting and cultivation techniques. This knowledge enabled them to express their informed opinions about shellfish management within the *cofradías* and to collaborate productively with marine biologists. At the time, the regional fisheries administration was implementing the Productive, Professional and Organizational Development Program for Shellfishing, focused on the cultivation of commercially valuable shellfish. European funds were also used in this project to finance the biologists needed for resource management.

Biologists and shellfishing organizations collaborated every year to design a proposal for the annual shellfish exploitation plan within the specific territory of each *cofradía*. The annual plan for shellfish gathering had been obligatory since 1993 (Law 6/93), but had not been immediately enforced. Each *cofradía* submitted their annual exploitation plan to the fisheries administration for validation. Regulating shellfish resources by limiting access and reducing fishing effort were the main objectives stipulated by the fisheries administration. Each exploitation plan defined extraction and rest zones, and stated the number of *mariscadoras* and harvesting days, taking into account the annual maximum tonnage of capture based on stock evaluation. The exploitation plan restricts the harvesting effort in the shellfish beds with the objective of maintaining good prices for shellfish catches. In fact, daily harvesting is reduced during seasons of low demand and increased during peak periods, such as Christmas, New Year and other dates when consumers are ready to pay a high price for shellfish. The *cofradía* authority aids the monitoring of compliance, but the regional fisheries administration supports these efforts and develops specific surveillance programs in the shellfish beds.[9]

---

[9] Since 2008, due to the economic crisis in Spain and Galicia, poaching in shellfish beds has been on the rise. Surveillance in these areas is usually difficult, and organized groups of poachers are known to use violence occasionally against enforcers.

Due to the difficulty of eradicating poaching, it is an important concern for both shellfish gatherers and the fisheries administration, and still dominates debate at local and regional levels; the fact of the matter is that anyone can gather shellfish. Decree 419/1993, which instituted the rules for selling fresh seafood, aimed to put an end to illegal harvesting. As a result, fishmongers are obliged to buy shellfish at auction, which must then accompanied by an invoice issued by the *cofradía* managing the sale.

During the training courses, women learned that they could improve the capacity of shellfish beds by restocking overexploited areas. Restocking shellfish beds or the extensive aquaculture of clams is an activity requiring strong collective organization between women within the community. For Marugán-Pintos, the development of extensive aquaculture contributed to enhancing women's organizations and empowerment. Extensive aquaculture requires well-organized collective work: when the clam seed arrives, farming zones must be ready, and the seed needs to be stocked into bags and brought to the shore. Throughout the year, women monitor seed growth, remove dead clams, clean the bags, and, finally, seed them on the shore (Marugán Pintos 2004). Semi-cultivation and extensive shellfish farming reinforce women's organizations, because members combine efforts to ensure all tasks are covered. Women work alongside biologists to define and tend to areas of the shore allocated to shellfish farming.

These developments have increased the *mariscadoras'* income. In 1996, women earned an average of 1,174 euro per year, while in 2000, average income reached 3,285.50 euro (Marugán Pintos 2004). The *mariscadoras'* earnings are a complementary source of income for their households; their income does not, however, give them complete financial independence. In 2000, the minimum legal wage in Spain was 424.80 euro per month[10] (40 h week). Although the figure earned by the *mariscadoras* was far lower, they harvested fewer than 15 days a month and for only 3 h a day, rendering their activity comparatively profitable on an hourly basis.

The downside of the new organizational model is the reduction in the number of *mariscadoras.* The exact number of women initially involved in the fishery's activity is unknown; numbers differ according to the consulted source. "Just before 1987, the number of people engaged in shellfishing was about 26,000. The first published statistics of the Consellería cited 14,520 in 1989–1990, and 16,355 in 1990–1991 with more complete data" (Frangoudes et al. 2008b, 225). After the 1993 regulation that established the license system (*permex*), the number of women fell to 7,852 in 1997. This number decreased further in 2000, because the "renewal of the permex required working at least 70% of authorized days, certificates of having paid social security of the sea and sales data" (Frangoudes et al. 2008b, 225). In 2003, only 5,693 *mariscadoras* held permits and in 2005, that number was further reduced to 5,490. This process has, however, led to disagreements related to the management

[10] See http://www.boe.es/boe/dias/1999/12/31/pdfs/A46461-46462.pdf. The minimum wage in Spain for 2011 is 641,40 euro, see http://www.boe.es/boe/dias/2010/12/31/pdfs/BOE-A-2010-20150.pdf

of the licenses or *permex,* controlled since its inception by the regional government's fisheries administration. Local *mariscadoras* organizations may demand an increase in the numbers of local licenses, but the decision remains with the regional administration. Many of these requests are frequently rejected. Entry into the activity and the renewal of the *mariscadoras*' licenses have both been sources of conflict between the regional administration and the local *mariscadoras association*. The first prefers to maintain a stable—or decreasing—number of *mariscadoras*, while the second typically wants to increase these numbers. The exclusion of many potential new *mariscadoras* from the activity due to this fact may have exerted some influence on the surge of poaching in this area, and some of the related conflicts.

Women's Organizations or *Agrupación de Mariscadoras*

Women working in shellfish harvesting needed organizations to ensure that free access areas would be transformed into managed territories and in order to gain professional recognition. In 1995, when the first *mariscadoras'* regional meeting was held, there were seven local organizations or *agrupaciones de mariscadoras* in Galicia, a figure that rose to 21 by 2001. The 188 organized training courses and the four regional meetings played an important role in the capacity building of *mariscadoras*, because women understood that proper organization would enhance their work.

Mariscadoras' associations soon became full and equal participants in *cofradías*, on par with other organizations already present. The *mariscadoras'* self confidence was boosted by their success in shellfish management and in being able to improve their incomes; their collective participation in the *cofradías* gave them the chance to be heard by governing bodies. "The involvement of mariscadoras in cofradía management had increased democracy, clarified memberships and responsibilities, and contributed to improved functioning as information was distributed to all members" (Marugán Pintos 2005, 5). However, it must be said that the acceptance of women by the *cofradías* left no one indifferent and did not occur without confronting the negative reactions of some men. The first *mariscadoras* to join *cofradías* struggled alone, with no support from their colleagues. Their actions had a positive impact for the collective interests of women, because they took responsibility in male-oriented organizations. Women decided to become board members of *cofradías*, because they had something to say about the allocation of shellfish licenses (*permex*) and realized they could benefit from the services of a biologist, whom the *cofradías* could employ. The legal framework of the *cofradía* was required, because the *agrupación* lacked the legal structures to employ biologists on their own accord. The women developed the capacity to discuss their own problems and interests, as well as those of artisanal fisheries.

In 1995, only two women were chairs of *cofradías*. In 2005, 351 women were involved on the boards of *cofradías*: 4 were *patronas mayores* (chairs), 120 held a position within the *cabildo* (governing bodies), and 227 were members of the *junta general* (general assembly). By joining official male-dominated fishers' organizations, women succeeded in obtaining formal professional recognition and gained access to decision-making bodies. The active role of women within male organizations also gave them power.

Discussion

Women are the main stakeholders in shellfish harvesting. As a marginalized group, however, they were largely unaware that what they did was real work and that the resource could disappear. Women in fisheries often fail to view their contribution as true employment and fail to realize that family-based enterprises would struggle to survive without this contribution. The lack of gender-specific statistics in Europe (Frangoudes et al. 2008a) makes women invisible in fisheries and reinforces an image that minimizes their role.

In the case of Galicia, the participatory approach chosen by the fisheries administration, which targeted women, proved empirically successful. By helping them articulate their needs through training and express their hopes for the future, they succeeded in building a social group. This, in turn, enhanced women's empowerment and resource management. The process, which aimed at improving governability in the sector, enabled women to better cope with their most urgent concerns (Jentoft 2007). The *mariscadoras* understood the difficulty of their situation and realized that action was needed if they were to retain a source of income. The other stakeholders were the *cofradías* and fisheries authorities. The fisheries authorities elevated the empowerment of women to a position of importance and provided new tools for resource management. The *cofradías* were required to participate in a licensing system that allocated property rights, which was a huge change for them. The sexism that held sway in Spanish society was also present in the *cofradías*. Their attitude towards female activity was not always positive and, in some cases, women wanting to join certain governing bodies were met with considerable resistance. A number of *cofradías* chairs were strongly opposed to ceding power to women, a common feature of many other organizations in Spain at the time, and tried to maintain traditionally sexist ways of operation. However, some *cofradías* pioneered the change by placing women in roles of importance, such as chairs of organizations, well before the administration imposed shellfishers' organizations on *cofradías* in Galicia. The mere existence of *cofradías* made the transformation easier and, despite the shortcomings of these institutions, their presence on a local scale paved the way for future scenarios.

National government had long disregarded the role of women shellfishers, perhaps because they were so unusual in such a male-dominated sector. Neither artisanal fisheries nor *mariscadoras* were considered significant within the context of deep-sea fleet development that absorbed the attention of the Spanish fisheries administration. The regional governments formed after the Spanish Constitution of 1978 were more sensitive to local challenges, such as the *mariscadoras* in Galicia. The principle of subsidiarity had a significant effect on changing the situation. Certain shrewd politicians and public servants, mainly women, openly supported the process, and European funding helped create an opportunity for change (Mahou Lago 2008). In this case, funding was invested in human capital and empowerment, with positive long-term consequences for the sector.

From a governability perspective, taking into account the system-to-be-governed, interaction between the social and natural systems improved with the implementation of local regulations concerning shellfish resources. In the case of the *mariscadoras*, we observe that modifications within the social system impact positively on the natural system and that resource control is also modified. The establishment of *mariscadoras'* organizations contributed to changes in both the natural system and the institutions. Women were gradually accepted into *cofradías*, a process for which the duration depended on the *cofradía* in question. It was usually sped up once the women had shown their capacity for public speaking within fisheries organizations and assumed new leadership positions in both the *mariscadoras'* organizations and *cofradias*. Training courses further contributed to leadership-building at local and regional levels.

The shellfisher organizations within *cofradías* had an opportunity to play a decisive role in the governance of resources. Partnerships with the regional government, which conceived a licensing system to avoid the *tragedy of the commons* in shellfish beds, were a key factor. The *cofradías* that pioneered the process experimented with some of these measures early on. However, the support of the regional government for implementing the model throughout Galicia was decisive, a process that was far from easy. Changing a centuries-old system of free access to shellfish beds, a major resource open to all and a source of food for the underprivileged, was met with considerable resistance. The governability of the sector increased radically, but was threatened by poaching, as some of the population did not accept the restrictions and consistently challenged the rules. As a direct result of the recent economic crisis and the extremely high rate of unemployment in Spain, these norms have been strongly challenged by large numbers of poachers throughout Galicia since 2008. Such circumstances have served to further aggravate an already difficult situation for *mariscadoras'* organizations, the regional government, and the coastguard service.

Another issue that characterized *mariscadoras'* groups at the end of the process in the mid 1990s was their acceptance of scientific knowledge and the employment of biologists. The combination of scientific knowledge and the empirical knowledge of the *mariscadoras* contributed to the sustainable management of the resource. Male-dominated fisher organizations were less interested in the support of biologists in artisanal fisheries. In some cases, the male leaders of cofradías considered an educated outsider a threat. For women, however, scientific knowledge helped them establish annual exploitation plans, which in turn contributed to better management of the harvesting effort. Scientific knowledge represented a guarantee for the compulsory validation of annual plans by regional fisheries authorities. The use of empirical and scientific knowledge positively impacted on resource management and on the development of extensive aquaculture. Restocking also contributed to enhancing the natural system. The evolution of the social, institutional, and natural systems had an impact on women's economic independence; resulting in a relatively high income bearing in mind the number of hours they worked each month. Recently, some *mariscadoras'* groups have taken on other economic activities related to their main tasks. For instance, some have become guides to the shore ecosystem: during the tourist season they organize visits to the shoreline, informing tourists about the marine ecosystem and their work. These are the positive aspects of female empowerment and its impact on resource governance.

Financial security has been a source of uncertainty for *mariscadoras'* associations. In some cases, funding may depend on the affinity between *cofradías* or *mariscadoras'* groups and the ruling political party in the region. As individuals, *mariscadoras* have increased their economic independence, but they have yet to increase their independence as a group. They remain vulnerable to political power, which can build or destroy collective movements devoid of strong roots in civil society.

Within 15 years, the image of *mariscadoras* has undergone a considerable change in regional society thanks to their capacity to organize themselves and to manage shellfish resources. They were able to transform their informal activity into a profession that gives them access to their own social benefits. Their image, recast in the media, has also changed in their own eyes. The understanding of image here reflects its use in governance literature: images can be understood as judgments and assumptions on fundamental matters, such as the relation between society and nature or the role of government (Kooiman et al. 2005a). They are relevant because people tend to act on their images of how the world works or should work. Frequently, "it is the assumption of how the system should be that determines what are perceived as the problems" (Jentoft et al. 2010, 1315). This consequently shapes the approach to solutions. In the wake of all of the measures that were introduced, *mariscadoras* understand the world and their role in the community in a different way. Their understanding of their capacity for managing the activity and the resources has changed dramatically. Nowadays, that capacity is threatened by pervasive poaching practices that have arisen as a direct result of the economic crisis and jeopardize the entire management system. As mentioned before, the difficult entry into the *permex* system, under the control of the regional government, may have had some consequences in this area.

The differences in governability between the early 1990s and the present day are striking. A system of resource use originally characterized by open access has been transformed into a licensing system largely co-managed by *mariscadoras'* associations and *cofradías*, with the support of regional government. The contrast in the capacity to respond to the most urgent concerns since this transformation has been overwhelming. This constitutes one of the main elements of governability (Jentoft 2007). The *mariscadoras* have changed their status, become professionalized, and begun to pay social security and receive social benefits. The changes have transformed the system-to-be-governed, streamlining the *mariscadoras'* group, creating new institutional arrangements and professionalizing its activity. Perhaps the most salient feature of this selective process was the requirement to make social security contributions; as a result, many part-time participants abandoned the shore. This event helped to clarify the boundaries of the group of *mariscadoras*, providing them with a shared interest to preserve their resources. They could thus improve productivity and income by planning when to gather and by selling under the best market conditions. However, the system's exclusion of some traditional shore users was not easily accepted during the implementation process.

These transformations (summarized in Table 13.1) were linked to the dynamics of the governing system. *Cofradías* developed new roles in the process, assuming the management of the new licensing system, and integrating new actors with increased power: *mariscadoras* and their organizations. As it was occasionally met with considerable resistance, the transformation was not easy, but the role of regional government and certain managers was decisive. By encouraging the transformation, they emphasized the need to improve the capacity of women to manage their own challenges. Training was one of the key factors in the process, encouraging previously uneducated women to think of themselves as capable of organizing and managing concerns. To a degree, it involved a change of image; an internal change in self-esteem that made women believe in their political abilities and their chance to succeed as managers of organizations. The strong emphasis placed by regional government on increasing *mariscadoras'* capacity for co-management through training and stakeholder or women empowerment enhanced the general governability of the sector. Some general lessons can be learnt from this process regarding governability. Inclusiveness constitutes a general principle for co-governance (Kooiman et al. 2005b). In this case, including women and gender considerations in institutional building processes proved a key element for improving the previous situation. When women constitute a particularly relevant user group, as in this case, the recognition of their activity and of the legitimacy of their voice implies their integration in the governing institutions. Introducing a gender lens in governance also contributes to improved equity, as the differential access to information, resources, or power has, in many cases, withheld women from participating in decisions that directly affected them. These three elements: legitimacy, inclusiveness and equity (second-order governance principles) constitute key aspects for the gender and governability perspective (Bavinck et al. 2005). In this specific case, *mariscadoras* overcame a situation of institutional and public marginalization, and gained, through their own efforts and the support of the government, a space in the institutions that manage the resources on which they depend.

Table 13.1 Challenges related to women shellfishers (*mariscadoras*) in Galicia

System-to-be-governed			
Natural system	Socio-economic system	Governing system	Governing interactions
Before mid 1990s			
Extremely rich shellfishing areas in many places, some of them overharvested. Very strong biodiversity, some species introduced to increase productivity. In some cases the activity was similar to aquaculture, tending to beds and seeding shellfish in order to increase productivity, with high success rates. Productivity is very high, with many species of commercial interest in the shellfish beds. Pollution constitutes the main external risks, apart from overharvesting.	Before the 1960s shellfish on foot was gathered mainly by women and children for home consumption: considered marginal food. Subsequently, demand and prices increased rapidly with thousands of women (and some men) working in the shellfish gathering fields, disorganized in the majority of cases. Strong conflicts. After the 1960s shellfishing areas sustained considerable pressure to harvest as a consequence of canning companies interested in new products and later because new markets for fresh shellfish were accessible. Some closed periods and minimal sizes, but strong poaching contrary to rules.	*Cofradías*, fisher organizations with a long tradition in Spain, were important in Galicia, as a sort of co-management institution. *Cofradías* differed considerably in power, depending on location. Although women played a significant role in some, especially from the early 1990s, in most they were marginalized. Shellfisher associations were rare. Few shellfishing areas under specific management till 1990s. Conflicts with the areas already under management from other *cofradías*. Increased interest in administrative regulation of the sector. Decisive new regulations in 1993. Regional government began to have more influence over *cofradías*. Control over *cofradías* was transferred to regional government in 1980s. Shellfisher groups were rare and women were habitually marginalized inside *cofradías*.	Women were not significant in the governing board of most *cofradías* before the 1990s. Critical stance of some chairs of *cofradías* against the inclusion of women and against giving them significant roles in the *cofradías*. Other men were supportive of a greater role for women. Some *cofradías* pioneered changes, entering into conflict with others; in some cases the conflict over inclusion of women was so important that whole communities were involved. Little interaction between women shellfishers and their groups within the region before 1995. The traditional open access of these resources was not easily transformed by the new regulatory process of the licensing system and limited entry. *Cofradías* were very important in coastal communities; in some cases, they were the only institutions representing the local population. Sometimes these processes were the cause of internal conflict within Cofradías and local populations.

Some areas overexploited after the 1960s boom of shellfishing. Strong menaces also from poaching, coastal urbanization and pollution.	Women shellfishers mainly from fishing families. No region-wide organization. No control over the marketing of shellfish outside (and within) the local community. Women commanded very low prices for their product. Until the 1990s, in most cases, shellfishing was practically open access, conflictive, and of low economic value. No serious government attempts to organize the sector till 1990s.		Organized in associations, women began to manage resources, prepare exploitation plans, and voice opinions on marketing plans. Important backing from the regional government.
After mid 1990s			
Menace of pollution increased. Management of fields by shellfisher associations increased productivity in many areas, with potential for additional increases. Menace from poaching significant, especially in periods of crisis.	Licensing system, with a maximum number of licenses: strong reduction in number of shellfishers. Shellfisher associations begin to manage exploitation plans, deciding what to extract each day, seeding, and tending to the fields. Increased control of marketing and prices. Incomes much higher.	Regional government pressure on *cofradías* for the integration of women and shellfisher associations. Some *cofradías* pioneered change, entering into conflict with others. Role of women in *cofradías* much more important, stronger involvement in decision-making and political positions. Region-wide organizations of shellfishers set up, with stronger voice in public scenario and media. Significant change of image of *mariscadoras* within sector and publicly.	Support from regional government and European funding was very important. Regional government provided venues for regional meetings and training; women learned rapidly, taking into account experience in other places. Training and organization for women was promoted by regional government, making women more capable of intervening in *cofradía* policies, and organizing themselves in associations. Change in self-esteem and political abilities were the key to success. Some women (but also some men) in key positions in regional government supported the process. Women taking leadership in more *cofradías*. Region-wide organization-building, some cases used as successful examples that could be imitated in other communities.

References

Alegret, J.L. (1996). Ancient institutions confronting change: The Catalan fishermen's confradies. In K. Crean & D. Symes (Eds.), Fisheries management in crisis (pp. 92–98). Oxford: Blackwell Science/Fishing New Books.

Alegret, J.L. (1998). Property rights, regulatory measures and strategic responses among the fishermen of Catalonia. In D. Symes (Ed.), Property rights and regulatory systems in fisheries (pp. 175–187). Oxford: Blackwell Science/Fishing New Books.

Alegret, J.L. (1999). Space, resources and history: The social dimension of fisheries in the northwest Mediterranean. In D. Symes (Ed.), Europe's southern waters: Management issues and practice (pp. 55–65). Oxford: Blackwell Science/Fishing New Books.

Bavinck, M., Chuenpagdee, R., Degnbol, P., Pascual-Fernández, J.J. (2005). Challenges and concerns revisited. In J. Kooiman, M. Bavinck, S. Jentoft, R. Pullin (Eds.), Fish for life: Interactive governance for fisheries (pp. 303–323). Amsterdam: Amsterdam University Press.

Bennett, E. (2005). Gender, fisheries and development. Marine Policy, 29(5), 451–459.

Bennett, E., Rey Valette, H., Maiga, K.Y., Medard, M. (Eds.) (2004). Room for manoeuvre – Gender and coping strategies in the fisheries sector. Report of a workshop held in Cotonou, Benin, 1–4 December 2003. INCO Contract No. ICA4-2002-50034. Portsmouth: IDDRA.

Consellería de Pesca Marisqueo e Acuicultura. (1993). Plan de ordenación de los recursos pesqueros y marisqueros de Galicia. Santiago de Compostela: Consellería de Pesca, Marisqueo e Acuicultura.

de Beauvoir, S. (1990 [1949]). Le deuxième sexe (Vol. 2). Paris: Gallimard.

Erkoreka Gervasio, J. (1991). Análisis histórico institucional de las cofradías de mareantes del país vasco. Vitoria: Gobierno Vasco.

Frangoudes, K., Escallier, C., Pascual-Fernández, J. (2005). Women in private and public spaces: Synthesis, analogies and comparisons. In Proceedings of the Santiago de Compostela Workshop Thematic network FEMMES 8–9 November 2004. Brest: CEDEM, University of Bretagne Occidentale. www.fishwomen.org/IMG/pdf/santiago_en.pdf

Frangoudes, K., Carroll, M., Holmyard, N., Marciniak, B., Moço, C., Pascual Fernández, J.J., Marugán Pintos, B., Rönn, C., Quist, C. (2008a). The role of women in the sustainable development of European fisheries areas. Brussels: European Parliament, Directorate General for Internal Policies of the Union, Policy Department B: Structural and Cohesion Policies, Fisheries (IP/B/TRAN/IC/2007_033, PE 389.586).

Frangoudes, K., Marugán-Pintos, B., Pascual-Fernández, J.J. (2008b). From open access to co-governance and conservation: The case of women shellfish collectors in Galicia (Spain). Marine Policy, 32(2), 223–232.

Jentoft, S. (2007). Limits of governability: Institutional implications for fisheries and coastal governance. Marine Policy, 31(4), 360–370.

Jentoft, S., & Chuenpagdee, R. (2009). Fisheries and coastal governance as a wicked problem. Marine Policy, 33(4), 553–560.

Jentoft, S., Chuenpagdee, R., Bundy, A., Mahon, R. (2010). Pyramids and roses: Alternative images for the governance of fisheries systems. Marine Policy, 34(6), 1315–1321.

Kooiman, J., & Chuenpagdee, R. (2005). Governance and governability. In J. Kooiman, M. Bavinck, S. Jentoft, R. Pullin (Eds.), Fish for life: Interactive governance for fisheries (pp. 325–349). Amsterdam: Amsterdam University Press.

Kooiman, J., Bavinck, M., Jentoft, S., Pullin, R. (Eds.). (2005a). Fish for life: Interactive governance for fisheries. Amsterdam: Amsterdam University Press.

Kooiman, J., Jentoft, S., Bavinck, M., Chuenpagdee, R., & Sumaila, R. (2005b). Meta-Principles. In J. Kooiman, M. Bavinck, S. Jentoft, R. Pullin (Eds.), Fish for life: Interactive governance for fisheries (pp. 265–284). Amsterdam: Amsterdam University Press.

MacAlister, E. (2002). The role of women in the fisheries sector. Final Report, 1443/R.03/C. Brussels: EC/DG Fisheries.
Mahou Lago, X.M. (2008). Implementación y gobernanza: La política de marisqueo en Galicia. Santiago de Compostela: Escola Galega de Administración Pública.
Marugán Pintos, B. (2004). Y cogieron ese tren... Profesionalización de las mariscadoras gallegas. Santiago de Compostela: Xunta de Galicia, Consellería de Pesca e Asuntos Marítimos.
Marugán Pintos, B. (2005). The Galicia shell fishers: A women silent revolution. (Paper presented at the MARE Conference, Amsterdam).
Meltzoff, S.K. (1995). Marisquadoras of the shellfish revolution: The rise of women in co-management on Illa de Arousa, Galicia. Journal of Political Ecology, 2, 20–38.
Meltzoff, S.K., & Broad, K. (1996). The rise of women in fisheries management: The marisquadoras of Illa da Arousa, Galicia. In R.M. Meyer, C. Zhang, M. Windsor, B. McCay, L. Hushak, R. Muth (Eds.), Fisheries utilization and policy. Proceedings of the World Fisheries Congress, Theme 2 (pp. 349–360). Oxford: Lib. Pub. Co., Pvt, Ltd.
Mikalsen, K.H., Hernes, H.-K., Jentoft, S. (2007). Leaning on user-groups: The role of civil society in fisheries governance. Marine Policy, 31(2), 201–209.
Pascual Fernández, J. (1999). Participative management of artisanal fisheries in the Canary Islands. In D. Symes (Ed.), Europe's southern waters: Issues of management and practice (pp. 66–77). Oxford: Blackwell's Science, Fishing New Books.
Pascual-Fernández, J.J. (Ed.). (2005). La mujer en la pesca, la acuicultura y el marisqueo en el contexto comunitario. La Laguna: Asociación Canaria de Antropología.
Pascual-Fernández, J.J., Frangoudes, K., Williams, S.B. (2005). Local institutions. In J. Kooiman, M. Bavinck, S. Jentoft, R. Pullin (Eds.), Fish for life: Interactive governance for fisheries (pp. 153–172). Amsterdam: Amsterdam University Press.
Rubin, G.S. (1975). The traffic in women: Notes on the 'political economy' of sex. In R.R. Reiter (Ed.), Toward an anthropology of women (pp. 157–210). New York: Monthly Review Press.
Santasmarinas, P. (2003). La mujer mariscadora y el proceso de cambio. In K. Frangoudes, C. Escallier, J. O'Doherty (Eds.), Women in private space: Women's role within the fishing enterprise. Proceedings of the Brest workshop 20 and 21 February 2003 (pp. 73–80). Brest: CEDEM, University of Bretagne Occidentale. http://www.fishwomen.org/IMG/pdf/brest_en.pdf
Sequeiros, J.L. (Ed.). (1995). A despensa de area: cambio social e formas productivas no sector marisqueiro das Rías Baixas Galegas Vigo: Edicións Xerais de Galicia.
Suárez de Vivero, J.-L., & de Lara, M. F. (1997). Regions at sea: the administrative region as a base for an alternative fisheries management system for Spain. Ocean & Coastal Management, 35(2–3), 173–183.
Suárez de Vivero, J.L., Frieyro de Lara, M., Jurado Estevez, J. (1997). Decentralization, regionalization and co-management: A critical view on the viability of the alternative management models for fisheries in Spain. Marine Policy, 21(3), 197–206.
Vara Miranda, M.J., & Maquieira D'Angelo, V. (Eds.) (1996). El trabajo de las mujeres, siglos XVI–XX: VI Jornadas de Investigación Interdisciplinaria sobre la Mujer. Madrid: Universidad Autónoma de Madrid, Instituto Universitario de Estudios de la Mujer.

Part IV
Methodological Approaches

Chapter 14
Approaches and Tools for Examining Governability

Ratana Chuenpagdee and Robin Mahon

Abstract Several tools and approaches are available for investigating the characteristics of the system-to-be-governed, the capacity of the governing system and the quality of their interactions. Diversity and complexity of the fisheries ecosystem can be understood using, for instance, food web analysis. Similarly, stakeholder analysis helps describe diversity and complexity of the social system that is being governed, as well as of the governing system. These tools, along with many others, provide ways and means to gauge what and where in these systems governability issues may be pronounced. We present some examples of these tools to illustrate their utility in examining key aspects of the fisheries system that may give rise to governability. They also illustrate the diversity of tools that may be required when the perception of governance and governability expands beyond the conventional to encompass the full range of stakeholders, processes and interactions. This investigation is one of the main steps in the full governability assessment framework.

Keywords Assessment tools • System complexity • Social network analysis • Fish chain • Indicators • Governability • Interactive governance

R. Chuenpagdee (✉)
Department of Geography, Memorial University of Newfoundland, St. John's, NL, Canada
e-mail: ratanac@mun.ca

R. Mahon
Centre for Resource Management and Environmental Studies (CERMES),
University of the West Indies, St. Michael, Barbados
e-mail: rmahon@caribsurf.com

M. Bavinck et al. (eds.), *Governability of Fisheries and Aquaculture: Theory and Applications*, MARE Publication Series 7, DOI 10.1007/978-94-007-6107-0_14,

Introduction

System characteristics can be looked at from a variety of perspectives, including those prescribed in resilience thinking, commons theory, sustainable livelihood approach, as well as interactive governance. While these frameworks differ in their theoretical groundings, they are similar in their focus on understanding and explaining the intricacies of the systems. According to the interactive governance theory, diversity, complexity, dynamics and scale are key characteristics of the natural and social systems-to-be-governed, and are likely to cause governability problems. Similarly, these features affect the capacity of the governing system to perform its functions. The main focus of the governability assessment framework is to examine these system properties using existing approaches and tools, including those employed in other frameworks.

Jentoft and Chuenpagdee (2009) operationalize the governability assessment framework using three guiding questions: 'where to look,' 'what to look for,' and 'what to look at.' The question about 'how to look' is left largely for the researchers and practitioners to explore on their own. This is because numerous approaches and tools are already available to characterize the systems and to understand their function. In this chapter, we provide some examples of what these tools are. Rather than doing an exhaustive review of all existing tools and approaches, we select a number of them to highlight what they may bring to the understanding of the systems and to examining governability. In this way, we contribute to the operationalization of one of the main steps in the full governability assessment framework, presented by Chuenpagdee and Jentoft (2013).

In keeping with the three system analysis, we first present tools and approaches that can be used to examine the natural and social system-to-be-governed. Recognizing the linkages of these two systems, we explore some of the tools that can be used to understand their interconnectivity. This is followed by tools for analyzing the governing system and the governing interactions, respectively. Further, we underscore meta-governance as an area that has been less explored, and more difficult to understand.

Systems-to-Be-Governed

Natural System-to-Be-Governed

Tools for examining the natural component of the system-to-be-governed are undoubtedly the ones that are best known to fisheries managers. They have been the basis for conventional fisheries management and natural science research. They span the full range of tools for examining the physical, geological and biological aspects of aquatic ecosystems. Many texts and reference books offer comprehensive coverage of these tools. The following aims to illustrate the breadth of what they cover, and to emphasize some that may not be normally thought of as governance-related tools.

Oceanographic surveys from research vessels are conventionally used to collect in-situ data about the ocean. Through them, scientists have been able to learn about the bio-physical properties of the sea and diversity of ocean life. Today, with the use of satellite imagery, advanced observation systems and the availability of modern geo-referencing tools, much more can be learned and spatially displayed. These modern technologies can be used to describe coastal morphology, monitor oceanographic conditions, such as temperature, salinity and waves, and predict effects of phenomena like tsunamis. Similar technology is used to assess biological oceanographic characteristics of the oceans, such as primary productivity or red tide occurrence, and to track the movement of mega-fauna like sharks, marine mammals, sea turtles and tuna. For the bottom of the ocean, technology like multi-beam sonar and remotely operated vehicles are being employed for habitat mapping and the assessment of aquatic life. Findings from the various monitoring and assessment systems are often reproduced in the form of maps, many of which are interactive, real-time, and offer multi-dimensional visualization. Aided by these modern tools and techniques, our knowledge about the diversity, complexity, dynamics and scale of the physical dimension of the natural system-to-be-governed has greatly increased, especially in the highly industrialized parts of the world.

The ecological aspects of the aquatic ecosystems are generally well studied with the long-established methods such as fisheries surveys and biological sampling for population dynamics and fisheries stock assessment. Life history, trophodynamics and food web studies are common and are important for ecosystem analysis and for ecosystem-based management. These methods make it possible to know distribution and ecology of species in marine ecosystems. Modeling tools such as Ecopath and the dynamic variation Ecosim (Pauly et al. 2000) are employed, not only to describe the connectivity of these species, but also to explore policy options through scenarios analysis. Decision-support systems like Atlantis (Fulton et al. 2004) are tools that scientists employ to help understand system complexity. Both approaches and other ecosystem modeling exercises are particularly useful to account for both the unknown and the uncertainty embedded in the natural system (e.g. Bundy and Fanning 2005). Increasingly, however, it is being demonstrated that the inclusion of stakeholders in the collection of information on resources and their ecosystems can be valuable. The development of indicators has also helped facilitate the participation of resource users in scientific research. All these approaches and tools add to the understanding of the biodiversity, relationships and dependency among species and on habitats, the dynamism of the system, and the spatial extent of life in the oceans.

In addition to scientific research, alternative ways of gaining information about the natural system include the use of local and/or traditional ecological knowledge (LEK and TEK). It has become widely accepted that resource users know a great deal about the resource system they depend on, which can be of high value for management.

Johannes (1998) provides several examples in which the use of local knowledge and commonsense led to improved management systems. He coined the term 'data-less management', causing some controversy in fishery science circles. However, he points out that this does not mean management without information, but instead refers to the use of information that is readily available from resource users. There is increasing evidence to suggest that the use of LEK and TEK can greatly enhance the information required for management. For example, fishers contribute to the collection of both fishery-dependent and fishery-independent data of the San Diego red sea urchin fishery (Schroeter et al. 2010).

The debate about the compatibility of these different types of knowledge is ongoing. So too is the discussion of their merits as a source of information versus a tool to empower communities (Gilchrist and Mallory 2007). These and other topics related to the strengths and limitations of using fisher knowledge are explored by Haggan et al. (2006) and Davis and Ruddle (2010). The struggles with LEK and TEK are not strictly with respect to the utility of the knowledge but also the method of elicitation and integration with scientific knowledge. Basic rapid appraisal approaches that seek to obtain and use this knowledge, such as community mapping, may be considered inevitable at an early stage, when the method was first explored (Berkes 1999; Berkes et al. 2000). With the advancement of mapping technology, especially participatory GIS software, many different types of data can be integrated and more sophisticated, accurate, maps can be produced, stored and easily shared with stakeholders (Douvere and Ehler 2009; Dalton et al. 2010). The sensitivity around the mapping of local knowledge does, however, need to be recognized because of rights considerations. This is particularly so when concerning aboriginal peoples, as territorial user rights may also be an issue. Roles of resource users, like fishers, in scientific research may thus go beyond data acquisition or provision to involve co-creation of knowledge and agreed-upon rules regarding intellectual property rights.

Social System-to-Be-Governed

A range of tools exists to describe the social, cultural and economic characteristics of people involved in coastal and ocean social ecological systems; in terms of their diversity, complexity, dynamics and scale. For instance, household surveys and censuses are often used to obtain information about demographics, livelihood activities, and income and expenditure. When employed on a regular basis, these methods can provide statistics that enable the observation of change at the national and regional levels. Oral history, field observation, interviews and focus groups are used to obtain detailed and specific information about individuals, households and communities. Similar to their use in the natural system-to-be-governed, rapid assessment methods have been developed to collect snapshots of information for purposes such as regional planning or aid distribution. They are often used in conjunction with the natural assessment methods and with a range of socio-economic indicators, including common ones like gross domestic product and emerging ones like job satisfaction (Pollnac et al. 2001) and human well-being (McGregor 2007). The latter is employed to indicate the current state of persons in the community, and consequently to design appropriate policy interventions to improve their well-being (UNEP 2006). The well-being approach is being promoted as a way to help link environmental sustainability with societal objectives like poverty alleviation and improving quality of life (Coulthard et al. 2011; Daw et al. 2011). The development of methods and approaches to measure well-being is currently a subject of discussion among economists interested in explaining peoples' choices, preferences and behavior (Kahneman and Krueger 2006).

Participatory approaches are regularly employed to obtain individual, group and community level data. They are used not only to recognize the limiting grasp of scientists on certain topics, but also in acknowledgement of the importance of stakeholder involvement in resource management and governance. Participatory research used to examine the social system-to-be-governed extends from the rapid appraisal type to participatory rural appraisal, to action oriented approaches like participatory action research. Future scenario analysis, prioritization and visioning exercises, and causal loop diagrams are among the tools used to gain understanding of the social system, as well as catalyze collective action and participatory decision-making (Colfer 2007; Almerigi et al. 2013).

Numerous analytical tools are used to assess situations, conditions, change and issues occurring in the social system-to-be-governed. Notably, stakeholder analysis is used to determine who the stakeholders are and to differentiate between them according to certain criteria, such as urgency, legitimacy and power (Mitchell et al. 1997; Buanes et al. 2005). Such analysis acknowledges the diversity, complexity and dynamics of individuals and groups who have varying interests and dependency on marine and fisheries resources. Given the different roles that men and women play, sensitivity to gender issues is also required in the case of fisheries. Fish harvesting is a male-dominated activity, especially in industrial fisheries, while women are key actors in post-harvest activities. In small-scale fisheries, the division of labor is more complex, and in many instances, more possibilities are available for women to be active in fishing. As shown in Frangoudes et al. (2013), an understanding of gender roles is required in order to institute policy interventions that enable fair and equitable involvement of women in fisheries and equal access to resources. March et al. (1999) offers a good explanation of gender studies and provides several frameworks to examine gender issues, concerns and relations, in particular how power is distributed between men and women. Examples of these frameworks are the Harvard Analytical Framework (Overholt et al. 1984), which analyzes gender roles, and the Social Relations Approach (Kabeer 1994), which examines gender relations. These frameworks can be used as a stand-alone application, or in combination, and are useful to integrate into social research and planning.

Similar to the Social Relations Approach, Social Network Analysis (SNA) describes the relationships between social actors, identifying, for instance, those that are central to the network (Bodin and Crona 2009; Bodin and Prell 2011; Mahon and McConney 2013). SNA is a methodology used to describe and examine interpersonal, economic, political, or any other type of relations (Scott 1991; 1996). The retrieved relational data can be expressed as graphs (referred to as sociograms), in which actors (individuals or groups) are connected by lines that express the relationships among them. SNA can be used to assess communications, cooperation and other inter-organizational environments within the network (Borgatti et al. 2009). An attractive feature of SNA is the explicit representation of information sharing patterns within and across groups, in turn, revealing key actors and important networks within the social system. Like other approaches, SNA alone is not sufficient to provide a full understanding of the situations and conditions surrounding the social systems (Martinez et al. 2003), and should therefore be used in combination with other methods.

Closely related to SNA are social capital assessment and the sustainable livelihoods approach. Both are useful when gauging local level capacity for self-support, organization and pursuing common goals (Campbell 2008; IMM 2008; Cinner and Bodin 2010). Defined as the 'institutions and relationships, as well as the trust, norms and values, that govern interactions among people and contribute to economic and social development' (Grootaert and Van Bastelaer 2002, 2), social capital has parallel concepts in many social science disciplines. In an effort to reduce poverty and inequality, the World Bank supported the development of 'Social Capital Assessment Tool' (SOCAT), a mixture of qualitative and quantitative instruments (e.g., household surveys, questionnaires, and interviews) used to determine the likely changes in productive behavior at the household and community levels in response to policy change. Applications in fisheries have been made, for instance, by Bodin and Crona (2008), who examine the role of social capital in resource management in Africa, and by Adger (2003) to look at collective actions in the context of climate change in coastal areas, drawing on case studies from Southeast Asia and the Caribbean.

As with the natural system, modeling techniques are employed to help understand complexity in the social system and account for uncertainty. Fishers' behaviors and fishing strategies at individual and group levels are analyzed to recognize the complexity and dynamics of a fishery system. Salas et al.'s (2004) study of fisheries in the Yucatan coast of Mexico, for instance, reveals that fishers make daily decisions about what species to target, what gear to use, where to fish and whether they should cooperate with fellow fishers to secure high catches. For managers, modeling and simulation exercises are generally helpful when predicting what fishers' decisions may be and how they may behave in the context of new rules and regulations, irrespective of the roles uncertainty and un-expectancy may play in their behavior.

Linked Social-Ecological System

In order to recognize the link between the natural and social systems-to-be-governed, we require an assessment of the impact that change in the natural system has on the social system, and vice versa, as well as an assessment of how the social system adapts and copes with said change. Composite tools such as environmental and social impact assessment, along with measurement of social-ecological resilience, vulnerability and adaptation are particularly useful (Harris 2007). When associating the change in the socio-ecological system with environmental change, various approaches, ranging from modeling the relationships (e.g. Cinner et al. 2009) to developing scenarios, are possible. The Millennium Ecosystem Assessment and its sub-assessments provide an accessible example of how four scenarios were used at the global level to provide a picture of different possible futures that would depend on the response to the current state of the planet (MEA 2005). Such scenario exploration helps us understand the dynamics of the coupled socio-ecological system, and thereby facilitates better natural resource use planning and environmental policies.

Social and ecological systems are also linked through the fisheries production chain. Chain analyses focusing on supply or global values can therefore be employed to examine issues such as flows of commodities, movement of labor and influence of markets (Béné et al. 2009). Value chain analysis focuses on the links in the fishing industry and describes the full range of activities required to bring fishery products from harvest and different phases of production to different levels of consumers. Global value chain analysis, in particular, has been very useful to not only track commodities and flows of goods and services, but also to understand the power dynamics of actors along the chain (Ponte 2007). Recognizing the importance of this dimension in affecting governability, the interactive governance approach employs the concept of the 'fish chain' as an analytical lens to examine the role of states, markets and civil society in pre-harvest, harvest and post-harvest activities.

Governing System

Governing systems for fisheries vary according to their mode of operation: hierarchical; co-governance, or co-management; and self-governance. The appropriateness of each mode depends on the characteristics of the natural and social systems that the governing system is tasked with. There may be characteristics and qualities of the governing systems that make them highly effective in performing governing tasks. Given the emerging demands for governing systems to cope with environmental and global change, and the usual constraints such as budget and knowledge, they are required to be increasingly adaptive, robust and resilient. Stability, endurance and effectiveness are some of the key measures used to gauge the quality of the governing system and, thus, its contribution to governability. The characteristics of the governing system, however, can be examined using methods that are similar to those of the social system-to-be-governed.

The function and roles of the governing system at international, regional, national and local levels can be described using organization charts. These can explain, to a certain extent, the simple horizontal and vertical relationships between various components in the system. Methods such as content analysis of policy documents can be used to reveal the purpose, aims, mandate and responsibility of the governing system (Ekstrom et al. 2009), as well as its relationship to the natural system (Ekstrom and Young 2009). As with the system-to-be-governed, the full complexity of the governing system can be comprehended using higher level analysis such as social networks. Resilience, adaptability and robustness of the governing system can also be assessed in a variety of ways (Allison et al. 2009; Resilience Alliance 2010). These include using institutional analysis frameworks like those developed by Anderies et al. (2004) and Ostrom (2009), or discursive- institutional analysis (Arts and Buizer 2009). The emphasis of the latter is on understanding institutional dynamics through an examination of ideas, concepts and narratives.

Other measures of the governing system relate to how well they perform their roles and pursue their mandates. Certain criteria are set and performance indicators are employed to gauge the capability of the governing system. As previously noted, there are many frameworks that have been developed to provide a context for the assessment of governance systems. There is vast literature on public policy analysis that many managers of fishery systems overlook all too frequently (Fischer et al. 2007). Other frameworks developed specifically for the linked social-ecological system range from relatively simple constructs to conceptually complex models. Grafton et al. (2007), for instance, provide a set of 23 indicators for benchmarking and subsequently monitoring fisheries governance in five categories: accountability; transparency; incentives; risk assessment and management; and adaptability. In each case, the indicator is scored on the basis of five levels from missing to fully in place and operational. Garcia et al. (2008) provide a more complex assessment system, offering a guide that links assessment approach with system characteristics and type of governance needed.

Governing Interactions

One approach to assessing the governing interactions among stakeholders or their organizations is to view them and their interactions as a network, and to analyze them using SNA described above (Mahon and McConney 2013). This can provide insights into where interventions could be expected to have the greatest effect, as well as prescribing a baseline against which progress can be measured. Until recently, SNA has not been prominent in discussions of adaptive governance and resilience (Carlsson and Sandström 2008). Moreover, as Cross et al. (2002) suggest, when viewed together by managers, decision-makers and fisheries stakeholders, the SNA diagrams can generate discussion reflective of the existing relationships, and recommendations about governing interventions that can help strengthen the network.

One area that is essential for effectively dealing with socio-ecological systems is the capacity to learn and retain knowledge; to build 'learning organizations'. This aspect of adaptability is flagged by the interactive governance approach as requiring special attention (Bavinck et al. 2005; Mahon et al. 2005). There is a rich body of literature on building and assessing learning organizations in other sectors (e.g. Senge 1990; Senge et al. 1999; Collison and Parcell 2001). Berkes et al. (2001) provide a number of suggestions for how better management of information can improve institutional memory and adaptive capacity. A recent manual on information sharing and management by the FAO similarly addresses these issues (2009). Finally, with heightened recognition of the importance of stakeholders' participation in decision-making process, methods have been developed to enhance participation and improve quality of interaction among stakeholders. Deliberative multi-stakeholder process is one such method. It focuses on collaborative problem solving, with stakeholders jointly defining the problems and developing solutions (Varjopuro 2008). A systemic observation of such a process can help assess the quality of the governing interactions.

Meta-order Governance

As explained by Kooiman and Jentoft (2009), meta-order governance plays a critical role in fostering or inhibiting the governing ability of the governing system. Normative governance traits such as values and principles, as well as subjective attributes like images that individuals and institutions hold about the system that is being governed and the governing system, are key aspects that need to be understood, deliberated and communicated (Jentoft et al. 2010). This is because they underpin the way governing actors perceive, define and perform their roles.

Although applications to fisheries are rather limited, methods and approaches developed for valuation of natural resources and environmental services are valid. As de Groot et al. (2002) suggest, ecological values are related to resource sustainability, social and cultural values are associated with equity and cultural perceptions, and economic values are about efficiency and cost effectiveness. Monetary valuation methods to estimate economic values are well developed and have been used to inform policymaking. These include direct market valuation based on production function and indirect market valuation such as replacement cost, travel cost and hedonic pricing methods. The most commonly used, and perhaps the most debated, valuation method is contingent valuation, which attempts at estimating people's willingness to pay (Haab and McConnell 2002). Critics argue that market-based approaches do not capture the overall values of environmental resources well, because they are often not traded in markets and thus have no price tag. Creating hypothetical market situations (as is done in contingent valuation studies) can therefore lead to misleading outcomes (Knetsch 1994). In many instances, it is best not to employ monetary valuation methods, no matter how tempting it may be. Some alternative approaches to valuation are choice experiments, which are not based on market prices but can still include monetary amounts in the choice set (e.g., Holmes and Adamowicz 2003). They can also be conducted using non-monetized units, as is done in the damage schedule approach (Chuenpagdee et al. 2001). Other deliberative methods, such as verbal protocols (Vatn 2009) and citizen workshops (Timotijevic and Raats 2007), are also used to obtain values, especially those related to the social and cultural aspects of the environment.

Unlike values, images are not as well studied because they are often considered unnecessary for governance. As Jentoft et al. (2010) argue, however, the current problems with fisheries are due in part to the lack of explicit discussion about the images that governors use to arrive at their decisions and whether these images are shared with other governance actors. Fisheries management is riddled with images that have led mostly to negative consequences for the resources and fishing communities. For instance, Thomas Huxley's famous image of the inexhaustible sea may have erased any concern anyone would have had about fisheries development through the modernization of fleet and gear. Retrospective analysis of images and the role they play in governance is easier than trying to determine what images currently influence decisions and behaviors. For the latter, surveys using different types of questionnaires, such as paired comparisons (David 1988) and Q-sort (Ten Klooster et al. 2008), which may contain statements, photographs or symbols, can be used.

With respect to principles, they are what fishers and other resource users are familiar with, as something determined by the governments. Insufficient attention is paid to questions about where principles come from, how they are deliberated, and what purpose they have. Deliberation about principles needs to take place, not only within governments but also with other key stakeholders, especially to differentiate between universal (e.g., human rights and the rights of indigenous peoples as specified in UN conventions and declarations) and contextual (e.g., local self-determination) principles (Kooiman and Jentoft 2009). Similar methods to those used to study values and images can be used to examine principles.

Conclusions

Virtually all tools that relate to assessing or evaluating the social and ecological components of social-ecological systems may have a role to play when assessing governability. What we have tried to achieve in this chapter is to demonstrate the extent to which past assessments have focused on the ecological, social and governing subsystems, and to emphasize the diversity of new frameworks, approaches and tools that are being brought to bear on the assessment of these systems, especially the social component and, within that, the governing system.

Many of these approaches have only recently been adapted from other disciplines and are being tested in fisheries contexts and other situations. We would like to highlight that the conceptual and methodological boundaries of governance assessment are expanding rapidly. This is being driven by the emerging appreciation for the complexity of governance in contrast to conventional management. There is an increasing understanding of governance that encompasses, for instance, the entire fish chain, and all processes from setting principles and visions through institution building to day-to-day operations of management bodies. Those interested in pursuing the assessment of governability will require a transdisciplinary perspective and successful 'end to end' assessments will almost certainly require multidisciplinary research teams.

References

Adger, W.N. (2003). Social capital, collective action, and adaptation to climate change. Economic Geography, 79,(4), 387–404.

Allison, E.H., Perry, A.L., Badjeck, M.C., Adger, W.N., Brown, K., Conway, D., Halls, A.S., Pilling, G.M., Reynolds, J.D., Andrew, N.L.. Dulvy, N.K. (2009). Vulnerability of national economies to the impacts of climate change on fisheries. Fish and Fisheries, 10, 173–196.

Almerigi, S., Fanning, L., Mahon, R., McConney, P. (2013). Creating a principle-based vision at the regional scale – The Caribbean example. In M. Bavinck, R. Chuenpagdee, S. Jentoft, J. Kooiman (Eds.), Governability of fisheries: Theory and applications. Amsterdam: Springer.

Anderies, J.M., Janssen, M.A., Ostrom, E. (2004). A framework to analyze the robustness of social-ecological systems from an institutional perspective. Ecology and Society, 9(1), 18. Retrieved from http://www.ecologyandsociety.org/vol9/iss1/art18/. Accessed 9 Jan 2013.

Arts, B., & Buizer, M. (2009). Forests, discourses, institutions: A discursive-institutional analysis of global forest governance. Forest Policy and Economics, 11, 340–347.

Bavinck, M., Chuenpagdee, R., Diallo, M., van der Heijden, P., Kooiman, J., Mahon, R., Williams, S. (2005). Interactive governance for fisheries: A guide to better practice. Delft: Eburon Academic Publishers.

Béné, C., Lawton, R., & Allison, E.H. (2009). Trade matters in the fight against poverty: Narratives, perceptions, and (lack of) evidence in the case of fish trade in Africa. World Development, 38(7), 933–954.

Berkes, F. (1999). Sacred ecology. Traditional ecological knowledge and resource management. London: Taylor & Francis.

Berkes, F., Colding, J., Folke, C. (2000). Rediscovery of traditional ecological knowledge as adaptive management. Ecological Applications, 10, 1251–1262.

Berkes, F., Mahon, R., McConney, P., Pollnac, R., Pomeroy, R. (2001). Managing small-scale fisheries: Alternative directions and methods. Ottawa: International Development Research Centre.

Bodin, Ö., & Crona, B.I. (2008). Management of natural resources at the community level: Exploring the role of social capital and leadership in a rural fishing community. World Development, 36(12), 2763–2779.

Bodin, Ö., & Crona, B.I. (2009). The role of social networks in natural resource governance: What relational patterns make a difference? Global Environmental Change, 19, 366–374.

Bodin, Ö., & Prell, C. (Eds.). (2011). Social networks and natural resource management: Uncovering the social fabric of resource governance. Cambridge: Cambridge University Press.

Borgatti, S.P., Mehra, A., Brass, D.J., Labianca, G. (2009). Network analysis in the social sciences. Science, 323, 892–895.

Buanes, A., Jentoft, S., Karlsen, G.R., Maurstad, A., Søreng, S. (2005). Stakeholder Participation in Norwegian Coastal Zone Management. Ocean & Coastal Management, 48(8–10), 658–669.

Bundy, A., & Fanning, P. (2005). Can Atlantic cod (Gadus morhua) recover? Exploring trophic explanations for the non-recovery of the cod stock on the eastern Scotian Shelf, Canada. Canadian Journal of Fisheries and Aquatic Sciences, 62(7), 1474–1489.

Campbell, J. (2008). Systematic approaches to livelihoods enhancement and diversification: A review of global experiences. Gland: IUCN.

Carlsson, L., & Sandström, A. (2008). Network governance of the commons. International Journal of the Commons, 2, 33–54.

Chuenpagdee, R., & Jentoft, S. (2013). Governability assessment: Moving forward. In M. Bavinck, R. Chuenpagdee, S. Jentoft, J. Kooiman (Eds.), Governability of fisheries: Theory and applications. Amsterdam: Springer.

Chuenpagdee, R., Knetsch, J.L., Brown, T.C. (2001). Environmental damage schedules: community judgments of importance and assessment of losses. Land Economics, 77(1), 1–11.

Cinner, J. E., & Bodin, O. (2010). Livelihood diversification in tropical coastal communities: A network-based approach to analyzing 'livelihood landscapes'. PLoS ONE, 5(8), 1–13.

Cinner, J.E., McClanahan, T.R., Daw, T.M., Graham, N.A.J., Maina, S., Wilson, S.K., Hughes, T.P. (2009). Linking social and ecological systems to sustain coral reef fisheries. Current Biology, 19, 206–212.

Colfer, C.J.P. (2007). Simple rules for catalyzing collective action in natural resource management contexts. Jakarta: Center for International Forestry Research.

Collison, C., & Parcell, G. (2001). Learning to fly: Practical lessons form one of the world's leading knowledge companies. Chichester: Capstone Publishing Ltd.

Coulthard, S., Johnson, D., & McGregor, J.A. (2011). Poverty, sustainability and human wellbeing: A social wellbeing approach to the global fisheries crisis. Global Environmental Change, 21, 453–463.

Cross, R., Borgatti, S., Parker, A. (2002). Making invisible work visible: Using social network analysis to support strategic collaboration. California Management Review, 44, 26–46.

Dalton T., Thompson, R., Jin, D. (2010). Mapping human dimensions in marine spatial planning and management: An example from Narragansett Bay, Rhode Island. Marine Policy, 34, 309–319.

David, H.A. (1988). The method of paired comparisons. London: Charles Griffin & Company.

Davis, A., & Ruddle, K. (2010). Constructing confidence: Rational skepticism and systematic enquiry in local ecological knowledge research. Ecological Applications, 20(3), 880–894.

Daw, T., Brown, K., Rosendo, S., Pomeroy, R. (2011). Applying the ecosystem services concept to poverty alleviation: The need to disaggregate human well-being. Environmental Conservation, 38, 370–379.

De Groot, R.S., Wilson, M.A., & Boumans, R.M.J. (2002). A typology for the classification, description and valuation of ecosystem functions, goods and services. Ecological Economics, 41, 393–408.

Douvere F., & Ehler, C. (2009). Ecosystem-based marine spatial management: An evolving paradigm for the management of coastal and marine places. Ocean Yearbook, 23, 1–26.

Ekstrom, J.A., & Young, O.R. (2009). Evaluating functional fit between a set of institutions and an ecosystem. Ecology and Society, 14, 2. Retrieved from http://www.ecologyandsociety.org/vol14/iss2/. Accessed 9 Jan 2013.

Ekstrom, J.A., Young, O.R., Gaines, S.D., Gordon, M., McCay, B.J. (2009). A tool to navigate overlaps in fragmented ocean governance. Marine Policy, 33, 532–535.

FAO (2009). Information and knowledge sharing. FAO Technical Guidelines for Responsible Fisheries. No. 12. Rome: FAO.

Fischer, F., Miller, G.J., & Sidney, M.S. (2007). Handbook of public policy analysis: Theory, politics and methods. New York: CRC Press.

Frangoudes, K., Marugán-Pintos, B., Pascual-Fernandez, J.J. (2013). Gender in Galician shell-fisheries: Transforming for governability. In M. Bavinck, R. Chuenpagdee, S. Jentoft, J. Kooiman (Eds.), Governability of fisheries: Theory and applications. Amsterdam: Springer.

Fulton, E.A., Fuller, M., Smith, A.D.M., & Punt, A.E. (2004). Ecological indicators of the ecosystem effects of fishing: Final report. Australian Fisheries Management Authority Report, R99/1546.

Garcia, S.M., Allison, E.H., Andrew, N.J., Béné, C., Bianchi, G., de Graaf, G.J., Kalikoski, D., Mahon. R., Orensanz, J.M. (2008). Towards integrated assessment and advice in small-scale fisheries: principles and processes. FAO Fisheries Technical Paper. No. 515. Rome: FAO.

Gilchrist, G., & Mallory, M.L. (2007). Comparing expert-based science with local ecological knowledge: What are we afraid of? Ecology and Society 12(1), r1. Retrieved from http://www.ecologyandsociety.org/vol12/iss1/resp1/. Accessed 9 Jan 2013.

Grootaert, C., & Van Bastelaer, T. (Eds.). (2002). Understanding and measuring social capital: A multidisciplinary tool for practitioners. Washington, DC: The World Bank.

Grafton, R.Q, Kompasa, T., McLoughlin, R., & Rayn, N. (2007). Benchmarking for fisheries governance. Marine Policy, 31, 470–479.

Haab, T.C., & McConnell, K.E. (2002). Valuing environmental and natural resources. Cheltenham: Edward Elgar.

Haggan, N., Neis, B., & Baird, I.G. (Eds.). (2006). Fishers' knowledge in fisheries science and management. Paris: UNESCO Publishing.

Harris, G. (2007). Seeking sustainability in an age of complexity. Cambridge: Cambridge University Press.
Holmes, T., & Adamowicz, W. (2003). Attribute-based methods. In K.J. Boyle & P.A. Champ (Eds.), A primer on non-market valuation (pp. 171–219). Boston: Kluwer Academic Publishers.
IMM (2008). Sustainable livelihood enhancement and diversification – SLED: A manual for practitioners. Geneva: IUCN.
Jentoft, S., & Chuenpagdee, R. (2009). Fisheries and coastal governance as a wicked problem. Marine Policy, 33, 553–560.
Jentoft, S., Chuenpagdee, R., Bundy, A., Mahon, R. (2010). Pyramids and roses: Alternative images for the governance of fisheries systems. Marine Policy, 34, 1315–1321.
Johannes, R.E. (1998). The case for data-less marine resource management: Examples from tropical nearshore fisheries. Trends in Ecology and Evolution, 13, 243–246.
Kabeer, N. (1994). Reversed realities: Gender hierarchies in development thought. London: Verso. Institute for Development Studies (IDS).
Kahneman, D., & Krueger, A.B. (2006). Developments in the measurement of subjective well-being. Journal of Economic Perspectives, 20, 1, 3–24.
Knetsch, J.L. (1994). Environmental valuation: Some problems of wrong questions and misleading answers. Environmental Values, 3, 351–368.
Kooiman, J., & Jentoft, S. (2009). Meta-governance: Values, norms and principles, and the making of hard choices. Public Administration, 87, 818–836.
Mahon, R., & McConney, P. (2013). A network perspective on governing interactions. In M. Bavinck, R. Chuenpagdee, S. Jentoft, J. Kooiman (Eds.), Governability of fisheries: Theory and applications. Amsterdam: Springer.
Mahon, R., Bavinck, M., Roy, R. (2005). Chapter 17: Fisheries governance in action. In J. Kooiman, M. Bavinck, S. Jentoft, R. Pullin (Eds.), Fish for life: Interactive governance for fisheries (pp. 353–378). Amsterdam: University of Amsterdam Press.
March, C., Smyth, I., Mukhopadhyay, M. (1999). A guide to gender analysis framework. Oxford: Oxfam Publication.
Martinez, A., Dimitriadis, Y., Rubia, B., Gomez, E., de la Fuente, P. (2003). Combining qualitative evaluation and social network analysis for the study of classroom social interactions. Computers and Education, 41(4), 353–368.
McGregor, J.A. (2007). Researching wellbeing: From concepts to methodology. In I. Gough & J.A. McGregor (Eds.), Wellbeing in developing countries: From theory to research (pp. 316–350). Cambridge: Cambridge University Press.
MEA. (2005). Millennium ecosystem assessment – Ecosystems and human well-being: Synthesis. Washington, DC: Island Press.
Mitchell, R.K., Agle, B.R., Wood, D.J. (1997). Toward a theory of stakeholder identification and salience: Defining the principle of who and what really counts. Academy of Management Review, 22(4), 853–888.
Ostrom, E. (2009). General framework for analyzing sustainability of socio-ecological systems. Science, 325, 419.
Overholt, C., Anderson, M., Cloud, K., Austin, J.E. (1984). Gender roles in development projects: A case book. West Hartford: Kumarian Press.
Pauly, D., Christensen V., Walters, C. (2000). Ecosim and ecospace as tools for evaluating ecosystem impact of fisheries. ICES Journal of Marine Science, 57, 697–706.
Pollnac, R.B., Pomeroy, R.S., Harkes, I.H.T. (2001). Fishery policy and job satisfaction in three southeast Asian fisheries. Ocean and Coastal Management, 44(7), 531–544.
Ponte, S. (2007). Greener than thou: The political economy of fish ecolabeling and its local manifestations in South Africa. World Development, 36(1), 159–175.
Resilience Alliance. (2010). Assessing resilience in social-ecological systems: Workbook for practitioners. Version 2.0. Retrieved from http://www.resalliance.org/3871.php. Accessed 9 Jan 2013.
Salas, S., Sumaila, U.R., Pitcher, T. (2004). Short-term decisions of small-scale fishers selecting alternative target species: A choice model. Canadian Journal of Fisheries and Aquatic Sciences, 61(3), 374–383.

Schroeter, S.C., Gutierrez, N.L., Robinson, M., Hilborn, R., Halmay, P. (2010). Moving from data poor to data rich: A case study of community-based data collection for the San Diego red sea urchin fishery. Marine and Coastal Fisheries: Dynamics, Management, and Ecosystem Science, 1, 230–243.

Scott, J. (1991). Social network analysis: A handbook. Newbury Park: Sage Publications.

Scott, J. (1996). Software review: A toolkit for social network analysis. Acta Sociologica, 39, 211–216.

Senge, P.M. (1990). The fifth discipline. The art and practice of the learning organisation. New York: Doubleday.

Senge, P.M., Kleiner, A., Roberts, C., Ross, R., Roth, G., Smith, B. (1999). The dance of change: The challenges to sustaining momentum in learning organisations. New York: Doubleday.

Ten Klooster, P.M., Visser, M., De Jong, M.D.T. (2008). Comparing two image research instruments: The Q-sort method versus the Likert attitude questionnaire. Food Quality and Preference, 19, 511–518.

Timotijevic, L., & Raats, M.M. (2007). Evaluation of two methods of deliberative participation of older people in food-policy development. Health Policy, 82, 302–319.

UNEP. (2006). Marine and coastal ecosystems and human well-being: A synthesis report based on the findings of the Millennium Ecosystem Assessment. Nairobi: UNDP.

Varjopuro, R. (2008). Addressing multiple goals of ecosystem approach through deliberation and technology development. Marine Policy, 32, 212–222.

Vatn, A. (2009). An institutional analysis of methods of environmental appraisal. Ecological Economics, 68(8–9), 2207–2215.

Chapter 15
The Damage Schedule Approach

Andrew Song and Ratana Chuenpagdee

Abstract Illegal fishing is a problem widely observed in fisheries around the world and Lake Malawi is no exception. The long alleviation attempts of the central government of Malawi based on the strategy of enforcement and sanctions have proved largely ineffective leading to the persistence of this governability challenge. An alternate perspective is sought in this chapter by emphasizing people's fundamental notions such as values, images and principles. In order to better understand what they look like and how they may differ amongst stakeholders, the damage schedule was employed. The results display a significant disparity in what governors and resource users regard as a value priority, with the former group judging conservation to be a top concern, while the latter strongly favors the advancement of economic wellbeing. This finding demonstrates socio-economic diversity in people's underlying views about the fishery, which provides partial but important insights towards the alleviation of illegal fishing in Lake Malawi. Such diversity poses a certain limit to the governability of this fisheries system, and must be made aware and genuinely acted upon by all those involved in governance.

Keywords Governability • Diversity • Values-images-principles • Illegal fishing • Lake Malawi • Damage schedules

Introduction

Illegal fishing and non-compliance of regulations pose serious problems for fisheries around the world. They are governance issues that have wide implications at all fishery scales, from an inland artisanal fishery in a developing country to a large

A. Song (✉) • R. Chuenpagdee
Department of Geography, Memorial University of Newfoundland,
St. John's, NL, Canada
e-mail: amsong@mun.ca; ratanac@mun.ca

M. Bavinck et al. (eds.), *Governability of Fisheries and Aquaculture: Theory and Applications*, MARE Publication Series 7, DOI 10.1007/978-94-007-6107-0_15,

industrial-scale one in the high seas. Manifested in various ways, such as poaching, use of destructive fishing methods and zoning violation, the overall effect of illegal fishing and non-compliance behaviors threatens the integrity and health of the ecosystem, as well as the socio-economic basis of those who depend on the resources.

The study of compliance had an initial point of departure in the criminal behavior of economic individuals. Inspired by the work of Adam Smith and Jeremy Bentham, who reasoned that individuals in pursuit of economic self-interest could yield criminal behavior necessitating deterrence to reduce crime, there followed numerous studies that linked crime and economic circumstances (Hønneland 1999). In the 1960s, a formal theoretical framework that views criminals as any other individuals attempting to maximize personal utility was established by Becker's (1968) economic analysis. Stemming from this neoclassical thinking that underpins the economic models of regulatory compliance, the prevailing framework has regarded fishers as utility maximizing individuals driven by self-interest whose decision to engage in illegal fishing is primarily determined by expected payoffs and penalties (cf. Sutinen and Gauvin 1989; Sutinen et al. 1990; Furlong 1991). Thus, the overarching policy response to this issue has been the promotion of deterrence by enhancing enforcement and posing threats of severe sanctions or expensive fines (Hatcher et al. 2000). Such deterrent policies do, however, have severe limitations given that they are costly and difficult to implement (Hatcher et al. 2000). They are also coercive in nature, which can engender bitterness or hostility towards government authority (Sutinen and Kuperan 1999).

Contrary to what the neoclassical deterrence model prescribes, in reality, the probability of getting caught is usually low and the penalties are generally not large relative to the illegal gains (Kuperan and Sutinen 1998). In addition, considerable empirical evidence has shown that a large majority of fishers normally complies with regulations despite such shortcomings (Sutinen and Gauvin 1989; Sutinen et al. 1990). The examples of high compliance despite relatively low enforcement and deterrence are also commonly found outside the fisheries sector, such as the low rate of tax evasion (Elster 1990). Realizing that the neoclassical perspective alone is not adequate to explain the compliance behavior, several studies have embarked on accounting for this 'irrationality' by incorporating other factors into the compliance framework (Kuperan and Sutinen 1998; Sutinen and Kuperan 1999; Charles et al. 1999; Hatcher et al. 2000). These other factors include legitimacy, morality and socialization, which are emphasized through social norms and the social capital of small groups, for instance, in common property theory (Ostrom 1990).

Tyler (1990) introduces the terminology of 'instrumental perspective' and 'normative perspective' in distinguishing between these two streams of arguments. The instrumental perspective is synonymous with Becker's (1968) framework that assumes individuals as rational agents driven by self-interest and responding to incentives and sanctions. The normative perspective, on the other hand, emphasizes that individuals are influenced by what is just, fair, appropriate and morally right. These are in a sense 'priceless' and thus defy a benefit-cost calculation. It involves human values such as being moderate as well as environmental values such as existence value. This perspective also accounts for people acting as a matter of principle, even in the face of imminent inconveniences or opportunities to seek easy rewards. In addition, certain images held by people may direct their action in a manner that others with different 'ways of seeing the world' may not easily comprehend. Hence, broadly speaking, values, images and principles form an intricate part of this so-called normative perspective. The arising consensus in the fisheries circle, as well as elsewhere, is that the normative perspective plays a crucial role in influencing compliance behavior among fishers, and therefore theoretical elaboration, as well as policy intervention, should be shifted towards paying greater attention to this aspect.

Diversity in Values, Images and Principles

Despite much deliberation on the topic of illegal fishing and compliance and on ways to improve the situation, its widespread nature is an ever-persistent threat (e.g. FAO 2001; Flewwelling et al. 2002; Crawford et al. 2004; Hauck and Kroese 2006; Sumaila et al. 2006; Agnew et al. 2009). Why is illegal fishing so difficult to eradicate or at least manage at a level that is not detrimental to ecological integrity and social functioning? It becomes an especially acute conundrum if violating fishers are typically shown to understand the rationale (and good intentions) of the restrictive fisheries regulations, and are generally aware of its long-term consequences.

We submit that the issue of illegal fishing can exemplify the component of *diversity* inherent in fisheries governance as argued by, and emphasized in, interactive governance (Kooiman et al. 2005). More specifically, a socio-economic sub-system represented by various individuals and groups involved in a particular fishery setting can be characterized as having a diverse set of interests and ways of doing things that may complicate the issue of illegal fishing by putting people in a number of unique circumstances. On a deeper level, this implies that people's underlying normative and cognitive concepts, such as values, images and principles, are at the root of the diversity that makes compliance a thorny undertaking. Values tend to be incommensurable, implying the impossibility of comparison (Kooiman and Jentoft 2005). 'Hard choices' in fisheries arise from this value-ridden characteristic forcing any decisions to be painful and controversial. In this regard, Kooiman and Jentoft (2009) argue for the importance of making governance values as coherent and explicit as possible for the betterment of governability. Images and principles also frequently compete and contrast with each other, creating an antagonistic reality. Hence, as values, images and principles guide, shape and inspire people's decisions and actions, including fishing practice and strategy, illegal fishing may be imbued with different connotations, meanings and rationales among the various stakeholder groups. A pertinent example of this is when a person values both ecosystem conservation and secure livelihoods. These two particular values are certainly not mutually exclusive and can be promoted in accordance with one another. A value conflict could arise, however, when in pursuit of securing livelihoods, a fisher forgoes ecological conservation by catching fish beyond the legal guidelines. This would represent a weaker standing of conservation value in light of the more desirable conception of secure livelihoods. Likewise, an image of poverty in his/her household may pose more painful/fearful than the image of the barren sea. Such configurations would affect his/her principles so that one's action and decisions are first and foremost guided by the principles involving the fulfillment of human welfare and economic viability rather than those concerning nature/resource conservation. Acknowledging the diverse ways people's values, images and principles interact to produce different behavior and decisions, it may be of benefit to examine them and gain a more deep-seated understanding of why illegal fishing takes place and persist despite the various deterrence measures.

Illegal Fishing as a Governability Challenge

We reason that the varying views that exist about the compliance of fishing rules, as well as the real social and environmental consequences that may ensue from illegal fishing, all lead to governability challenges. Governability here can be rephrased as the capacity of fisheries as a whole to self-organize and maintain themselves with due consideration of the inherent and constructed qualities embedded in nature and society. Referred to in simpler terms as a propensity for successful governance (Kooiman 2008), implications for governability have several dimensions in the context of illegal fishing. First, the depletion of fish stocks and deterioration of ecosystem services pose a rather obvious difficulty in maintaining the governability of an overall fisheries system. This is because it results in the reduction of integrity and self-generating capacity of the ecosystem. Overfishing, combined with much suspected illegal fishing by foreign fleets in the Grand Banks, has led to the northern cod collapse in the waters around Newfoundland. The recovery has yet to fully occur, while the abundance of large cod prey such as snow crab and shrimp has greatly increased (Bundy and Fanning 2005). This complex trophic shift has made governing the fishery in the region a more contentious affair and thus less governable. Looking at the social aspect, illegal fishing contributes to a loss of social cohesion and sense of cooperation, as well as a stimulation of hyper-competitive spirit, hostility and even physical confrontation. These factors make governance an extremely delicate and tense subject, and reduce the ways in which governing can reach its potential. In the Gulf of California, for instance, poaching by outsiders was shown to jeopardize otherwise well-observed local cooperation. Leading to a rapid cascading effect on fisheries resources and locally-designed rule compliance, accountability was eroded among fishers and the fishery quickly became a free-for-all (Cudney-Bueno and Basurto 2009). Thirdly, people's varying views pose a challenge to the upkeep of governability in and of themselves. Incommensurable values may act as a limit to how governable a fishery can be (Jentoft 2007). If governors and those-being-governed hold highly polarized or otherwise vastly different views of an issue such as rule compliance or which fishery projects to give priority, ensuring stakeholder compromise and cooperation becomes a greater challenge. The point is that illegal fishing may proliferate or is difficult to eradicate due to the existence of diverse and conflicting stakeholder views encompassing value, images and principles. Understanding this diversity in values, images and principles would thus become an important step towards alleviating illegal fishing and increasing the level of compliance. Insights that stem from this alternate, but fundamental, viewpoint are welcomed and may contribute to finding ways to improve governability.

The following section proceeds with the aim of discussing how the diversity in stakeholder views can be systematically examined. By bringing in the case study accounts of illegal fishing in Lake Malawi as an empirical context, we highlight one of the tools, the 'damage schedule,' that can be employed to elicit people's judgments and preferences as a way to explore their underlying values, images and principles.

Studying Values, Images and Principles

Elicitation of values, images and principles does not present itself as a clear-cut exercise. There are a variety of ways to understand what people value and how much they do so based on one's approach and schools of thought. Satterfield and Kalof (2005) remind us of a useful way of categorizing values – axiomatic vs. relativistic traditions. An axiomatic approach operates on the premise that certain values are better, more important and intellectually defensible than others. Typically expert-driven from the fields of ethics, philosophy and ecological economics, the values under this tradition are formulated based on argument (e.g. Kellert 1993; Rolston 1994) and/or measurement (e.g. Costanza et al. 1998). The relativistic approach assumes that there are no right or wrong values, only different ones. Abiding by the principle of 'value-neutral,' researchers in this tradition rely on expressed preferences to monitor or elicit public opinion and conduct valuation exercises for policy and management purposes. This chapter focuses on the relativistic approach due to its heavy influences on policy and management process through the collection of stakeholder judgments and preferences (Satterfield and Kalof 2005).

Implicitly aligning with the relativistic tradition, Gregory (1999) offers a useful list of tools for identifying (environmental) values. Among the tools that comprise the category of economic valuations are travel costs – assigning economic value to natural resources based on visitation; hedonic pricing – capturing values in the prices of marketed goods; contingent valuation – willingness-to-pay and willingness-to-accept; and damage schedules – estimates of the relative seriousness of adverse impacts on natural resources. Non-economic valuation that involves non-monetary expressed preferences includes attitudinal and opinion surveys, and small-group discussions such as focus groups.

Damage Schedules

The survey method highlighted in this chapter is the damage schedule (Gregory et al. 1996; Chuenpagdee et al. 2001). A damage schedule is envisioned as a set of policy instruments similar to payments and sanctions that could be used to discourage damaging activities and compensate for resource losses. It collects public judgments on the relative importance of resource losses or the relative harmfulness of certain activities causing the losses. The assessed preferences or judgments are presented in the form of an interval ranking scale, which could work as a non-monetary indicator of the severity of resource losses or the impacts of the damaging activities. The result can aid policy makers in developing appropriate policy strategies to prevent certain activities, create a compensation scheme for resource damage, and deter incidents such as accidental oil spills and discharge pollution. Further, the developed damage schedules offer policy makers a platform to involve local communities in the management of resources and directly incorporate their inputs in policy design, since the schedules are based on the knowledge of resource users and on people's preferences and judgments about resources and their importance, as well as those of scientists and managers.

The damage schedule approach has been applied to several fisheries and coastal related studies over the years. An earlier application was to examine coastal development issues surrounding shrimp farming and tourism in Southern Thailand (Chuenpagdee et al. 2001). Chuenpagdee et al. (2002) surveyed community members in Mexico to reveal local judgments about the severity of damages to coastal habitats and the impact of activities that may cause the damages. A more elaborated set was developed to assess the relative severity of collateral impacts of the fishing gears commonly used in the United States (Chuenpagdee et al. 2003). Environmental damages in the urban coastal setting of Singapore were the subject of the study by Quah et al. (2006), who then used the resulting scale to derive willingness-to-accept compensation amounts for relinquishing top environmental concerns. As demonstrated by these applications, this quantitative survey method affords flexibility in design that, we believe, will allow modifications to be made to get at one's underlying values and principles.

Paired Comparison

The damage schedule relies on the use of paired comparison, which is a simple method frequently used to attain a ranking scale. Its basic unit is the comparison of two objects, and the comparison is presented to one or more judges. The term 'object' is used to cover what is being compared such as treatment or stimuli, while judges mean survey respondents (David 1988). This method has proven useful in situations where subjective judgments may play a role in people's evaluation, such as in taste tasting, personnel evaluation, or social values. It is particularly applicable in situations "when it is impossible or impractical to make relevant measurements in order to decide which of the two objects is preferable" (David 1988, 1). Initially employed in psychometry (e.g. Thurstone 1927), its fields of application have more recently been expanded to include, among others, acoustics, animal ecology, economics, epidemiology, food science and sports. Furthermore, its use in eliciting public preferences and judgments in an environmental study setting has been justified by a number of studies that employed this method with a similar intention (Peterson and Brown 1998; Rutherford et al. 1998; Chuenpagdee et al. 2001; Rudd 2001; Wattage and Mardle 2005; Quah et al. 2006). The method begins by establishing a set of objects under the theme of a particular study, whether it is resource losses, damaging activities or community programs. The objects are presented in pairs to each respondent, who then is asked to make a choice between them. This will continue one after another until all possible pairs are exhausted. Standard notation denotes N as the total number of objects, while the total number of respondents are denoted as k. For each respondent, the total number of all possible pairs for comparison is $N(N\text{-}1)/2$. Under normal circumstances, each object has the same probability of being selected as all objects are paired an equal number of times.

There are at least three key advantages of using paired comparisons. First, a fine judgment can be better achieved in a binary setting, especially when objects are deemed to have subtle differences. The usual difficulty that faces a simultaneous ordinal ranking of all N objects can be lessened. Secondly, the paired comparison method can be used to produce an interval scale in which the numerical differences between the objects have an arithmetic meaning. It shows the spread of the objects on a scale, and is useful in explaining the extent to which one object is preferred over others in numerical terms. The third advantage of the paired comparison exercise is that it is simple to conduct, and thus can be repeated or modified to accommodate changes in the systems that we are interested in. An obvious drawback of the approach is the limited number of objects that can be included in the comparison, at least in the complete design. When the number of pairs presented is too large, respondents may become fatigued and tend to believe that there are repetitive pairs, even if there were not.

Case Study: Southeast Arm Fishery in Lake Malawi

Like many fisheries around the world, illegal fishing is widespread in Lake Malawi. Towards the southeastern area called the Southeast Arm (SEA), illegal fishing is thought to contribute significantly to species decline, adding to existing concerns over stock depletion, as well as the loss of livelihood opportunities and sources of animal protein (Banda et al. 2005; Bulirani 2005). The concept of illegal fishing first became a topic of importance in Lake Malawi in the 1930s. British colonial conservationists, concerned with the lake's ecology and fish conservation, indignantly viewed traditional fishing methods, such as weirs and traps, as destructive and primitive. Determined to restrict these traditional practices, they introduced a series of early fishing regulations (Chirwa 1996). Through the establishment of the Department of Fisheries (DoF) in 1946, deterrence through enforcement of regulations and the threat of sanctions continued to be the main form of policy response to illegal fishing in the decades that followed, persisting beyond independence from the colonial government in 1964. Despite fishers' prolonged acquaintance with the regulations (Hara 2001), however, the ineffectiveness of this approach was evident. The process was often treated with mistrust and disdain, and the ensuing violent confrontations with resource users not only proved to be dangerous, but also demoralizing for the government enforcement personnel (Hara 2006a). With the rising eminence of co-management regimes as a democratic and cost-effective way of achieving compliance, a participatory management initiative was put into practice in the SEA in the mid-1990s. Unfortunately, as Njaya (2007, 2008) and Hara (2006b) point out, the management initiative's implementation has been beseeched by several challenges, and it has only been marginally successful in raising the support of the fishing communities needed in order to resolve illegal fishing. There is a need for an alternative measure that has a different focus and that recognizes the limited availability of financial and human resources in management in order to better address the illegal fishing problem in the SEA and the growing concern for the conservation of the lake's ecosystem.

A multi-species and multi-gear small-scale fishery is the main form of fishing activity in the SEA to this date (Smith 1998; Ngochera 2001). It typically uses small vessels such as planked boats and dug-out canoes to operate gillnets, longlines and open-water seine nets called *chilimira*. Fishing is not only boat based, however. It can also be done on land through, for example, beach seining. Much of the catch is sold for cash income, signifying the commercial importance of the fishery.

There is a great diversity too in the stakeholders involved in the SEA fishery. Gear owners commonly refer to those who own fishing gear, but may not necessarily participate in fishing. Crewmembers, on the other hand, are those who provide manpower and technical know-how in the actual fishing operation. Fish processors and traders are also highly visible in fishing communities. They set up their operation near landing sites to ensure a steady supply of fish and maintain a close relationship with fishers. Though DoF holds the ultimate authority in overseeing fisheries matters, the Traditional Authority governs day-to-day fishing matters at the village level. The Traditional Authority is upheld by three hierarchical levels of traditional leaders – the chief, the group village head and the village head. Other relevant governing institutions include the Members of Parliaments that represent the area at the national level and the Commercial Fishermen's Association, which exists to represent the preferences and judgments of a handful, but powerful, large-scale fishing owners and operators.

Study Design for Elicitation of Judgments and Preferences

In the context of illegal fishing and the diverse group of stakeholders involved in the Lake Malawi fishery, this case study was designed to assess the extent to which fishery stakeholders value conservation through elicitation of judgments and preferences.

Table 15.1 Objects for each paired comparison set (in Set B, a value associated with each community program is shown in brackets)

Set A: Fishing activities
Catching juvenile fish
Fishing using mechanized gear
Fishing using gears that disturb lake bottom
Fishing in offshore deep water
Fishing using non-selective gear
Too many people fishing in one area
Fishing in spawning area
Set B: Community programs
Protect fish habitat and fish species (conservation)
Promote scientific research on lake fisheries ecosystem (precaution)
Provide micro-credit loans to expand fishing-related work (economic wellbeing)
Help reduce fish spoilage during catching and processing (frugality)
Promote small-scale community fish cage culture (innovation)
Ensure fishing access for local fishers and communities (social justice)
Provide ownership of resources to local communities (subsidiarity)

By examining how much they value conservation as it competes with other interests, we may acquire a deeper understanding of what motivates illegal fishing and find support for focusing on values, images and principles in dealing with illegal fishing.

Two sets of paired comparisons were developed, as listed in Table 15.1. In order to gauge the level of conservation awareness, one measures the judgment of respondents in relation to which fishing activities are more damaging to the fisheries resources in the SEA. We asked respondents "In your opinion, which of these two activities do you consider more damaging to the fishery resources in the SEA?" All of the objects can be deemed potentially damaging by the respondents, albeit with different degrees of severity. For instance, it may be entirely possible that one sees 'fishing using mechanized gear' just as damaging as 'catching juvenile fish' depending on his/her unique fishing circumstances. Although the question refers to opinion, it is acknowledged that the responses may reflect local, specialized knowledge of the respondents, not just opinions or preferences. The second set assesses respondents' preferences towards conservation-oriented community fisheries programs by asking the question: "If a program were to be implemented in your community, in your opinion, which of these two programs, A or B, do you prefer?" There are seven objects included in each set, giving a total number of 21 pairs per set. They represent site-specific fishing activities and the community programs that are relevant in the SEA. They were developed based on existing literature, including the fisheries regulation, direct observations during field visits, informal interviews with key informants, and the results of several rounds of pre-tests. The lists were also verified with a group of fishery managers in the SEA to ensure that these activities and programs indeed best reflect the concerns of the stakeholders in the region.

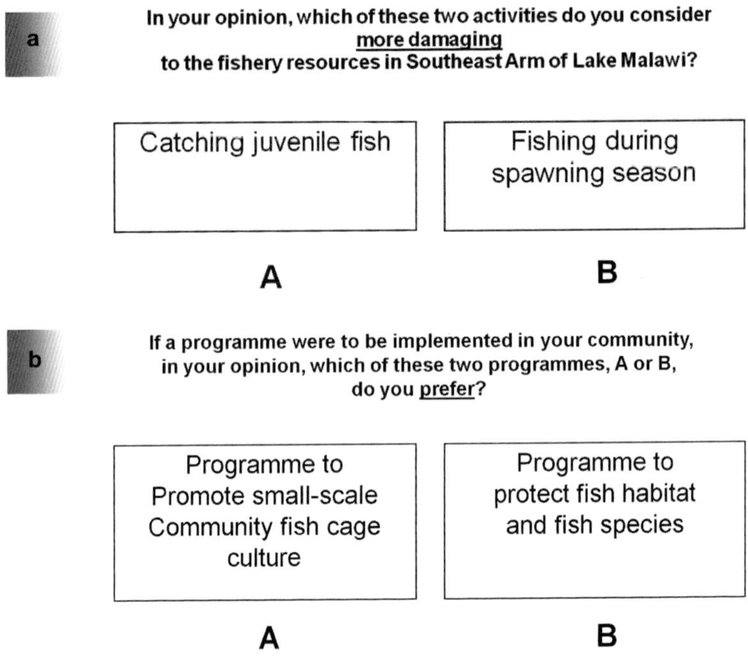

Fig. 15.1 Sample paired comparison questions drawn from (**a**) Set *A* and (**b**) Set *B*

Responses to the first set, Set A, produce a scale, from most damaging to least damaging fishing activity, on the basis of respondents' perceptions. It is an indication of their level of understanding about the lake's fisheries and ecosystems, and may enable voluntary engagement in conservation-oriented fishing practices. The assumption is that, for any principle to influence one's fishing decisions and behavior (both legal and illegal), s/he must first be equipped with sufficient understanding of which activities promote conservation and should thus be encouraged, and which activities jeopardize conservation and should therefore be made illegal. As shown in Table 15.1, the objects are void of any specific details. For example, there is no number indicating the degree of mechanization in 'fishing using mechanized gear', and any mention of a specific fishing gear is avoided. This was to minimize strategic voting of the respondents by basing the comparisons on the concept – or the *image* – of the fishing activities and not on the specifics that may conjure up certain attachments to their own fishing activities. An example of paired comparison used in Set A is displayed in Fig. 15.1a.

The second set, Set B, results in a scale that reveals the extent of respondents' inclination for fisheries conservation. Even if the stakeholders hold sufficient understanding on how to proceed with conservation, it is necessary to confirm that they do in fact value conservation. The set was designed such that two community programs that directly promote conservation are included in the choice pairs. 'Programme [sic] to protect fish habitat and fish species' is a scenario that has a direct attachment to conservation, while ‘programme [sic] to promote scientific research on lake fisheries ecosystem’, which draws on the well-known precautionary principle, also deems conservation highly relevant. Valuing these over the other community programs, which may largely be driven by other interests and motivations that show little compatibility with the conservation objective, can be interpreted as a fair indication of their genuine inclination towards conservation. Each of the community programs has an associated value, which is suggested in brackets in Table 15.1. The programs are presented to the respondents in a hypothetical sense as something that could be implemented, but without the promise of implementation. This was to prevent immediate expectation from influencing their choices. An example of the paired comparison used in Set B is displayed in Fig. 15.1b.

Survey Information

The survey was directed at seven groups of respondents and involved multiple sites, as shown in Fig. 15.2. Active fishing villages on the eastern shore of the SEA were chosen to be the main location for surveying the resource-dependent groups, that is, gear owners, crew members, fish processors/traders and community members. This side of the water body is believed to yield better catch than the stock-depleted western shore (Njaya 2008). Also, due to its relative remoteness, coupled with the shortage of infrastructure and tourism development, fishing still remains a key economic activity supporting people’s livelihoods on the eastern shore. To investigate any potential disparity that may arise from the east-west geographical distinction, two more resource-dependent groups, gear owners and crew members on the western shore, were added to the survey. Thus, together with the managers/scientists group, seven respondent groups were formed. The survey was conducted with the assistance of a local person, who is native to the area and, in addition to being proficient in English, has fluency in two of the most widely spoken local languages, Chichewa and Chiyao. The survey with the scientists/managers group, comprising of various government officials such as planners, researchers, statisticians, enforcement officers, lecturers, and also scientists from several nongovernmental organizations (NGOs) working in the field of fisheries, took place in various locations around the lake. These were conducted without the assistance of the local translator, because of their proficiency in the English language.

All 21 pairs from each set were included in the survey booklet, resulting in a total of 42 pairs. The sequential order of the pairs in the booklet and the left-right position of the two choices in each pair were both randomly generated to ensure the uniqueness of each booklet, as well as avoiding any possible order-related bias. Due to concerns about the availability and reliability of a census database, random sampling of respondents was not feasible. Instead, quota sampling was employed to obtain the total number of respondents, 144. Approximately 20 respondents were surveyed for each group. As shown in the demographic breakdown of the survey respondents (Table 15.2), only modest differences exist in the number of respondents, their average age and the average years of fishery experience. One may think that the small sample size of each

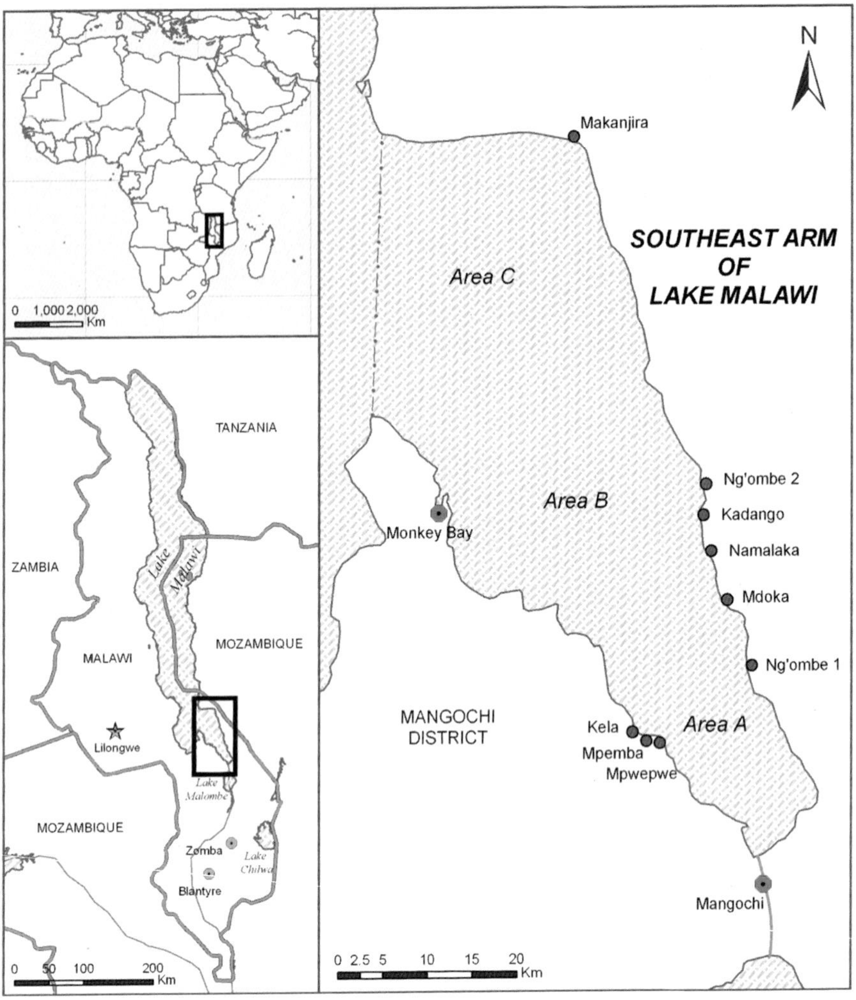

Fig. 15.2 Map of the Southeast Arm of Lake Malawi (Survey sites are shown as *dark circles*; *double circles* indicate town centres) (Source: Song 2009)

group could reduce the reliability of the results of this study. However, as other studies of similar methodologies and intent have shown (c.f. Chuenpagdee et al. 2003; Quah et al. 2006; Bose and Crees-Morris 2009), in-group consistency was quickly reached with the number of respondents obtained in the study. Thus, increasing the number of respondents would be of no pragmatic value from the view of both cost-effectiveness and time-efficiency (Bose and Crees-Morris 2009). Whenever possible, a one-on-one setting was preferred when conducting the survey in order to minimize any strategic bias that may arise from social pressure or fear of reprisal. Hence, most surveys were conducted in a quiet, sheltered environment in the absence of other community members. Each survey took an average of about 25 minutes to complete.

Table 15.2 Demographic information of survey respondents

	Gear owners (East)	Crew members (East)	Processors/ Traders	Community members	Gear owners (West)	Crew members (West)	Manager/ Scientists	Total
Total number of respondents	20	20	20	20	21	17	26	144
Male	20	20	7	8	19	17	21	112
Female	0	0	13	12	2	0	5	32
Age[a]	38	30	37	34	38	34	40	–
Years of fishery experience[a]	10	9	9	–	6	9	13	–
Years in education[a]	1	5	6	3	3	3	13+[b]	–

[a]Denotes average value
[b]Indicates tertiary education, which typically surpasses 13 years of schooling

Results

Figure 15.3 is a simplified version of a damage schedule that illustrates the rankings of conservation awareness and conservation value based on the results of the two paired comparison sets. According to the Kendall's tau rank correlation analysis ($p = 0.05$), the survey results revealed no significant difference among the respondent groups as to which fishing activities are more or less damaging to the fisheries resources in the SEA. Hence, the judged responses of all seven groups were combined into a single scale as shown in Fig. 15.3. 'Fishing in spawning area' was unanimously identified to be the most damaging practice with the normalized score of 82 out of 100. This was followed by 'fishing using gears that disturb lake bottom' and 'fishing using non-selective gear', which shared the identical score of 65. 'Catching juvenile fish' was the next most damaging activity. As for activities with moderate damage, 'too many fishers in one area' was deemed as damaging as 'fishing using mechanized gear'. With a score of 8, 'fishing in offshore deep water' was overwhelmingly judged to be the least damaging form of fishing activity.

Understanding the ecological impact of pertinent fishing activities forms an essential aspect of realizing fisheries conservation. Not only are the results attained here agreed on by all surveyed groups in the SEA fishery, they are also judged to be consistent with the commonly-held knowledge about fishing gears and their impacts in other fisheries worldwide. For instance, utmost attention on fishing in spawning areas observed in the SEA corresponds with the high emphasis placed on the role of marine reserves in enhancing spawning stocks and protecting juvenile production (Murawski et al. 2000; Manríquez and Castilla 2001). Also, the relatively severe damage from disturbing the lake bottom perceived by the respondents in the SEA is a contentious issue globally. with special regard given to bottom-trawling (Watling and Norse 1998; Chuenpagdee et al. 2003). Hence, the general correspondence of the survey result with prevailing ecological issues in global fisheries suggests that the stakeholder groups in the SEA hold a moderately high level of understanding of fisheries conservation.

Unlike the awareness portion, the survey results demonstrate the existence of a significant divergence of judgments and preferences between the resource-dependent groups and the manager/scientists group when it comes to the preference of fisheries-related programs. While the resource-dependent groups unequivocally preferred the program that provides micro-credit loans for the expansion of their fishing-related work, the same inclination was not found within the managers/scientists, who ranked the program near the bottom. The second notable difference between the two groups lies in the preference ranking of the program that provides ownership of resources to local communities, which was ranked relatively low for the resources-dependent groups compared to a high ranking among the managers/scientists. The third disparity concerns the program that helps reduce fish spoilage during catching and processing. This was by far the least preferred program in the eyes of resource-dependent groups with the score of 7, and in stark contrast to the score provided by the managers/scientists, 50. Two programs that have direct relevance for

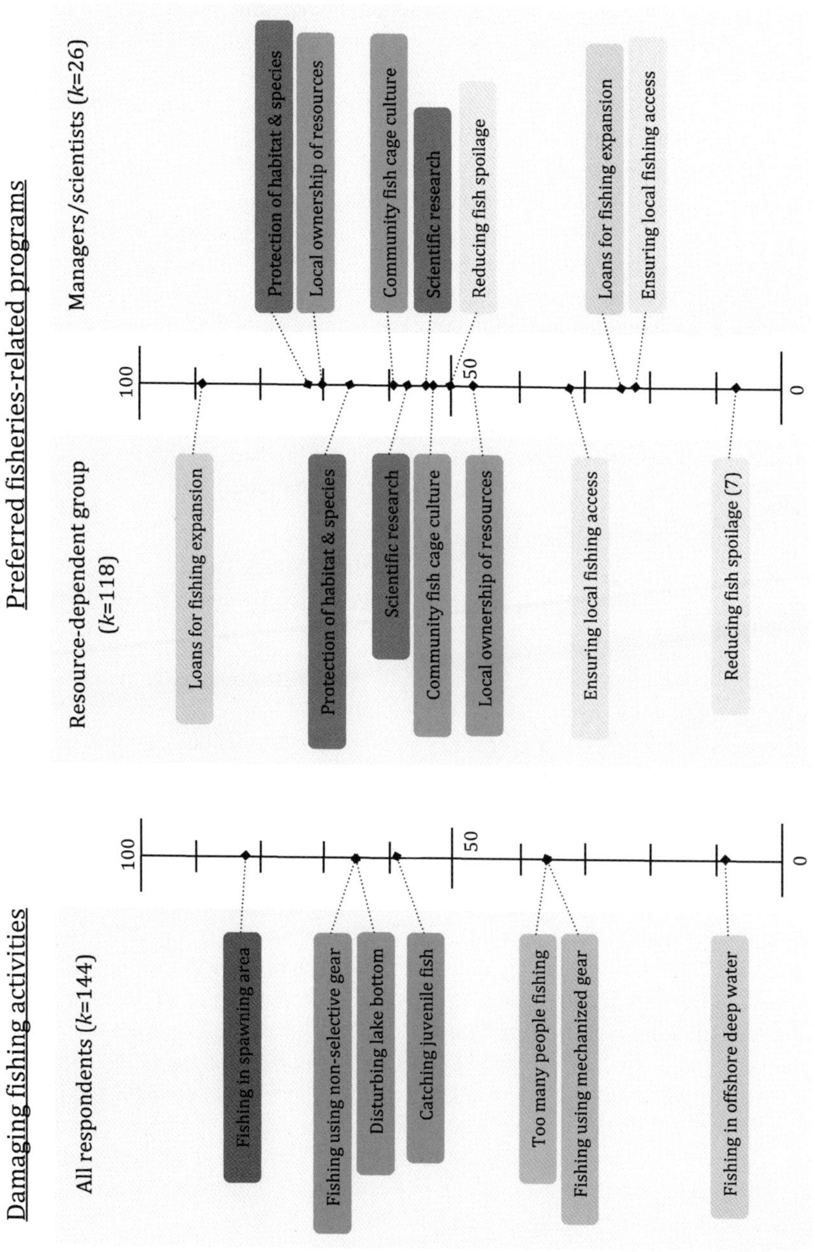

Fig. 15.3 Damage schedule of damaging fishing activities and preferred community programs

conservation, protecting fish habitats and species, and promoting scientific research on the lake ecosystem, were ranked comparatively high by both groupings. An agreement was also found in the mid-ranking of the promotion of small-scale community fish cage culturing. Finally, the program that ensures fishing access to local fishers and communities was, in comparison to the other six hypothetical programs, generally the least preferred one by both groups.

As Fig. 15.3 highlights, the two conservation-oriented programs are positioned mid-to-high in both groups' preference ranking scales. This can be loosely interpreted as the respondents valuing conservation in a moderate sense – in other words, they care about conservation to some degree. Based on this, and together with the presence of the shared understanding of damaging fishing activities, which can help create a common vision for the fisheries, we can expect moderate potential in advancing conservation goals in the SEA fishery. An important challenge must, however, be noted when looking at the top inclinations of the two groups. While the managers/scientists group were partial to the program that protects fish habitats and species, the resource-dependent group showed a clear preference for the provision of loans that would enable the expansion of one's fishing work. We believe that this difference must be duly recognized and reconciled if the conservation potential and the resolution of the illegal fishing situation is to be realized.

Discussion: Conservation, Economic Wellbeing and Illegal Fishing

The resource-dependent group's leading preference was shown to be the expansion of one's fishing work by acquiring capital through loans. In other words, the resource-dependent group's preference of economic expansion and development seems to clearly outpace other inclinations, including conservation. In fact, this empirical finding corresponds to the general, prevailing line of thinking in Malawi, which, whether it concerns fisheries or other sectors, revolves around the development agenda. Poverty alleviation and raising the standard of living through development, modernization and rapid economic growth are the main themes that reverberate in all sectors of government affairs (GoM 2009). According to the 2001 National Fisheries and Aquaculture policy, the major policy goals in fisheries are aimed at "maximizing the sustainable yield…to improve the efficiency of exploitation, processing and marketing…to promote investment in the fishing industry, rural fish farming units and exploit all opportunities to expand existing and develop new aquatic resources" (GoM 2001, 5). Furthermore, persistent attempts to expand fisheries further offshore, a production-oriented modernization agenda and examples of policy support for the industrial sector exemplify the government's pro-development position in steering the fisheries sector. It is worth noting, however, that some inconsistencies and ambivalence have been observed in fisheries development policy over the years (Chirwa 1996; Allison et al. 2002). The theme of poverty alleviation and development is also prominent in rural fishing villages at the community level, as well as among resource users at the individual level. The economic motive involving the expansion of one's fishing work therefore appears to be an important factor hindering a more resilient expression of conservation value in the SEA fisheries.

The findings indicate that the short-term enhancement of standard-of-living is one of the resource-dependent group's main conceptions of 'what is desirable', and that this competes with conservation values. From the perspective of policy-makers and governors, a compromising solution is likely required when promoting conservation and rule compliance in order to respond to the real demands of resource-users. We submit that a policy recommendation that does not lead to enhanced standards-of-living would not be well-received and would therefore bring little positive improvement to the governability of the system. Consequently, initiatives that supply the resource-dependent group with an economic incentive to engage in conservation-oriented fishing practices are strongly recommended.

One possible approach that directly utilizes resource users' economic-minded preferences is conservation payments (or conservation performance payments). Conservation payments compensate people for their role in looking after fisheries resources (Simpson and Sedjo 1996; Ferraro 2001; Ferraro and Simpson 2002). Conventional development interventions attempt to reduce pressures on ecosystem by steering the economic development process towards a path that is compatible with ecosystem protection through initiatives such as eco-tourism and aquaculture. However, this indirect way of encouraging conservation is often observed to be ill-suited for the proper alignment of economic incentives and conservation goals. This is mainly due to the complexity of development interventions vis-à-vis the temporal and spatial scales at which conservation objectives must be achieved. This difficulty often creates little effect on conservation-related household behavior (i.e. in fishing practices) (Ferraro 2001). The premise of the payment scheme is to offer a far more cost-effective way of ensuring conservation results than conventional development projects by directly linking explicit payments to conservation progress. Although the direct payment system is not without its own set of shortcomings – ones that necessitate careful program design and implementation – past and ongoing examples can be found in several developing countries in the tropics, where they have been employed to protect ecosystems and promote stewardship of forest resources (Ferraro and Simpson 2002). For instance, Ferraro and Simpson (2002) report that Guatemala's example delivers direct payments to forest stewards through the Forest Incentives Program (World Bank 2000), while, in Costa Rica, institutional mechanisms were established to allow local, national, and international beneficiaries of ecosystem services to compensate those who protect ecosystems (Castro et al. 2000). As with these examples, such programs can be made feasible through financial support garnered from national and international donors, NGOs and various interest groups around the world who share a keen interest in protecting particular ecosystems.

Various forms of conservation subsidies and direct payments schemes should garner meaningful attention in meeting the economic/development demand of the resource-dependent group in the SEA. This would be an important policy addition aimed at raising the level of inclination towards fisheries conservation over time and encouraging self-driven restraint in illegal fishing, especially given that the resource users surveyed in this study are already well-equipped with adequate conservation knowledge. In the process, we hope that the overall illegal fishing problem in Lake Malawi will move towards a resolution.

Conclusion

In this chapter we argued that illegal fishing may be exacerbated by the diversity of values, images and principles that people hold. An examination of such underlying concepts could therefore help generate meaningful insights about the ways illegal fishing could be lessened. Taking a cue from the relativistic nature of values, images and principles – conceived as each competing against many other values, images and principles one holds in a given situation – ecosystem conservation value was examined in detail to determine its standing vis-à-vis other values deemed to be associated with the practice of illegal fishing. The emphasis was laid on showcasing one possible tool that can be employed to gauge people's values through elicitation of judgments and preferences. The case study of a Lake Malawi fishery, in which illegal fishing and non-compliance has been a pressing issue for many years, has shown that despite the general concurrence in the conservation knowledge of all stakeholder groups interviewed – especially between resource users and governors – the degree to which they attach importance to conservation and other values is significantly different and, subsequently, indicates a plausible source for the persistence of illegal fishing. Such a result has an influence on governability. It requires empathy and higher appreciation of each other's standpoints in minimizing controversy and dissatisfaction. It also puts a limit on the overall governability of the fisheries by narrowing the range of workable agendas and demanding more creativity in governance. It is thus important that governors are made sensitive to the levels of governability and its potential sources. This chapter highlighted that one of the confounding elements giving rise to the governability problem is the deep-seated diversity of the socio-economic system founded on more durable and fundamental values, images and principles. Focusing on these underlying concepts to explore and improve governability presents an intriguing direction worthy of further elaboration and empirical testing. With the right tools and their continuing refinement, however, the effort should receive timely assistance.

Acknowledgements Support for this research was provided by the Social Sciences and Humanities Research Council of Canada (SSHRC) and a partnership between the Association of Universities and Colleges of Canada (AUCC) and the Canadian International Development Agency (CIDA) (Project No. S61268-528/I). Authors would like to acknowledge numerous in-kind support of the Marine Institute (MI) and the Department of Fisheries (DoF) of Malawi.

References

Agnew, D.J., Pearce, J., Ganapathiraju, P., Peatman, T., Watson, R., Beddington, J.R., Pitcher, T.J. (2009). Estimating the worldwide extent of illegal fishing. PLoS One, 4(2), e4570. doi:10.1371/journal.pone.0004570.

Allison, E.H., Mvula, P.M., & Ellis, F. (2002). Conflicting agendas in the development and management of fisheries in Lake Malawi. In K. Geheb, M.-T. Sarch (Eds.), Africa's inland fisheries: the management challenge (pp. 49–73). Kampala: Fountain Publishers.

Banda, M.C., Kanyeree, G.Z., Rusuwa, B.B. (2005). The status of the chambo in Malawi: Fisheries and biology. In M. Banda, D. Jamu, F. Njaya, M. Makuwila, A. Maluwa (Eds.), The chambo restoration strategic plan. Proceedings of the national workshop, Mangochi, Malawi, 13–16 May 2003. WorldFish Center Conference Proceedings 71 (pp. 1–7). Penang: WorldFish Center.

Becker, G.S. (1968). Crime and punishment: an economic approach. Journal of Political Economy, 76, 169–217.

Bose, S., & Crees-Morris, A. (2009). Stakeholder's views on fisheries compliance: An Australian case study. Marine Policy, 33, 248–253.

Bulirani, A. (2005). Observations on the factors behind the decline of the chambo in Lake Malawi and Lake Malombe. In M. Banda, D. Jamu, F. Njaya, M. Makuwila, A. Maluwa (Eds.), The chambo restoration strategic plan. Proceedings of the national workshop, Mangochi, Malawi, 13–16 May 2003. WorldFish Center Conference Proceedings 71 (pp. 8–11). Penang: WorldFish Center.

Bundy, A., & Fanning, L.P. (2005). Can Atlantic cod (Gadus morhua) recover? Exploring trophic explanations for the non-recovery of the cod stock on the eastern Scotian Shelf, Canada. Canadian Journal of Fisheries and Aquatic Sciences, 62, 1474–1489.

Castro, R., Tattenbach, F., Gamez, L., Olson, N. (2000). The Costa Rican experience with market instruments to mitigate climate change and conserve biodiversity. Environmental Monitoring and Assessment, 61, 75–92.

Charles, A.T., Mazany, R.L., Cross, M.L. (1999). The economics of illegal fishing: a behavioral model. Marine Resource Economics, 14, 95–110.

Chirwa, W.C. (1996). Fishing rights, ecology and conservation along southern Lake Malawi, 1920–1964. African Affairs, 95, 351–377.

Chuenpagdee, R., Knetsch, J.L., & Brown, T.C. (2001). Coastal management using public judgments, importance scales, and predetermined schedule. Coastal Management, 29, 253–270.

Chuenpagdee, R., Fraga, J., Euán-Avila, J.I. (2002). Community perspectives toward a marine reserve: a case study of San Felipe, Yucatán, México. Coastal Management, 30, 183–191.

Chuenpagdee, R., Morgan, L.E., Maxwell, S.M., Norse, E.A., Pauly, D. (2003). Shifting gears: Assessing collateral impacts of fishing methods in US waters. Frontiers in Ecology and the Environment, 1, 517–524.

Costanza, R., Andrade, F., Antunes, P., van den Belt, M., Boersma, D., Boesch, D.F., Catarino, F., Hanna, S., Limburg, K., Low, B., Molitor, M., Pereira, J.G., Rayner, S., Santos, R., Wilson, J., Young, M. (1998). Principles for sustainable governance of the oceans. Science, 281, 198–199.

Crawford, B.R., Siahainenia, A., Rotinsulu, C., Sukmara, A. (2004). Compliance and enforcement community-based coastal resource management regulations in North Sulawesi, Indonesia. Coastal Management, 32, 39–50.

Cudney-Bueno, R., & Basurto, X. (2009). Lack of cross-scale linkages reduces robustness of community-based fisheries management. PLoS ONE, 4(7), e6253. doi:10.1371/journal.pone.0006253.

David, H.A. (1988). The method of paired comparisons. London: Charles Griffin & Company.

Elster, J. (1990). Selfishness and altruism. In J.J. Mansbridge (Ed.), Beyond self-interest (pp. 44–52). Chicago: The University of Chicago Press.

FAO. (2001). International plan of action to prevent, deter and eliminate illegal, unreported and unregulated fishing. Rome: FAO.

Ferraro, P.J. (2001). Global habitat protection: Limitations of development interventions and a role for conservation performance payments. Conservation Biology, 15, 990–1000.

Ferraro, P.J., & Simpson, R.D. (2002). The cost-effectiveness of conservation payments. Land Economics, 78, 339–353.

Flewwelling, P., Cullinan, C., Balton, D., Sautter, R.P., Reynolds, J.E. (2002). Recent trends in monitoring, control and surveillance systems for capture fisheries. FAO Fisheries Technical Paper, No. 415. Rome: FAO.

Furlong, W.J. (1991). The deterrent effect of regulatory enforcement in the fishery. Land Economics, 67, 116–129.

GoM (Government of Malawi). (2001). National Fisheries and Aquaculture Policy. Government Printer, Lilongwe, Malawi.

GoM (Government of Malawi). (2009). The Government of Malawi, Lilongwe, Malawi. Retrieved September 1, 2008, from http://www.malawi.gov.mw

Gregory, R. (1999). Identifying environmental values. In V.H. Dale & M.R. English (Eds.), Tools to aid environmental decision making (pp. 32–58). New York: Springer.

Gregory, R., Brown, T., Knetch, J. (1996). Valuing risks to the environment. In H. Kunreuther & P. Slovic (Eds.), The annals of the American Academy of Political and Social Science (Vol. 545, pp. 54–63). Thousand Oaks: Sage Publications.

Hara, M.M. (2001). Could co-management provide a solution to the problem of artisanal fisheries management on the Southeast Arm of Lake Malawi? Dissertation, PhD thesis, University of Western Cape.

Hara, M.M. (2006a). Restoring the chambo in Southern Malawi: Learning from the past or re-inventing the wheel? Aquatic Ecosystem Health and Management, 9, 419–432.

Hara, M.M. (2006b). Nesting participatory fisheries management within district decentralisation: Case of Mangochi district, Malawi. (Paper presented at the Eleventh Conference of the International Association for the Study of Common Property, Bali, Indonesia).

Hatcher, A. Jaffry, S., Thébaud, O., Bennett, E. (2000). Normative and social influences affecting compliance with fishery regulations. Land Economics, 76, 448–461.

Hauck, M., & Kroese, M. (2006). Fisheries compliance in South Africa: a decade of challenges and reform 1994–2004. Marine Policy, 30, 74–83.

Hønneland, G. (1999). A model of compliance in fisheries: Theoretical foundations and practical application. Ocean & Coastal Management, 42, 699–716.

Jentoft, S. (2007). Limits of governability: institutional implications for fisheries and coastal governance. Marine Policy, 31, 360–370.

Kellert, S.R., (1993). The biological basis for human values of nature. In S.R. Kellert & E.O. Wilson (Eds.), Biophilia hypothesis (pp. 42–69). Washington, DC: Island Press.

Kooiman, J. (2008). Exploring the concept of governability. Journal of Comparative Policy Analysis, 10, 171–190.

Kooiman, J., & Jentoft, S. (2005). Hard choices and values. In J. Kooiman, M. Bavinck, S. Jentoft, R. Pullin (Eds.), Fish for life: Interactive governance for fisheries (pp. 285–299). Amsterdam: Amsterdam University Press.

Kooiman, J., & Jentoft, S. (2009). Meta-governance: Values, norms and principles, and the making of hard choices. Public Administration, 87, 818–836.

Kooiman, J., Bavinck, M., Jentoft, S., Pullin, R. (Eds.). (2005). Fish for life: Interactive governance for fisheries. Amsterdam: Amsterdam University Press.

Kuperan, K., & Sutinen, J.G. (1998). Blue water crime: Deterrence, legitimacy, and compliance in fisheries. Law & Society Review, 32, 309–338.

Manríquez, P.H., & Castilla, J.C. (2001). Significance of marine protected areas in central Chile and seeding grounds for the gastropod Concholepas concholepas. Marine Ecology Progress Series, 215, 201–211.

Murawski, S.A., Brown, R., Lai, H.-L., Rago, P.J., & Hendrickson, L. (2000). Large-scale closed areas as a fishery-management tool in temperate marine systems: The Georges Bank experience. Bulletin of Marine Science, 66, 775–798.

Ngochera, M.J.R. (2001). Status of small scale fishery in Malawi. In O.L.F. Weyl & M.V. Weyl (Eds.), Proceedings of the Lake Malawi fisheries management symposium, Lilongwe, 4–9 June 2001 (pp. 95–104). Lilongwe: Organizing Committee of the Lake Malawi Fisheries Management Symposium.

Njaya, F. (2007). Governance challenges for the implementations of fisheries co-management: Experiences from Malawi. International Journal of the Commons, 1, 137–153.

Njaya, F. (2008). Participatory Fisheries Management revisited. Samudra, Triannual Report of International Collective in Support of Fishworkers, 49, 29–34.

Ostrom, E. (1990). Governing the commons: The evolution of institutions for collective action. Cambridge: Cambridge University Press.

Peterson, G.L., & Brown, T.C. (1998). Economic valuation by the method of paired comparison, with emphasis on tests of the transitivity axiom. Land Economics, 74, 240–261.

Quah, E., Choa, E., Tan, K.C. (2006). Use of damage schedules in environmental valuation: The case of urban Singapore. Applied Economics, 38, 1501–1512.

Rolston, H., III. (1994). Conserving natural value. New York: Columbia University Press.

Rudd, M.A. (2001). The non-extractive economic value of spiny lobster, Panulirus argus, in the Turks and Caicos Islands. Environmental Conservation, 28, 226–234.

Rutherford, M.B., Knetsch, J.L., Brown, T.C. (1998). Assessing environmental losses: Judgments of importance and damage schedules. Harvard Environmental Law Review, 22, 51–101.

Satterfield, T., & Kalof, L. (2005). Environmental values: an introduction – Relativistic and axiomatic traditions in the study of environmental values. In L. Kalof & T. Satterfield (Eds.), The Earthscan reader in environmental values (pp. xxi–xxxiii). London: Earthscan.

Simpson, R.D., & Sedjo, R.A. (1996). Paying for the conservation of endangered ecosystems: a comparison of direct and indirect approaches. Environment and Development Economics, 1, 241–257.

Smith, L.W. (1998). Use of traditional practices and knowledge in monitoring a Lake Malawi artisanal fishery. North American Journal of Fisheries Management, 18, 982–988.

Song, A.M. (2009). Dealing with uncertainty in governance outcomes: Illegal fishing and conservation in the Southeast Arm fishery of Lake Malawi (pp. 248). Masters dissertation. St. John's: Department of Geography, Memorial University of Newfoundland.

Sumaila, U.R., Alder, J., & Keith, H. (2006). Global scope and economics of illegal fishing. Marine Policy, 30, 696–703.

Sutinen, J.G., & Gauvin, J.R. (1989). An economic study of regulatory enforcement and compliance in the commercial inshore lobster fishery of Massachusetts. In P.A. Neher, R. Arnason, & N. Mollet (Eds.), Rights based fishing (pp. 415–428). Dordrecht: Kluwer.

Sutinen, J.G., & Kuperan, K. (1999). A socio-economic theory of regulatory compliance. International Journal of Social Economics, 26, 174–193.

Sutinen, J.G., Rieser, A., Gauvin, J.R. (1990). Measuring and explaining non-compliance in federally managed fisheries. Ocean Development and International Law, 21, 335–72.

Thurstone, L.L. (1927). The method of paired comparisons for social values. The Journal of Abnormal & Social Psychology, 21, 384–400.

Tyler, T. (1990). Why people obey the law. New Haven: Yale University Press.

Watling, L., & Norse, E.A. (1998). Disturbance of the seabed by mobile fishing gear: A comparison to forest clearcutting. Conservation Biology, 12, 1180–1197.

Wattage, P., & Mardle, S. (2005). Stakeholder preferences towards conservation versus development for a wetland in Sri Lanka. Journal of Environmental Management, 77, 122–132.

World Bank. (2000). Project proposal: Guatemala Western Altiplano natural resources management project. Project ID GTPE64883. Latin America and Caribbean Region.

Chapter 16
A Network Perspective on Governing Interactions

Robin Mahon and Patrick McConney

Abstract The interactive governance approach views the fisheries system or 'fish chain' as consisting of a system-to-be-governed, a governing system and governing interactions between them. This system can also be viewed in a complementary way as a network of nodes or actors with links among them. Some nodes will belong to the system-to-be-governed, some to the governing system and some belong to both system-to-be-governed and governing system. The linkage between any pair of nodes can take any or several of four forms: goods, services, payment or a governing interaction. A network perspective on the interactive governance approach offers the opportunity to analyze the fisheries system using network analysis tools. These analyses can reveal system characteristics such as where links are weak or lacking, or how power and centres of organization are distributed within the system. These findings can guide interventions to create or strengthen governing interactions. This information can be used to determine where interventions can be designed to improve governability of the system, for example by improving network connectivity to increase capacity for learning, or creating new pathways to distribute power more equitably.

Keywords Social network analysis • Governing interactions • Resilience • Social ecological systems • Identification • Learning systems

---

R. Mahon (✉) • P. McConney
Centre for Resource Management and Environmental Studies (CERMES), University of the West Indies, St. Michael, Barbados
e-mail: rmahon@caribsurf.com; patrick.mcconney@cavehill.uwi.edu

M. Bavinck et al. (eds.), *Governability of Fisheries and Aquaculture: Theory and Applications*, MARE Publication Series 7, DOI 10.1007/978-94-007-6107-0_16,

Introduction

The Interactive Governance Approach to Fisheries

The interactive governance approach emphasizes the importance of considering the entire fish chain from the resource in the water to the consumer and all of the interconnections (Kooiman et al. 2005). Its definition of governance emphasizes interactions, "*Governance is the whole of public as well as private interactions...*" (Kooiman et al. 2005). It views the fisheries system or 'fish chain' as consisting of a system-to-be-governed, a governing system and governing interactions that connect them, including feedback (see Chap. 2, Fig. 2.1). The reality of this perspective is rather complex. The system-to-be-governed and governing system overlap, as many members are common to both groups and switch roles. They are both givers and receivers of governing interactions, as well as being involved in non-governing transactions that consist of goods, services and various forms of payment with pervasive positive and negative feedback.

Dealing with this complexity of interactions in a way that is useful for assessing and ultimately influencing governance is a critical question that must be addressed in the operationalization of the interactive governance approach if it is to become a basis for useful interventions to improve governance. In order to pursue this aim, the concept of governability has been introduced in concert with the interactive governance approach (Kooiman et al. 2008; Mahon 2008). This concept aims to encourage the assessment of fishery systems from the interactive governance approach perspective by considering the full spectrum of interactive governance approach components, i.e. taking a holistic approach to the extent to which the governing system and the governing interactions have the capacity to provide governance given their characteristics.

In this paper we suggest that one way to operationalize the interactive governance approach may be to view the entire system as a network. Operationalizing the interactive governance approach in practical terms and techniques that are easily accessible to interested parties strengthens its utility as a conceptual framework. Accessibility encourages the development of a body of knowledge needed to test and learn from the application of the interactive governance approach. In the following sections, networks and analysis are introduced before discussing their application to the interactive governance approach. We especially invite resource managers and researchers to test this perspective.

Networks and Network Analysis

Networks are entities linked by relationships that tie them together into a web-like structure. Networks are found everywhere in nature and human society, and they are conceptualized as analytical tools in many disciplines and professions, ranging from mathematics to public health (Barabasi 2002; Cross and Parker 2004). In social systems they consist of nodes, which may be individuals, groups of individuals or organizations, and the ties between them. The ties may be unidirectional or bidirectional. They can be described in many ways, depending on what aspect of the interaction is of interest. For example, they may consist of information, advice, cooperation in activities and exercise of authority. The use of networks in the analysis of policy-making and governance is very much an emerging field in which there is a need to develop both theory and a body of case studies that will allow concepts and theories to be tested (Raab and Kenis 2007).

Social network analysis can be used to examine the extent and nature of the ties among network nodes. Social network analysis seeks to describe, understand and explain the social relationships among nodes by examining the interactions within the network (Brandes and Wagner 2004). It can be used to assess communication, cooperation and other inter-organizational environments of the network (Jorgensen 2004). Until recently, however, it has not been prominent in discussions of adaptive governance and resilience (Carlsson and Sandström 2008; Bodin and Crona 2009). Although governability from a network perspective involves understanding the nature of both the nodes and the ties, emphasis is placed on the latter as these represent the interactions in the interactive governance approach. Conventional analyses of node attributes alone, which do not include the relationships among nodes, run the risk of missing much that can inform our emerging understanding of the nature of governability. We can draw conceptual or mental maps of the relationships that we think exist, or ought to exist, among entities engaged in governance. This use of 'network' as more of a metaphor is often the initial stage in an analysis. However, the complexity of interactions and structural arrangements, even in a network with a just few handfuls of nodes, is such that quantitative analysis is required to reveal patterns and uncover the hidden properties of the network that qualitative analysis cannot fully describe.

Various software packages are available for social network analysis, with UCINET being one of the most widely used (Borgatti et al. 2002). Network diagrams or sociograms used to visualize (graphically represent) the network are created with NetDraw, a component of UCINET 6.0. Many important network metrics, particularly of large and complex structures, require computational power to be derived and analyzed statistically. Computer applications greatly assist visualization by using a suite of algorithms to provide options for how nodes and ties among nodes can be made to appear in sociograms. For very small networks, the analysis and graphics can be done manually. For large ones, the use of software is essential to convert the data tables into diagrams that are meaningful. The size, shape and color of the symbols used to depict nodes can represent various attributes. Relationship type, strength and direction can be shown by line color, thickness and the use of arrowheads. For very large networks, graphics may not be useful and analysis must be based on metrics alone.

Social network analysis uses a variety of network metrics to describe network structure. For purposes of illustration, we focus on two network measures that relate to the nature of and limitations to interaction in the network: density and centrality.

Density is the proportion of potential ties that actually occur among the nodes in the network, and measures the richness of connections among those nodes (Mohrman et al. 2006). In other words, the more actors that have ties to one another, the denser is the network and the more interactions there are among the actors in the network (Scott 1991). The implications of high or low network density are often situation specific. For example, a dense network in one fishery may result in shared conservation-oriented norms and values, whereas in another fishery high density may serve to consolidate a group of illegal and irresponsible fishers. Centrality (which has several sub-types: degree, closeness and betweenness) measures the extent to which a node occupies a critical location between others in the network and, hence, may exercise control. This measure assists in the identification of key players (the more central nodes) within the network. Density and centrality can also be used to study the power relations and resource flows within the network that have implications for governability.

A Network Perspective on the Interactive Governance Approach

The Fish Chain as a Network

In fisheries, network models such as ECOPATH are commonly used to depict natural components of the system-to-be-governed, but they typically ignore the social components and other governing system interactions. There are many ecosystem and social system interactions in the fish chain. Some of these and their linkages to the ecological part of the chain are illustrated in Fig. 16.1. All of these need to be accounted for and modeled in the interactive governance approach.

From the perspective of the interactive governance approach, which emphasizes interactions, the entire fish chain of fisheries with its governing system, system-to-be-governed and governing interactions can also be viewed as a network of nodes or actors with links among them. Some nodes will belong to the system-to-be-governed, some to the governing system and others to both the system-to-be-governed and governing system. The interaction or linkage between any pair of nodes can take any of four forms: goods, services, payment or a governing interaction. A link between any two nodes can consist of more than one of these four linkage types. Among the nodes that comprise the system-to-be-governed, the interactions are solely goods, services or payment in money or kind. This is essentially a resource use network that corresponds to the combined 'Resource Systems and Users' in Ostrom's (2009) general framework for analyzing sustainability in Social Ecological Systems. It is worth mentioning that the network approach has been noted as having potential for operationalizing that framework as well.

When the fish chain (Fig. 16.2a) is 'unpacked' into a network, it will probably resemble the image shown in Fig. 16.2b. This network is intended to illustrate a relatively simple system and is based on categories of individuals. There will be

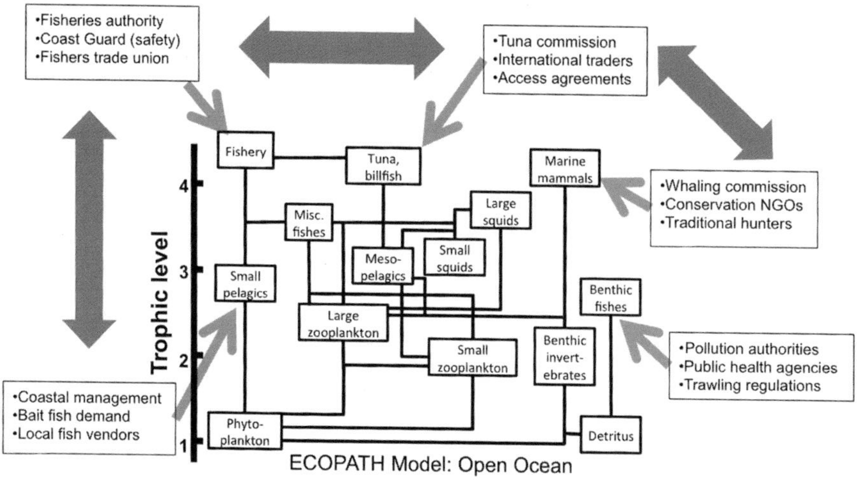

Fig. 16.1 Interactions between parts of a classical food web as modeled by ECOPATH and some examples of the human parts of the fish chain and governing system

a flow of fish along various pathways that take the resource from the sea to the consumer. Each category of actor is comprised of several types.

The network diagram in Fig. 16.2b also shows a number of government departments and an NGO that, while part of the network, are not in the marketing/distribution fish chain. The leftmost government department (20) is a fisheries department working with fishers and on resource and habitat problems. The central one (21) is a market oriented or local trade department working with middlemen and retailers. At the bottom is a foreign trade department (22) working with an exporter. In the upper right hand corner is a tourism department (23) concerned with retailers, including restaurants and consumers. The NGO shown (24) interacts with fishers and the resource. These nodes that are not directly involved in the distributive fish chain may wield considerable power over how the chain and its institutions function, creating enabling or constraining environments for various actors. For example, government departments exercise control through legislation, regulations and the allocation of funds. NGOs exercise control directly through allocation of funds or by influencing public opinion.

It is important to recognize that although Fig. 16.2b shows groups of actors, these groups are comprised of individuals. Each group will therefore have a subnetwork within the overall network (Fig. 16.3). There may be many links among the individuals within each category, as well as links between individuals in different categories. In taking a network approach to the fish chain, it is necessary to decide on the level of unpacking required in order to address a particular governability issue. For example, when fishers are to be fully engaged in governance processes through representatives, it is useful to understand the network structure at the individual level to determine whether the representative is legitimate and truly capable of representing the group.

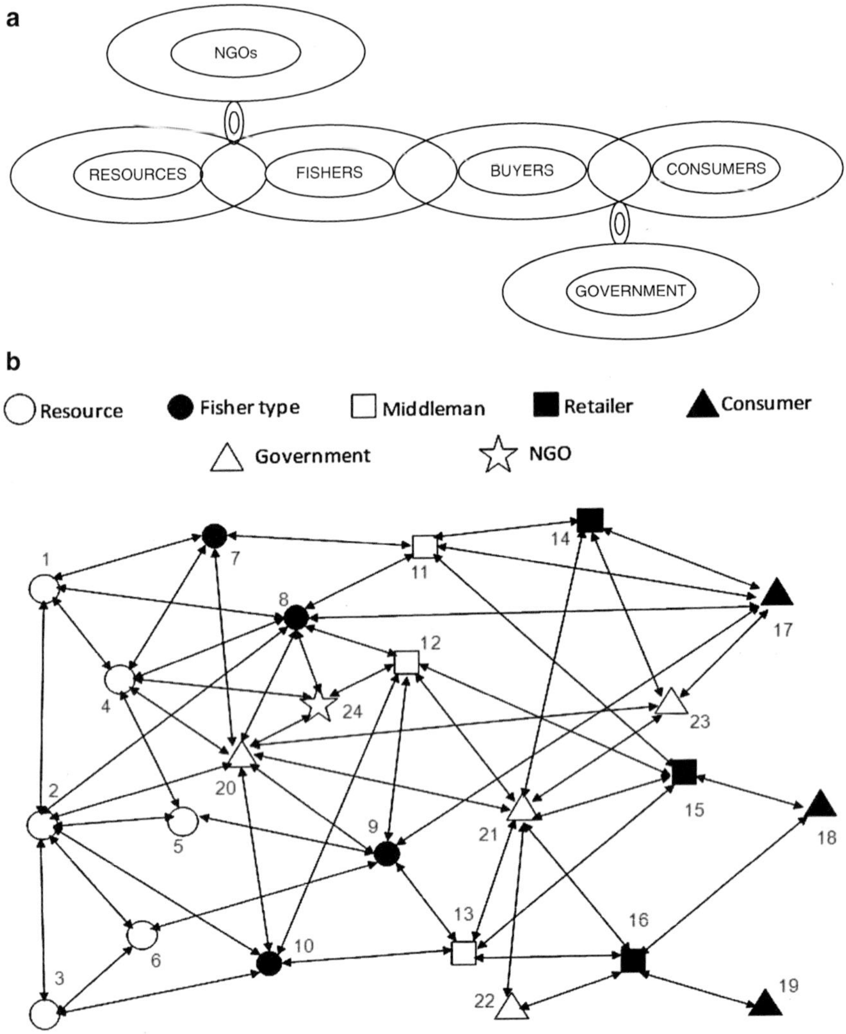

Fig. 16.2 (**a**) The main categories of actors in the fish chain and two external governing actors (**b**) An illustrative fisheries network depicting possible interactions among multiple sub-categories of the actors in the fish chain (proceeding from *left* to *right*) as well as two sub-categories of the governing actors inserted into the chain. There are six types of resource (1–6), four types of fisher (7–10), three types each of middleman (11–13), retailer (14–16) and consumer (17–19), four government departments (20–23) and an NGO (24)

All the interactions among the nodes that comprise the governing system and between these nodes and the nodes that comprise the system-to-be-governed are by definition governing interactions. These governing interactions are aimed at trying to control the ways things are done by actors in order to maximize flows within the resource use network in favor of some actor or actor group (not necessarily oneself). Separating this network into a system-to-be-governed and governing system is

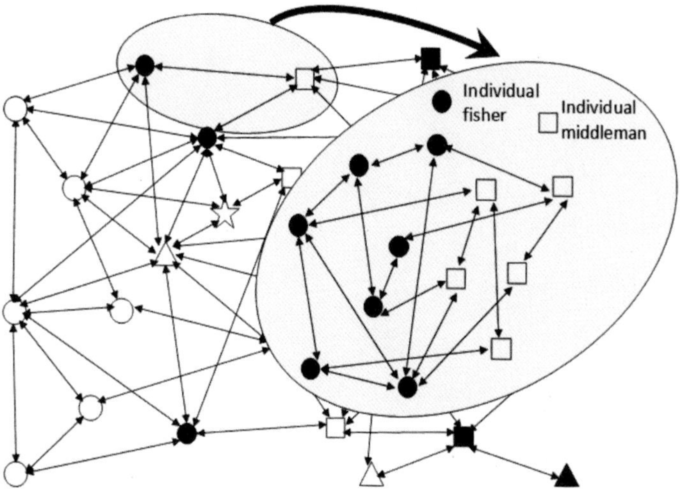

Fig. 16.3 Each category of actor in the network shown in Fig. 16.1 comprises individuals and has its own internal and intergroup network structure. The *larger ellipse* is an expansion of the *smaller* one, which contains a fisher type and a middleman type. The expansion shows that each category type consists of several individuals; in this case seven fishers and five middlemen comprise the sub-network (Symbols as in Fig. 16.2b)

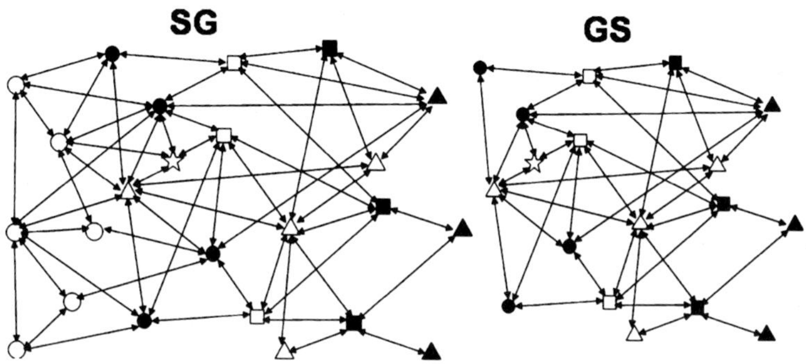

Fig. 16.4 An attempt to subdivide the network into system-to-be-governed (*SG*) and governing system (*GS*)

challenging if one adheres strictly to the definition of governance provided by Kooiman et al. (2005). By definition, any governing interactions will be from the governing system to the system-to-be-governed, thus members will at times be in the system-to-be-governed and at times in the governing system. Figure 16.4 is an example of how the network might be divided into the system-to-be-governed and governing system.

As can be seen by comparing Fig. 16.4 with Fig. 16.2a, the system-to-be-governed is essentially the entire network, because any node in the network can and usually does receive governing interactions. Some might argue that the government

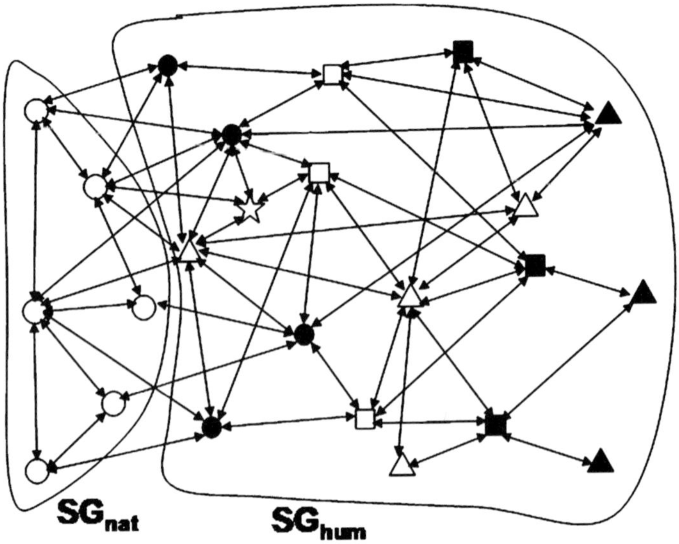

Fig. 16.5 Subdivision of the system-to-be-governed into its natural and human components. Note that SG_{hum} and the governing system comprise the same actors depending on the direction of the governing interactions

departments and the NGO do not belong in the system-to-be-governed, but in reality, these entities can and should be the recipients of governing interactions from each other and from the other social components of the system. Otherwise, how are they to learn and adapt to changing circumstances in the system? Learning and adaptation are critical aspects of governability, and social network analysis assists our understanding of the nature of these processes.

The system-to-be-governed can be seen as a socio-ecological system comprised of two intricately linked parts: the social/human component (SG_{hum}) and the ecological/natural part (SG_{nat}) (Fig. 16.5). SG_{hum} is contained entirely within the governing system, because it is difficult to conceive of a governance node in SG_{hum} that is totally neutral. SG_{nat} is outside the governing system, because although there may be interactions that affect the SG_{nat}, these are usually not intentional and are therefore not governing interactions. A special type of intentional interaction can take place directly between the governing system and SG_{nat} in the form of habitat enhancement or restoration, stocking and the like. These can therefore be considered to be governing interactions. They are not the key matters that the interactive governance approach seeks to address, but cannot be ignored. Feedback can occur from the SG_{nat} to SG_{hum}. However, while these may play a significant role in determining the dynamics of the system-to-be-governed and the way in which the system is governed, they cannot be perceived as governing interactions. They are system constraints or conditions with which the governing system must deal.

Why Take a Network Perspective?

From a reductionist viewpoint, taking a network perspective facilitates addressing a variety of questions that are important for assessing governability. The first of these is: "Who are the nodes?" This requires a process of stakeholder identification that is fundamental to any governance initiative (Mahon et al. 2005). Asking, "How do these interactions relate to actors' characteristics?" requires that the characteristics of these stakeholders be assessed. Most importantly, a network approach within a governability perspective facilitates the development of a detailed picture of the types of interactions among the stakeholders or groups of stakeholders. It can provide a picture that identifies which ties represent the transfer of goods and services; which ones are payments, whether in money or in kind, and which ones are governing. In this way, it can help reveal where the power in the governing system lies and how it is used.

The ultimate aim is to assess and design interventions that will improve governance. The interactive governance approach to fisheries (Bavinck et al. 2005; Mahon et al. 2005) has suggested fisheries governance can be improved through:

- Promoting inclusivity;
- Shared principles and values;
- Enhancing learning systems.

The network approach allows for the formal assessment of points at which interventions might be useful and most effective, because, for example:

- If stakeholders are weakly engaged, inclusivity is difficult to achieve;
- If interaction pathways are absent or controlled, sharing will be constrained;
- If information flow/feedback is weak, there will be a low probability of learning.

A simple, small-scale pelagic fishery on the west coast of Grenada is illustrated in Fig. 16.6. It shows the difference between the resource use network and the governing network. It also shows where governing interactions are expected to occur, but are currently lacking. Depending on the interactions that are considered, these gaps may, in network terms, be missing links or 'holes'. In Fig. 16.6, for example, although you would expect them to be linked in order to complete the fish marketing chain, retailers are outside the network of governance exchanges. Perhaps they are excluded because they operate mainly 'under the radar' of regulation in the informal economy. These insights often provide valuable information about the specific target region and the key nodes to be explored in further research or addressed by interventions. It may not be necessary or desirable to repair these gaps, but it helps to understand why they exist. This in part because constraints on connectivity may be due, among other things, to the exercise of power elsewhere in the network. For example, indebtedness to a specific buyer may prevent fishers from transacting business with other buyers, thus distorting the local market but also potentially creating the economies of scale and stability of supply needed for export marketing.

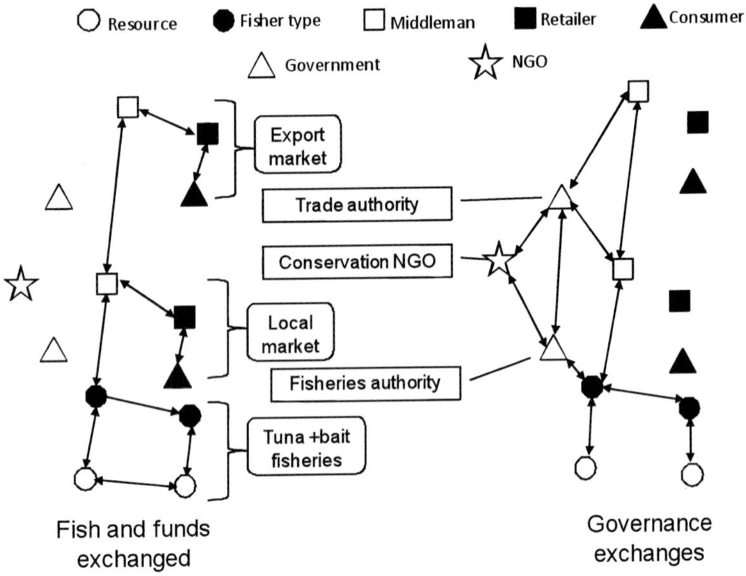

Fig. 16.6 Goods and services versus Governing Interactions in a small-scale export-oriented tuna fishery

Another example of how network structure can be interpreted to guide intervention is provided by an social network analysis of the cooperation interactions among organizations (NGOs, key government departments and schools) with an interest in sustainable development in the Grenadine Islands of St. Vincent and the Grenadines and Grenada (Mahon et al. 2010). The Grenadine islands lie on the Grenada Bank that extends some 120 km between the main islands of Grenada and St. Vincent. There are over 30 islands, of which 9 have permanent settlements with a total population of about 18,000. The largest islands have towns and communities with public (schools, clinics, utilities) and private supporting infrastructure. Others are resort islands – Palm I., Petit St. Vincent. The majority of the remaining islands are visited by yachters and fishers.

Marine-based activities are the mainstay of the region's economy. Tourism is a major source of employment and tourism development is proceeding apace. Private sector activities include: resorts, hotels, guest houses, restaurants, SCUBA diving operators, cruise ships, day and longer-term cruise operators, crafts and shops. There are also under-utilized land-based opportunities for earnings through cultural and heritage developments that would diversify the tourism sector. Fishing is the other major source of employment in the area and serves as a key source of exports to neighboring islands. Sustainable use of living marine resources is essential for the well being of the islanders. Due to their small size and disconnectedness from the main islands, sustainability remains a significant challenge.

The Sustainable Grenadines Project targets the promotion of sustainable use of living marine resources by building the capacity of civil society organizations, and the linkages between them and governmental development partners. The network

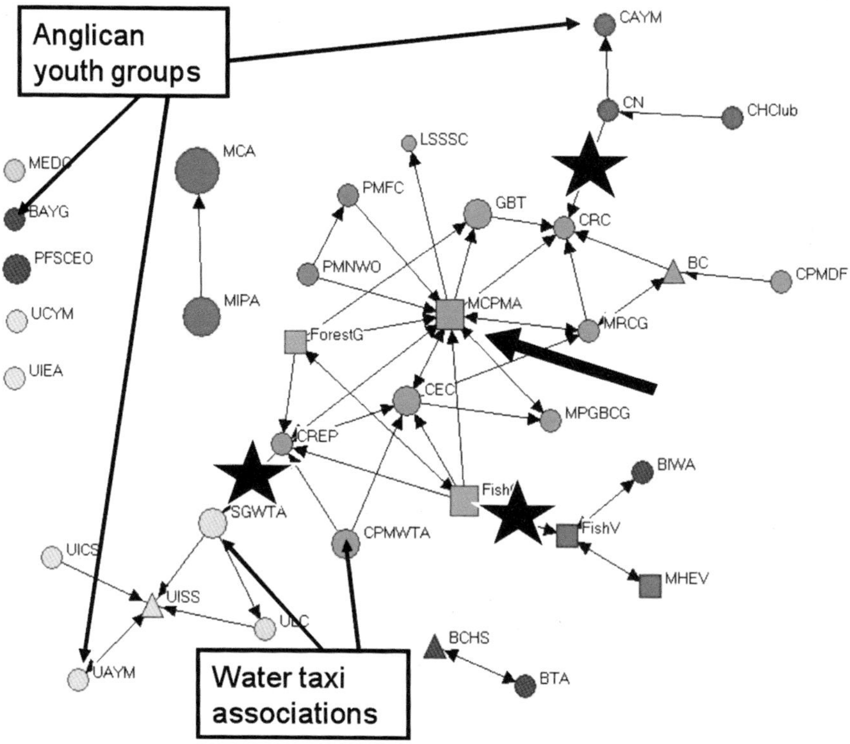

Fig. 16.7 Social network map of communication among small NGOs (*circles*), schools (*triangles*) and government departments (*squares*) in the Grenadine Islands (Mahon et al. 2010), showing areas where interventions can be made to improve the network interactions

shown in (Fig. 16.7) reflects the baseline situation that was found in the Grenadines at the outset of the project. It provides indications of where interventions might be useful to build a stronger network. For example, several NGOs are not connected to the larger network. Anglican Church groups that would likely have common interests are widely separated. Fostering communication among these church groups could strengthen them all. Likewise water taxi associations on adjacent islands have little interaction. Subsequent interventions have increased their cooperation in addressing common problems and accessing funds.

In Fig. 16.7, stars indicate where subgroups are only weakly connected to the network, with their interactions depending on only one link. Should either organization in those dyads fail, the subgroup would become disconnected. Targeted efforts can be made to bring these subgroups closer to the main network. Key actors can be identified as those having high centrality. In this case MCPMA, indicated by the large arrow, is a central government department that links many NGOs. Some might argue that this is not a healthy structure for an empowered civil society and that an emphasis on networking NGOs directly could increase their empowerment for independent action.

From a holistic perspective, taking a network approach that involves network metrics such as density and centrality facilitates addressing questions such as:

- Do some network characteristics promote good governance through
  - Sharing of principles? Such as the communication of values in dense networks;
  - Responsive institutions? For example, within modular networks that self-organize;
  - Effective actions? i.e. where a central powerful actor can mobilize resources.
- How do metrics such as density and different types of centrality relate to governance performance in key areas such as, responsiveness, adaptive capacity, resilience, transparency, accountability and tradeoffs among these characteristics? The evolving realm of dynamic network models has much potential to offer insight into changing relationships.
- Can action arenas (Ostrom 2005), within which targeted interventions will most likely be successful, be identified by examining networks? Shown as network hubs, cliques, etc.
- Do we have any expectations about network structure? E.g. normative versus actual linkages.
- Are there network characteristics that make the networks that possess them more adaptive or resilient? For example, the levels of redundancy required to achieve resilience but without rigidity.

Answering these questions will require a combination of conceptual development and analysis of empirical studies. The former, while progressing on several fronts, is mainly developed in connection with ideas of resilience (e.g. Carlsson and Sandström 2008). The latter are increasing in number to the extent that reviews and syntheses such as the one by Bodin and Crona (2009) can be attempted. Their review indicates that network structure can be related to several important features of natural resource governance and that the network approach to assessing governability may have considerable potential. They conclude that significant differences in governance processes and outcomes can be expected among networks experiencing structural differences in terms of the density of relations, degree of cohesiveness, subgroup interconnectivity and degree of network centralization. They observe a need to develop a body of empirical work on governance networks as a basis for further analyses of the relationship between network structure and governance.

Bodin and Prell (2011) build on this by emphasizing that network analysis is important for adding relational dimensions to existing social theories, such as those that underpin many aspects of governability, and for developing entirely new interdisciplinary conceptual frameworks and theories. One of the most prominent examples of a social theory to which network analysis could contribute is the dynamics of power. They also identify social learning, social movements and social arrangements such as adaptive co-management as being very amenable to elaboration from network perspectives.

We conclude that a network perspective of the interactive governance approach offers the opportunity to analyze fisheries and other social ecological systems using network analysis tools. These analyses can reveal a tremendous amount about system characteristics, identifying missing or weak links or showing how power is distributed within the system. These findings can guide interventions to create or strengthen governing interactions.

Acknowledgements This contribution has been developed through: (1) The MarGov Project at CERMES, Marine Resource Governance in the Eastern Caribbean, supported by the International Development Research Centre, Ottawa, Canada; and (2) The PROGOVNET Project at Dalhousie University, Halifax, Canada and CERMES, 'Strengthening Principled Ocean Governance Networks – Transferring Lessons from the Caribbean to the Wider Ocean Governance Community' Supported by The Nippon Foundation of Japan.

References

Barabasi, A.L. (2002). Linked. New York: Plume.

Bavinck, M., Chuenpagdee, R., Diallo, M., van der Heijden, P., Kooiman, J., Mahon, R., Williams, S. (2005). Interactive governance for fisheries: A guide to better practice. Delft: Eburon Academic Publishers.

Bodin, Ö., & Crona, B.I. (2009). The role of social networks in natural resource governance: What relational patterns make a difference? Global Environmental Change, 19, 366–374.

Bodin, Ö., & Prell, C. (Eds.). (2011). Social networks and natural resource management: Uncovering the social fabric of resource governance. Cambridge: Cambridge University Press.

Borgatti, S.P., Everett, M.G., Freeman, L.C. (2002). UCINET for Windows: Software for social network analysis. Harvard: Analytic Technologies.

Brandes, U., & Wagner, D. (2004). Visone - Analysis and visualization of social networks. In M. Jünger, P. Mutzel, Graph drawing software (pp. 321–340). New York: Springer.

Cross R., & Parker, A. (2004). The hidden power of social networks: Understanding how work really gets done in organizations. Boston: Harvard Business School Press.

Jorgensen, D.J. (2004). Networking to meet the demands of youth: An analysis of communications networks among Nueces County Youth Service Organisations. Southwestern Social Science Conference (17 pp). Corpus Christi, Texas.

Kooiman, J., Bavinck, M., Jentoft, S., Pullin, R. (Eds.). (2005). Fish for life: Interactive governance for fisheries. Amsterdam: University of Amsterdam Press.

Kooiman, J., Bavinck, M., Chuenpagdee, R., Mahon, R., Pullin, R. (2008). Interactive governance and governability: An introduction. The Journal of Transdisciplinary Environmental Studies, 7(1), 1–11.

Mahon, R. (2008). Assessing governability of fisheries using the interactive governance approach: Preliminary examples from the Caribbean. The Journal of Transdisciplinary Environmental Studies, 7(1), 1-12.

Mahon, R., Bavinck, M., Roy, R. (2005). Chapter 17: Fisheries governance in action. In J. Kooiman, M. Bavinck, S. Jentoft, R. Pullin (Eds.), Fish for life: Interactive governance for fisheries (pp. 353–378). Amsterdam: University of Amsterdam Press.

Mahon, R., Blackman, K., Lee, R., McConney, P. (2010). Assessing social network structure of stakeholder organisations in the Grenadine Islands. CERMES Technical Report No.31.

Mohrman, S.A., Galbraith, J.R., Monge, P. (2006). Network attributes impacting the generation and flow of knowledge within and from the basic science community. In J. Hage, & M. Meeus (Eds.), Innovation, science and industrial change: The handbook of research (pp. 196–216). London: Oxford Press.

Ostrom, E. (2005). Understanding institutional diversity. Bloomington: Princeton University Press.

Ostrom, E. (2009). A general framework for analyzing sustainability in social ecological systems. Science, 325, 419–422.

Raab, J., & Kenis, P. (2007). Taking stock of policy networks: do they matter? In F. Fischer, G.J. Miller, M.S. Sidney (Eds.), Handbook of public policy analysis: Theory, politics and methods. New York: CRC Press.

Scott, J. (1991). Social network analysis: A handbook. Newbury Park: Sage Publications.

Chapter 17
Working with Principles and Visions

Sharon Almerigi, Lucia Fanning, Robin Mahon, and Patrick McConney

Abstract The geo-political complexity and diversity within the Wider Caribbean Region call for cooperation in living marine resource governance. Most living marine resources are transboundary and require some level of international cooperation for sustainable use. Guided by the Interactive Governance Approach, this Wider Caribbean consultation on ecosystem-based management (EBM) employed facilitation methods to explore multi-level functionality, shared vision, guiding principles and priority actions. Inclusivity and participation were emphasized in the agenda, which began with presentations from participants on principled ocean governance and recent advances in marine EBM. These were followed by a visioning process facilitated in groups, with each group representing expertise in specific ecosystems and regional governance. Informal, small group conversations followed. Throughout the process, the set of governance principles identified by consensus earlier on were revisited and used as a lens through which to view subsequent outputs. A vision synthesis and map of strategic directions based on process outputs was compiled and shared with participants after the symposium. The process described in this chapter generated a rich collection of ideas and suggestions for EBM in the Wider Caribbean upon which further conversations can build. These types of processes are seen as fundamental to enhancing governability by engaging stakeholders in building and articulating a foundation of shared principles and visions for marine EBM.

Keywords Visioning • Shared vision • Shared principles • Participatory methods • Facilitation • Participatory strategic planning • Focused conversations

S. Almerigi (✉)
People Dynamics Associates, St. James, Barbados
e-mail: salmerigi@caribsurf.com

L. Fanning
Marine Affairs Program, Dalhousie University, Halifax, NS, Canada
e-mail: lucia.fanning@dal.ca

M. Bavinck et al. (eds.), *Governability of Fisheries and Aquaculture: Theory and Applications*, MARE Publication Series 7, DOI 10.1007/978-94-007-6107-0_17,

Introduction

The Interactive Governance Approach (IGA) includes "…the formulation and application of principles guiding [governing] interactions and care for institutions that enable them" (Kooiman et al. 2005, 19). As the move to more participatory methods in fisheries management organizations is relatively new within the Wider Caribbean, little attention has been paid to the installation of principles that are necessary for effective governance. When principles are mentioned, it has mainly been within the texts of multilateral agreements. The IGA also recognizes that effective governance requires functionality at multiple levels: working with images, developing instruments and taking action, which it refers to as the 'elements of governance'. Here again, the fisheries management literature provides little guidance on working with images and how these are converted into instruments that guide action. One form of imagery that can play an important role in fisheries governance is the vision that stakeholders have for a particular fishery. Mahon et al. (2005) emphasize three ways forward in pursuing interactive governance: (1) the development of principles and visions to guide governance; (2) the inclusivity and sharing of the responsibility for governance; and (3) the promotion of capacity to learn and adapt. The extent to which stakeholders in both the system-to-be-governed and the governing system are working from a common set of principles and have a shared vision is considered to be a critical factor in the governability of the system (Jentoft et al. 2010).

In this chapter we describe a facilitated process in which a diverse group of stakeholders from fishery governing systems and systems-to-be-governed in the Wider Caribbean Region came together to: (1) explore the extent to which they are working from a common set of principles; (2) to develop a shared vision for marine EBM (EBM) in the Wider Caribbean Region; and (3) to identify key strategic actions necessary to set the stage for implementing EBM. In this process, which took the form of a 3-day symposium, December 10–12, 2008, attention was paid to inclusivity and participation to ensure that the strategic directions developed would include the widest possible range of perspectives. The objective of the process was to produce outputs that other stakeholders around the region could identify with and choose to adopt.

The Wider Caribbean Region is one of the most geopolitically diverse and complex regions in the world (Chakalall et al. 2007; Fanning et al. 2009). It extends from the mouth of the Amazon River, Brazil, in the south, through the insular Caribbean, Central America, the Gulf of Mexico and north along the east coast of North America to Cape Hatteras (Fig. 17.1). The Wider Caribbean Region is a

---

R. Mahon • P. McConney
Centre for Resource Management and Environmental Studies (CERMES),
University of the West Indies, St. Michael, Barbados
e-mail: rmahon@caribsurf.com; patrick.mcconney@cavehill.uwi.edu

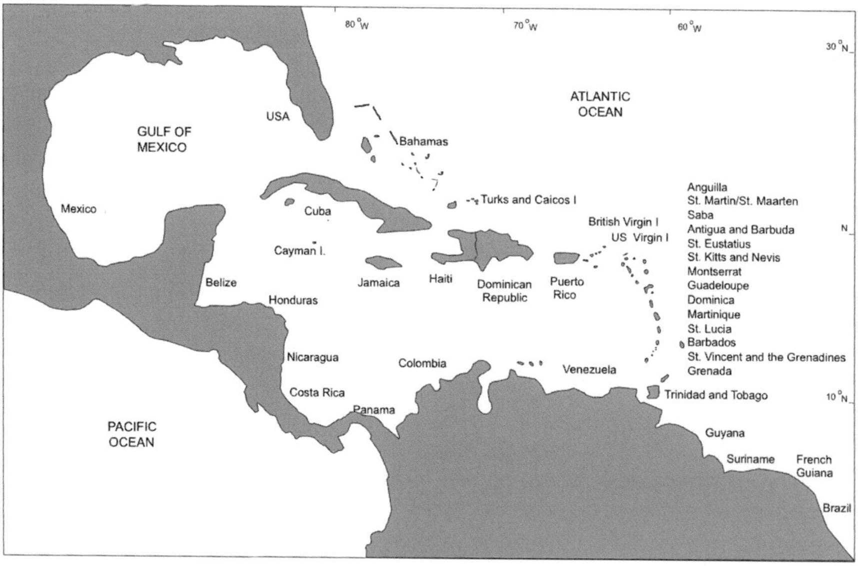

Fig. 17.1 The Wider Caribbean Region showing hypothetical Exclusive Economic Zones based on the equidistance principle (Credit: R. Mahon)

recognized geopolitical entity defined by the Cartagena Convention and is covered by the Caribbean Environmental Programme of the United Nations Environmental Programme Regional Seas Programme. It encompasses 32 countries, as well as 12 territories that are dependencies of France, United Kingdom, United States and The Netherlands. The Exclusive Economic Zones of the Wider Caribbean Region cover the entire region. Consequently, there is a high incidence of transboundary resource management issues, even at relatively small spatial scales.

The marine ecosystems of the region and the fisheries dependent on them further contribute to complexity (Mahon et al. 2011). There is great diversity in the number of mainly small-scale fisheries that operate in a variety of marine ecosystems. Most resources are shared in one way or another. This requires a great deal of cooperation in living marine resource governance. The nature of the resources is such that cooperation is required at multiple scales. Diversity also exists in the range of regional and subregional organizations that have an interest in and deal with living marine resources. This means that interactions are needed among the organizations, as well as among the countries (Fanning et al. 2007).

The diversity of the region increases the need to explore the extent to which stakeholders operate from common principles and are aligned in their vision for the governance of their living marine resources prior to embarking on the reform or establishment of institutions and actions. The Caribbean Large Marine Ecosystem Project is a multiyear Global Environment Facility-funded initiative to promote marine EBM in the Wider Caribbean Region (Fanning et al. 2009). It includes the marine ecosystems, resources and resource use systems in all of the 44 states in the

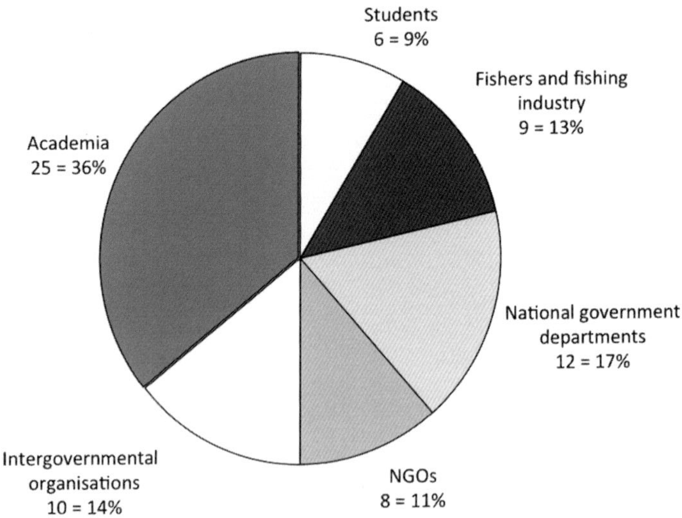

Fig. 17.2 The mix of participants in the symposium. The number and percentage of participants is shown under each label

region. It encompasses all of the diversity, dynamics and complexity described above. An attempt to create a shared vision or visions of marine EBM was deemed crucial from the outset of the project to ensure that the many stakeholders involved in the Caribbean Large Marine Ecosystem Project could use it as a common point of reference to guide their activities. It was decided to approach this by holding a symposium that brought a wide cross-section of stakeholders together to generate a vision for a shift to marine EBM that would be grounded in a set of agreed-upon principles. A secondary goal of the symposium was to promote networking among key collaborators who are in a position to steward the shift. Creating linkages among key players will help to keep the shared EBM vision alive and in sight, and will help to coordinate efforts regionally. Symposium participants came from the groupings illustrated in Fig. 17.2. They were selected based either on the role of their organization in regional living marine resource governance, their connection with the CLME initiative or their known expertise and activity in marine EBM.

Symposium Process Design and Organization

Facilitation services were engaged to plan and implement the symposium. Design meetings were held between the lead facilitator and symposium conveners to determine the most effective methods to achieve the desired outputs. What the conveners wished to achieve was the identification of a shared vision for EBM and its inherent principles and how stakeholders will work together, or individually, to achieve it.

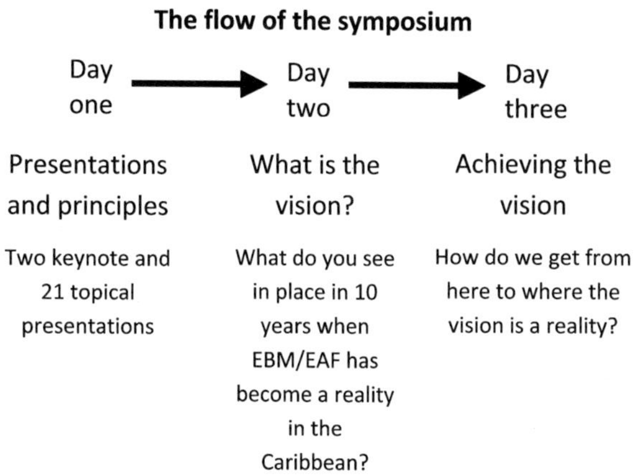

Fig. 17.3 The overall organizational flow of the symposium

The profession of facilitation, as it relates to whole systems change, includes a wide array of processes to encourage group effectiveness and participation (Holman et al. 2007). Open Space Technology, Bohmian dialogue, Appreciative Inquiry, Dynamic Facilitation, Future Search, Scenario Thinking are among the many processes that could have been considered for this symposium. However, a method adapted from a strategic planning model was selected because of its emphasis on developing a vision, values and action steps to reach the vision. A conversational method known as World Café was also chosen, because the informal process promotes networking among stakeholders that would ultimately be necessary to collectively make EBM a reality.

The visioning process of the Participatory Strategic Planning method developed by the Institute of Cultural Affairs[1] was used due to its ability to draw out individual ideas and synthesize these into a cohesive group consensus.

The process of World Café would provide small focused conversations in sequentially different configurations of participants. The intent of this approach was to put people at ease in a mock 'café situation', promote open conversation about issues important to them, and – as participants change their seats during the 'rounds' of conversation – allow new insights to emerge from listening to a variety of perspectives.

The facilitated processes discussed above accentuate the principles of inclusivity and participation, bringing out the best thinking on the issue of EBM with this particular group of regional stakeholders. The chosen methods had to be tailored to a large group and composition. The need for concurrent sessions and the use of additional facilitators was therefore included in the final design.

The overall flow of the symposium is depicted in Fig. 17.3. It was an integrated process that began with two keynote speakers and 21 topic-specific presentations, followed

[1] The Institute of Cultural Affairs in the U.S.A. (ICA-USA) is a private, non-profit, social change organization that promotes positive change in communities, organizations and individual lives. ICA Head Office: 4750 N. Sheridan Road, Chicago, Ill, 60640, http://www.ica-usa.org/

by the facilitated group work. The keynote presentations focused on principled ocean governance and recent advances in marine EBM, while the topic-specific presentations were solicited from regional experts to focus on the diversity of issues that must be considered in marine EBM. They covered various marine ecosystems, fisheries, economics, social considerations and institutional arrangements. These sessions were intended to provide the background information for the development of a shared vision, foundational principles and a set of actions for implementation.

Working with Principles

The first keynote presentation emphasized the importance of placing principles at the forefront of discussions about marine EBM. It was noted that making these explicit would ensure that all who are working in marine EBM in the Caribbean would be aware of the need for a common set of principles and share a willingness to work from them. The symposium participants selected a common set of principles through prioritization and discussion.

Using the method of dot prioritization participants used sticky dots to develop a list of key principles from a set of principles derived from various publications and multilateral agreements (Mahon et al. 2010). Dot prioritization is an established facilitation method for prioritizing ideas with a large number of people (Diceman 2010). Participants were each given ten adhesive dots to distribute among the set of principles displayed on wall posters. They were instructed to affix dots to their highest priorities – those principles they believed to be essential for successful EBM. They were not allowed to split dots or use more than one dot per item. The result of the process is shown in (Fig. 17.4).

The top ten principles identified by the dot voting exercise were adaptiveness, integration, accountability, conservation, precaution, sustainability, empowerment, equity, use of science and participation. This should not be interpreted as a suggestion that the remaining principles are not important, but instead as an indication of the relative importance of the top ten. The full set of principles gleaned from EBM agreements and publications is shown in Mahon et al. (2011). Participants once again focused on principles at the later stage of the visioning process.

Developing the Shared Vision

A vision is the articulation of a specific destination; a picture of a desired future. Visions also have an attractive function. To be effective, visions must be grounded in the current reality; a true picture of the way things are. The gap between the current reality and the desired vision is a source of energy. This ‘creative tension’ wants to resolve itself by pulling either in the direction of reality or the vision. Holding firmly to an articulated vision, while being scrupulously honest about the current

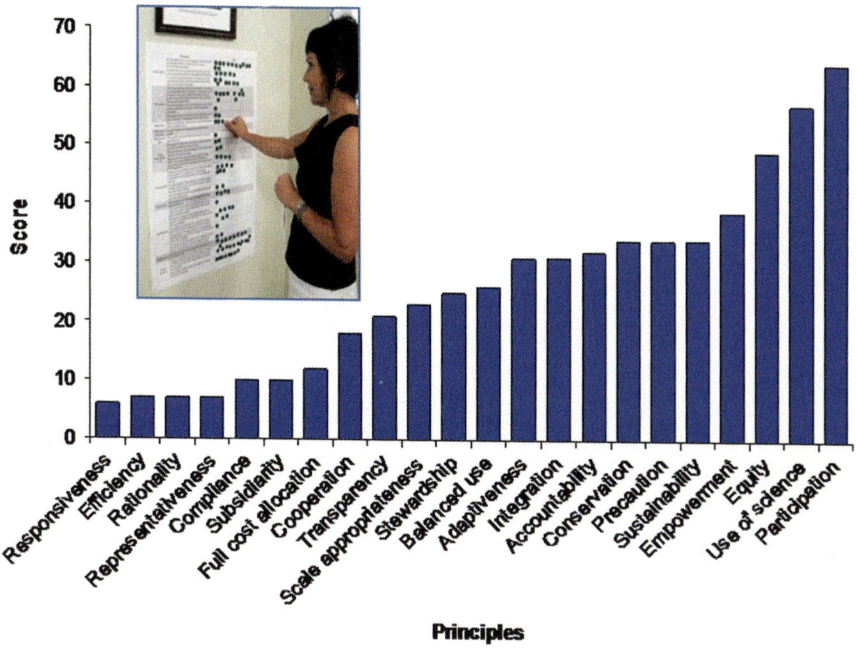

Fig. 17.4 The output from the exercise in which participants ranked principles by dot-voting (*inset*) (Credit photograph: D. Benskin)

reality, will therefore have the attractive effect of a 'rubber band' pulling toward positive change (Senge 1990; Harrison 1995).

The tension created between what is desired – the vision – and what currently exists – the current reality – becomes a creative tension that can make things happen (Wilson et al. 2003). Creative tension is present when the vision is easily seen and clearly stated and the current reality is also clear. The questions to ask are, "What would we see and hear, or what would be in place if we reached our vision?" rather than simply querying, "What do we want in the future?"

It is important that the tension between vision and current reality remains taught, steady and forward-thinking and that it is not allowed to slacken or oscillate between competing goals such as change and stability or short- and long-term goals. If there is advancement toward the vision, the tendency is to consistently move forward with each achievement serving as a foundation for further achievements. (Fisk 1999).

The ICA Participatory Strategic Planning shared vision process used in the symposium is designed to capture the emerging wisdom of the group (Spencer 1989; Stanfield 1995; Holman et al. 2007). The method presupposes that the vision is latent in the group in the form of hopes and desired outcomes for the future. These consist of various ideas, pictures and scenarios that individual group members have of what they hoped for. The process pulls these together into a synthesis that creates an appealing vision to the group.

The Consensus Workshop Method provides the structure for the ICA visioning process. This method was designed to "enable people to think together, to plan and to work together as teams" (Stanfield 2002, 5). As a step-by-step process, it begins with a focus question. In the case of visioning, the question usually regards what stakeholders would like to see 'in place' in the next 5–10 years with reference to a certain topic or arena. Participants are first asked to brainstorm ideas individually, and then repeat the activity in small groups with a focus on what will have changed after a specified time period if their personal vision has come about. In the small groups the best ideas are chosen and written on cards (one idea per card). These are then given to the facilitator who – guided by the entire group – organizes the ideas into clusters of similar ideas on a sticky wall.[2] Little-by-little, like-minded ideas are grouped on the wall until none are left.

The intent of clustering is to encourage inquiry within a group in contrast to what is often the case in groups where group members advocate and defend positions. A good advocate recommends a specific perspective, is convinced that one position is right and seeks support for that view from others. An inquirer comes to a topic with an open mind in search of a greater understanding of, or new meaning to an old established truth (Stanfield 2002). As inquiry, the Consensus Workshop invites intuition and associations to reveal connections between ideas that are not always immediately obvious. According to Stanfield (2002, 71), "It seeks a breakthrough of insight that is contained in the data but is also beyond the data." Therefore, in the 'naming step' that follows, the facilitator pushes the group to give a title to each of the clusters. In this step, the group reads the cards in the cluster and then tries to find the meaning suggested through the collection of cards. The ICA refers to the clustering and naming process as 'gestalting'. It is about achieving breakthrough insights that spring from the juxtaposing of many pieces of data. The titles given to the clusters become the groups' shared positive vision of the combined group's hopes and desired outcomes (Holman et al. 2007). The final step of the process includes a discussion in which the group reflects on the significance of the work that has been produced.

For the purpose of developing a vision for marine EBM in the Wider Caribbean, participants were assigned into four breakout groups, each representing an area of interest or responsibility. These sessions represented the following elements of the Caribbean Sea ecosystem: (1) the continental shelf; (2) offshore pelagic resources; (3) coral reef resources; and (4) regional governance. These topic areas were chosen because they are the major marine ecosystems of the Wider Caribbean Region within which EBM would need to be undertaken. Care was taken to ensure that there was a diversity of participants in each topic area. Participants were assigned to groups based on their background and experience. They were, however, offered the opportunity to change if they preferred to be in a group with a different topic. It was recommended that people find someone from the desired group to exchange with so as to not reduce the size of the groups. Each breakout session undertook a visioning exercise, led by a facilitator trained

[2] A nylon sheet sprayed with a non-permanent adhesive that allows for the repositioning of cards by a facilitator.

Fig. 17.5 Visioning breakout session with the ideas being grouped on the sticky wall (Credit photograph: D. Benskin)

in the aforementioned methodology. Figure 17.5 shows the grouping of ideas on a sticky wall in one of the breakout sessions, while Table 17.1 shows the vision completed by the continental shelf group. The vision elements were then named with title cards that reflected the consensus of the group regarding what that element was about.

After the visioning process, participants again used dot prioritization to indicate the individual ideas on the sticky wall they would like to see assigned the highest priority or that needed to be addressed first. Each person was given three red dots to select their top three priorities, and seven blue dots to indicate other areas of importance.

Next, the groups returned to the top ten principles to consider whether and how these relate to their vision. They did this by writing the numbers of the principles (shown on a flip chart sheet numbered 1–10) on each title card.

The final step in visioning was to combine the visions from the four separate groups into one overall vision. After the workshop, the organizers compiled the synthesized vision by extracting the common elements in each of the four visions. The combined vision depicting those elements that were aligned and those that were unique is shown in Table 17.2.

Advancing Towards the Vision

Assisting and Resisting Factors

The process of identifying assisting and resisting factors is adapted from the concept of force field analysis. This concept, developed by American psychologist Kurt

Table 17.1 An example of a vision session output – the vision for ecosystem based management for the continental shelf shrimp and groundfish fisheries in the Wider Caribbean

FOCUS QUESTION: What do you see in place in 10 years time when EBM/EAF has become a reality in the Caribbean?

Improved quality of life	Effective inclusive governance systems	Restored and maintained ecosystem integrity	Effective institutional networks	Value ecosystem assets	An engaged public
Secured livelihoods-happy faced Improved quality of life for stakeholders Healthy use of the ecosystem that benefits all users Balanced usage of freshwater including the coastal zone Sustainable benefits from ecosystem goods and services	Harmonized inclusive policy on EBM Harmonized governance Subsidiarity in decision-making and management Well developed legal framework Adequate enforcement measures Wider Caribbean coordinating body established	Native marine biota very close to natural numbers Ecosystem integrity being maintained Quantify habitats under extinction risk Climate change mitigation and adaptation measures	Capacity in place to deliver EAF/EBM A well managed co-coordinated ecosystem Improved trans-boundary linkages/information sharing	Ecosystem recognized and treated as natural and regional assets	Informed educated citizens Public awareness of the concept

Table 17.2 Combined vision for EBM/EAF for the Wider Caribbean Region

The elements that make up of our vision from the four groups					
Governance	Reef ecosystems	Pelagic ecosystems	Continental shelf ecosystems		Our overall vision
	Only clean water to the sea Healthy functional ecosystems	Healthy and productive social-ecological systems	Restored and maintained ecosystem integrity	⇨	**Healthy marine ecosystems in the Wider Caribbean**
Sustainable financing			Value ecosystem assets	⇨	**that are fully valued**
Comprehensive coherent consistent decision-making Coherent adequate legal frame work Efficient multi-level networks and institutions	Politicians serious about environment and with a will to manage Appropriate effective accountable governance Enforcement that works Caribbean wide marine space use management	Effective and responsible governance for ecosystem sustainability Effective, harmonized access and control Equitable practical regionalism	Effective institutional networks	⇨	**and protected through strong institutions at local national and regional levels providing effective governance**
Inclusive participation	Stakeholders fully involved Quality information accessible to all	Inclusive and participatory collaborative management	Effective Inclusive governance systems	⇨	**that involves everyone**
Individual motivation for stewardship	Environmentally educated public		An Engaged Public	⇨	**is fully understood and supported by the public**
Enhanced social capital	Secure and sustainable livelihoods	Secure, social and economic opportunities	Improved quality of life	⇨	**and enhances livelihoods and human well being**

Lewin (1976), looks at factors (forces) that are either driving movement toward a goal (helping forces) or blocking movement toward a goal (hindering forces) (Lewin 1976; Shankland 2003).

Following the completion of the shared vision exercises, the four groups discussed factors that assist movement toward the vision and those that resist it. The exercise was carried out in two parts. The first phase identified the assisting and resisting factors, while the second categorized the ideas into a SWOT analysis, a strategic planning method for identifying the internal and external factors that are favorable (strengths and opportunities) and unfavorable (weaknesses and threats) to achieve an objective.

Strategic Directions and Actions Using World Café

In the next stage of the Symposium, participants were asked to consider what actions would be necessary for marine EBM to become a reality in the Caribbean. To do this, they engaged in the World Café conversational methodology described earlier. The method is known to be useful in accessing the best thinking of groups (Creative Commons 2008). In the session, four to six people sat together at café-style tables to explore a question or issue related to EBM. Other persons seated at similar tables explored related questions. As participants talked to each other, they were encouraged to write down key ideas, or sketch them on paper tablecloths provided for that purpose. After a 20–30 min 'round of conversation' participants were invited to change tables, carrying the insights from their previous conversation to a newly formed small group. One participant, who volunteered to be 'host', remained at each table to share key ideas or questions from the previous conversation with the new arrivals. The host, while taking part in the conversation, was also responsible for recording the main ideas discussed at their table. After three rounds, the groups met as a whole to 'harvest' the most important ideas from the conversations.

The World Café is based on a set of 'integrated design principles' that are intended to foster authentic dialogue. These principles are as follows:

Setting the context – Defining the purpose for convening the Café plus the desired outcomes and range of perspectives that need to be included in the process.

Creating a hospitable space – Having a warm and friendly café setting alerts participants that this gathering is not a business-as-usual meeting. Additionally, meeting in small groups creates conversations that are quite different from those that might take place at larger tables. Every effort is made to provide natural light, flowers and refreshments to nourish good conversation.

Explore questions that matter – The questions to be considered by the group are those they care deeply about. In addition, participants are invited to explore possibilities rather than thinking about what went wrong or who is to blame.

Fig. 17.6 Small groups use the World Café process to generate key actions required to achieve the vision (Credit photograph: D. Benskin)

Encourage everyone's contributions – The process encourages all participants to contribute to the conversation. Each participant in the Café represents a part of the whole system's diversity and, as each person has the chance to contribute, more of the insights inherent in the group become accessible.

Connect diverse perspectives – As each person shares their perspective, new ideas may emerge. Tablecloths are used, plus paper and markers, to create a "shared visual space" through drawing the emerging ideas.

Listen Together and Notice Patterns – The quality of the listening is an important factor determining the success of a Café. Participants are encouraged to listen closely to each other and to try not to formulate their ideas while another is talking.

Share Collective Discoveries – Conversations held at one table reflect a pattern of wholeness that connects with the conversations at the other tables. The last phase of the Café involves making this pattern of wholeness visible to everyone. To do so, the facilitator holds a conversation with the individual tables and the whole group.

To connect the conversations to the visions for EBM in the Caribbean created earlier in the day, each breakout group was asked to discuss and put forward the necessary actions to realize each element of their vision. To do this, each table was assigned specific vision element(s) to work on. Figure 17.6 shows the small groups' discussion actions in the World Café process.

For example, the group addressing coral reefs discussed how to implement one of its vision elements related to creating 'healthy functional ecosystems', offering the following actions:

- Effectively enforced legislation;
- Adoption of a healthy reef indicator framework which is regularly made public;

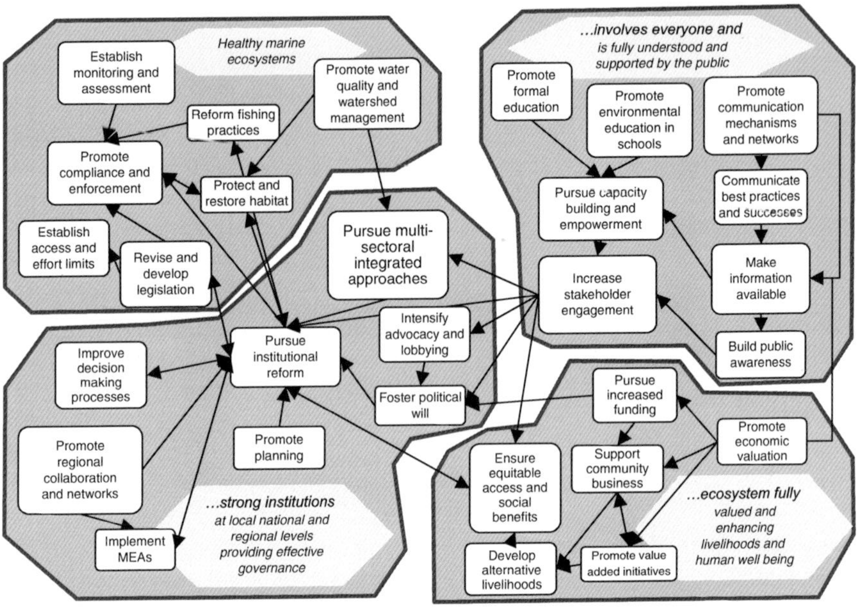

Fig. 17.7 After the session, the organizers combined the World Café outputs and the collective vision into this strategic directions map that relates to the key elements of the combined vision

- Pursue advanced modeling of multi-species, predator/prey climate and species/habitat interaction to accomplish management learning.

In another example from the group addressing governance, the group discussed their vision element related to 'inclusive participation' and suggested the following actions:

- Promote comprehensive public education programmes in the school curricula and to the general public;
- Target participatory workshops at the community level;
- Integrate coastal zone issues in school curriculum;
- Build and implement co-management relationships where appropriate.

To conclude the small group conversations, the facilitator led a whole group reflection to harvest any further insights supporting movement toward each vision element.

Following the Symposium organizers combined the action ideas from the four breakout sessions and produced a map illustrating strategic directions for moving forward with recommended priorities (Fig. 17.7).

Final Reflections

The Group Process Methodology

The suite of group process methods described in this paper generated a considerable amount of information from a large number of people in a relatively short period. A weakness of the process was that there was inadequate time for synthesis of the outputs on the final day. Therefore, the synthesis of the visions into an overall vision and the synthesis of the strategic direction map both took place after the symposium and were then shared with participants. This led some participants to conclude that the process had not reached a satisfactory 'closure point' during the symposium. This was more the case for participants who were unfamiliar with such methods and who expected the symposium to lead to conclusive recommendations for implementation.

Upon reflection, these expectations could probably have been tempered by sharing a clearer description of the process and expected outcomes before and during the process. This would have included an even greater emphasis on the symposium's role in the Caribbean Large Marine Ecosystem Project and the long-term process of developing regional ocean governance in the Wider Caribbean (Fanning et al. 2009, 2011). Had stakeholders appreciated the fact that the symposium was not an isolated event, but part of an ongoing process of enhancing ocean governance in the Wider Caribbean Region that may take a decade or longer to provide significant results, they may have been better equipped to make their input.

It is also important to be aware that the outcome of the process is dependent upon the information provided to participants in the building blocks session. While all participants are knowledgeable about marine ecosystems and their management in terms of their particular role or experience, the information provided sets the context for the process, and in some ways sets boundaries for its outcomes. This is necessary if a useful product is to be obtained within a specific time frame and budget, but must be borne in mind in interpreting the product.

Implications for Governability

Notwithstanding the process's design issues, the organizers maintain that the prioritization of principles process and the four separate group processes generated useful outputs that reflected the consensus of the participants. It also appears that the sequence of processes produced an overall result that can serve as a planning foundation for, or at least a basis for discussion of the principles and visions that might underlie an ecosystem approach to living marine resources in the Wider Caribbean Region. In broad terms, the overall process would appear to provide a means to both determine the extent of, and to work towards improving, governability at the scale of a large marine ecosystem.

Exploration and operationalization of the Interactive Governance Approach and governability will increasingly take action researchers in the direction of facilitated group processes such as this one. The Interactive Governance Approach emphasizes the Governing Interactions between the Governing System and the System-to-be-Governed. These are, for the most part, social interactions that should be carried out according to agreed procedural principles if they are to be effective (Mahon et al. 2011). The objectives of governance should also be guided by agreed fundamental principles (Mahon et al. 2011). Beyond the need for agreed principles, stakeholders must have a common vision of what governance is trying to achieve. This common vision can be expected to greatly enhance the governability of the system (Bavinck et al. 2005; Jentoft et al. 2010).

Given this need for interaction among stakeholders in the Governing System and System-to-be-Governed, methods that can facilitate the development of consensus on principles and visions are likely to be key components of Interactive Governance and initiatives to enhance governability. Group processes will be particularly important when: (1) attempting to engage multiple stakeholder types; (2) attempting to address governance at larger spatial and organizational scales, as is required for working with large marine ecosystems; and (3) working with intangible aspects of governance, such as principles and visions. We expect that working on marine governance issues in this way will be an iterative process in which participants will both contribute to governance and learn about it at the same time. Engagement of stakeholders in such processes is expected to also lead to an increase in their capacity to participate meaningfully. This should lead to an improvement in governability in the entire system. However, given the geopolitical and scale complexity of the region, such process will be necessary at several scales and organization scale levels if significant change is to be achieved.

References

Bavinck, M., Chuenpagdee, R., Diallo, M., van der Heijden, P., Kooiman, J., Mahon, R., Williams, S. (2005). Interactive governance for fisheries: A guide to better practice. Delft: Eburon Academic Publishers.

Chakalall, B., Mahon, R., McConney, P., Nurse, L., Oderson, D. (2007). Governance of fisheries and other living marine resources in the wider Caribbean. Fisheries Research, 87, 92–99.

Creative Commons (2008). Welcome to the World Café. Retrieved from http://www.theworldcafe.com. Accessed 9 Jan 2013.

Diceman, J. (2010). Dotmocracy handbook: A simple tool to help large groups find agreement. Retrieved from http://www.dotmocracy.org. Accessed 9 Jan 2013.

Fanning, L., Mahon, R., McConney, P., Angulo, J., Burrows, F., Chakalall, B., Gil, D., Haughton, M., Heileman, S., Martinez, S., Ostine, L., Oviedo, A., Parsons, S., Phillips, T., Santizo Arroya, C., Simmons, B., Toro, C. (2007). A large marine ecosystem governance framework. Marine Policy, 31, 434–443.

Fanning, L., Mahon, R., McConney, P. (2009). Focusing on living marine resource governance: The Caribbean Large Marine Ecosystem and Adjacent Areas Project. Coastal Management, 37, 219–234.

Fanning, L., Mahon, R., McConney, P. (Eds.). (2011). Towards marine ecosystem-based management in the wider Caribbean. Amsterdam: Amsterdam University Press.

Fisk, R. (1999). The path of least resistance for managers. San Francisco: Berrett Koehler Publishers.

Harrison, R. (1995). Consultant's journey: A dance of work and spirit. San Francisco: Jossey Bass.

Holman, P., Devane, T., Cady, S. (Eds.). (2007). The change handbook: The definitive resource on today's best methods for engaging whole systems. San Francisco: Barrett-Koehler.

Jentoft, S., Chuenpagdee, R., Bundy A., Mahon, R. (2010). Pyramids and roses: Alternative images for the governance of fisheries systems. Marine Policy, 34, 1315–1321.

Kooiman, J., Bavinck, M., Jentoft, S., Pullin, R. (Eds.). (2005). Fish for life: Interactive governance for fisheries. Amsterdam: University of Amsterdam Press.

Lewin, K. (1964/1976). Field theory in social science. New York: Harper & Row.

Mahon, R., Bavinck, M., Roy, R. (2005). Fisheries governance in action. In J. Kooiman, M. Bavinck, S. Jentoft, R. Pullin (Eds.), Fish for life: Interactive governance for fisheries (pp. 353–378). Amsterdam: University of Amsterdam Press.

Mahon, R., Fanning, L., McConney, P., Pollnac, R. (2010). Governance characteristics of large marine ecosystems. Marine Policy, 34, 919–927.

Mahon, R., Fanning, L., & McConney, P. (2011). Principled ocean governance for the Wider Caribbean Region. In L. Fanning, R. Mahon, P. McConney (Eds.), Towards marine ecosystem-based management in the Wider Caribbean (pp. 27–38). Amsterdam: Amsterdam University Press.

Senge, P. (1990). The fifth discipline: The art and practice of the learning organization. New York: Currency Doubleday.

Shankland, E. (2003). Assisting and resisting factors with the wave trends analysis. Unpublished Manuscript, 8 p.

Spencer, L.J. (1989). Winning through participation: Meeting the challenge of corporate change with the Technology of participation. Dubuque: Kendall/Hunt Publishing Company.

Stanfield, B. (1995). Transparent strategy. Edges, 7(2), 2–4.

Stanfield, B. (2002). The workshop book: From creativity to group action. Gabriola Island BC, Canada: New Society Publishers; Toronto ON Canada: The Canadian Institute of Cultural Affairs (ICA Canada).

Wilson, P., Harnish, K., Wright, J. (2003). The facilitative way: Leadership that makes the difference. Shawnee Mission: Team Tech Press.

Part V
Future Prospects

Chapter 18
Assessing Governability – What's Next

Ratana Chuenpagdee and Svein Jentoft

Abstract Issues, challenges and concerns in fisheries and aquaculture, and the problems associated with addressing them, must be examined through a comprehensive governability lens. Drawing from lessons learned in the application of the interactive governance perspective to assessing governability of various fisheries and aquaculture systems, we summarize what we believe should be the key elements of such an assessment framework. The first two parts focus on how to recognize and conceptualize the problems facing governance, and examine the main characteristics of the systems-to-be-governed, the governing system and the governing interactions that improve or inhibit governability, respectively. The next two parts of the framework outline the additional steps required to investigate: (1) goodness of fit, responsiveness and performance of the governing system in addressing the challenges found in the systems-to-be-governed; and, (2) how the system-to-be-governed and the governing system interact. Such assessment broadens the understanding of the limits and opportunities for improving governance, leading to a higher level of governability.

Keywords Governability assessment • Wicked problems • Interactive governance • Governing interactions • Fisheries

R. Chuenpagdee (✉)
Department of Geography, Memorial University of Newfoundland,
St. John's, NL, Canada
e-mail: ratanac@mun.ca

S. Jentoft
Faculty of Biosciences, Fisheries and Economics,
University of Tromsø, Tromsø, Norway
e-mail: svein.jentoft@uit.no

M. Bavinck et al. (eds.), *Governability of Fisheries and Aquaculture: Theory and Applications*, MARE Publication Series 7, DOI 10.1007/978-94-007-6107-0_18,

Introduction

The previous chapters of this book make the case for an examination of governability related to capture fisheries and aquaculture. It is argued that factors affecting governability may be found in the system-to-be-governed, the governing system and the governing interactions. The interactive governance approach offers a comprehensive lens through which governability challenges can be researched. In this chapter, we present an overview of the governability assessment framework and what it has to offer. We see such an assessment as a logical 'third' step following the descriptive analysis in *Fish for Life*, the preliminary exploration of the governability concept in the two special issues and the previous chapters of this book, which systematically analyze the basic elements and apply it to real life situations. Building on what we wrote in Chap. 3, we posit that a governability assessment must begin by recognizing the 'wickedness' of fisheries problems, which arises from various concerns and hard choices in fisheries. In other words, we argue that these problems, and the associated governance challenges, may be difficult to define and delineate from other governance issues. As wicked problems, they require a holistic approach that recognizes the traits (i.e., diversity, complexity, dynamics and scale) of the natural and social systems-to-be-governed, the governing system and the governing interactions (Jentoft and Chuenpagdee 2009).

The first step of the governability assessment begins with an examination of the nature of problems, the degree and nature of their wickedness, and the ways in which problems are perceived and understood by relevant stakeholders. These are questions pertaining to the dimension of images that, according to the interactive governance approach, plays a major role in setting the stage for governance (Kooiman 2003; Kooiman and Jentoft 2009; Jentoft et al. 2010). The next step (Step 2) is to perform a systematic assessment of how the properties of the governed natural environment and social system may contribute to lowering or enhancing governability in particular situations. The same assessment applies to the governing system. Governability also relates to the performance of the governing system and its capabilities (in other words, functional governability, see Jentoft and Chuenpagdee, Chap. 3, this volume). In Step 3, we therefore assess what elements make up and drive the governing system, how suitable each mode of governance is in responding to fisheries challenges and in facilitating appropriate and timely interactions, and how the different orders of governance foster or inhibit the governing system to produce desirable outcomes for systems-to-be-governed. Finally, governability rests upon the ways in which the system-to-be-governed and the governing system interact. This calls for Step 4, which is an examination of factors affecting various interactions and the degree to which these interactions are conducive to governability.

In the following section, we present the governability assessment framework, detailing the key concepts and steps involved. Learning from the applications presented in the book, we conclude with some reflections about what can be done to improve governability in fisheries and aquaculture.

Framework for Assessing Governability

Table 18.1 summarizes the targets, features and key measures involved in assessing governability. *Targets* are the various governance systems that we must explore in order to locate the governability problems. *Features* are the qualities and characteristics of the systems that provide the conditions for governability, and *measures* are the variables that may positively or negatively affect governability, depending on the degree of their existence, performance and response.

Table 18.1 Governability assessment framework

Assessment step	Targets (Where to look)	Features (What to look for)	Measures (What to look at)
Step 1	Fisheries governance problem	Degree of wickedness of the fisheries problem	Stakeholders' images of the problem Existence of stopping rules The embedded nature of problem Cost and reversibility of prescribed solutions
Step 2	Natural and social systems-to-be-governed Governing system Governing interactions	Prevalence of system properties (i.e., diversity, complexity, dynamics and scale)	Components Relationships Interactions Boundaries
Step 3	Governing system	Goodness of fits of elements (i.e. images, instruments and actions)	Behaviour, decisions, mental models, institutional arrangements, implementation
		Responsiveness of modes (i.e. self-, co-, and hierarchical)	Awareness, learning, sensitivity, conflicts
		Performance of orders (i.e. first, second and meta)	Consistency, effectiveness, transparency, justice
Step 4	Governing interactions	Presence and quality of interactions	Information sharing, co-learning, adaptiveness
		Enabling and restrictive role of power relations	Inclusiveness, representativeness, participation

Step 1: Identifying Problem Wickedness

The persistence and re-occurrence of certain fisheries and aquaculture governance problems suggest that there may be features related to aquatic ecosystems, social systems and governing systems, as well as their interactions, which make management particularly challenging. Deconstructing these challenges is therefore an essential starting point.

Fisheries governance typically follows a conventional planning paradigm. The process begins with the definition of the problem, the setting of goals, seeking and analysing information, outlining possible solutions, calculating their relative merits, making a decision, and finally evaluating outcomes (cf. Gilmore and Camilius 1996). Following Rittel and Webber (1973), the interactive governance theory questions the overly simplistic and unrealistic assumption of this paradigm, arguing that the planning process for some problems is not linear. They purport that such a linear approach would work for what they term "tame" (or benign) problems, but not for "wicked" problems. We hold that fisheries and aquaculture problems resemble the latter more than the former, and thus cannot be addressed by a methodology typical of, for instance, engineering or scientific inquiry.

In this assessment framework, we focus on some key features of wicked problems. First, fisheries and aquaculture problems are wicked in the sense that they can be looked at from several perspectives; there is no single way, right or wrong, to define them. Involved parties, decision-makers and stakeholders may disagree about what these problems are and what causes them (see for example, De la Cruz Modino and Pascual-Fernandez, Chap. 12, this volume). We therefore need to assess the extent to which images diverge or overlap in the way stakeholders rationalize, communicate and deliberate the problems (see also Jentoft et al. 2012). It is for this reason that Rittel and Webber (1973) argue that defining the problems ***is***, in and of itself, a wicked problem. As Schön (1983) argues, it follows that a problem must first be set, which can only happen through an interactive process involving stakeholders, before it can be solved. In fisheries and aquaculture, this process is particularly challenging, because stakeholders often do not agree on basic values and principles; they frame the problem differently and in accordance with their own individual images and interests (Jentoft et al. 2010). For this reason, it may be difficult to achieve full consensus about problem definitions on, for instance, what activities should be promoted (Liu et al., Chap. 11, this volume), what constitutes distributional justice (Jentoft, Chap. 4, this volume) or what actually makes livelihoods more sustainable (Johnson, Chap. 5, this volume). By the same token, solutions that will make everyone satisfied are hard to reach.

Second, Rittel and Webber (1973, 162) note that wicked problems "have no stopping rule" that informs us exactly when they are solved. It is therefore hard to recognize when stated goals are accomplished. They tend to reappear and must therefore be *re*solved, as is illustrated in this volume. Ecosystem health, for instance, is not determined by a single measure, assessed at a particular point in time (Pascual and Chuenpagdee, Chap. 7, this volume). Marine protected areas may help address ecosystem health, but they must be continuously monitored (De la Cruz Modino and Pascual-Fernandez, Chap. 12, this volume). The food supply of a particular region may be under ongoing threat (Pullin, Chap. 6, this volume). Similarly, gender inequality (Frangoudes et al., Chap. 13, this volume) and poverty (Onyango and Jentoft, Chap. 9, this volume) have yet to be solved once-and-for all. The wickedness of the latter problem is further illustrated by the fact that poverty in fisheries is not only pervasive, but it has also expanded from a predominantly income issue to a capability and entitlement issue (Jentoft and Eide 2011).

Third, Rittel and Webber argue that wicked problems are problems within problems. In the context of fisheries, this means that not all fisheries problems are contained within the fisheries sector. Some are instead embedded in problems that relate to societal concerns at a larger scale. Frangoudes et al.'s (Chap. 13, this volume) discussion of gender inequality and Johnson's (Chap. 5, this volume) elaboration of the wickedness associated with sustaining livelihoods in coastal communities illustrate this point well. Finally, solutions to wicked problems tend to leave traces, and mistakes are costly and irreversible. There is little room for experimentation because solutions create path dependency that may eventually lead to tipping points. For instance, Chuenpagdee and Jentoft (2007) demonstrate that mistakes made in the initial phase of co-management implementation have consequences that may inhibit the progression of this management scheme at later stages. The privatization of common pool resources through mechanisms such as individual transferable quotas (ITQs) is another example of irreversibility. Once established, it is practically impossible to revert, even when substantial costs to communities are likely to occur (see discussion in Archeson 2006; Sabau 2011). Another example is the management decision leading to the industrialization of fishing fleets at the national scale (as in India, Scholtens and Bavinck, Chap. 10, this volume), as well as globally (Bavinck 2011). These decisions affect the viability of small-scale fishing sector (Chuenpagdee 2011). Even if replacing small-scale fisheries with large-scale industries is regarded, in hindsight, as a mistake for reasons of food security, poverty alleviation, and ecosystem health, it may already be too late. Rebuilding small-scale fisheries by reallocating resources to them and reinstituting their rights from a justice perspective (Jentoft, Chap. 4, this volume) require interventions that, for political and other reasons, may not be practical, feasible or popular.

Step 2: Examining System Properties

The second step of the assessment deals with the properties of the natural and social systems-to-be-governed. At this step of analysis, the governing system is also seen as an entity that needs to be governed. The assessment involves an examination of the diversity, complexity, dynamics and scales associated with each of these systems. The hypothesis would be that the more diverse, complex, dynamic and scale extensive the system-to-be-governed and the governing system are, the lower their governability. This is due in part to the fact that these features give rise to the wickedness of governance. We therefore consider an industrialized aquaculture system, such as that described by Liu et al. (Chap. 11, this volume), to be more governable than, for instance, the more "chaotic" system of small-scale fisheries in Lake Victoria (Onyango and Jentoft, Chap. 9, this volume).

Here *natural* system-to-be-governed refers to the aquatic environment, including flora, fauna and other organisms contained within the system, the habitats in which they reside (including the water column), and oceanographic currents such as upwelling and storm surges. Coastal ecosystems connected to the aquatic environment are likewise considered part of the natural system-to-be-governed. The *social* system-to-be-governed is comprised of direct and indirect stakeholders such as small-scale and large-scale fishers, recreational fishers, shellfish gatherers, fish farmers, divers, tourists, developers, and oil and gas explorers. The governing system refers to the institutions and steering instruments and mechanisms that are established to deal with governability issues arising as wicked problems within the system-to-be-governed.

In the examination of system properties, we first look for *diversity* as it relates to resource units and relevant stakeholders. That is, we explore the heterogeneity and quantity of system elements. Investigations into this characteristic focus on system *components*; the elements that comprise the system; their characteristics and numbers. We need to ask what and who they are, how similar or different they are from each other, and how many of each exists. The hypothesis would be that the greater the diversity among system components, the lower the governability. For instance, diversity implies a broader range of knowledge, including "vertical knowledge" that enables a deep understanding of ecosystems (Kenyon et al. 2006), and "high resolution" data regarding, for example, particular habitats such as spawning grounds and biotopes (Degnbol 2004). With respect to social systems, there is a similar need to determine who benefits directly or indirectly from ecosystem services. Is the group heterogeneous or homogenous? How do the beneficiaries extract from the resource base, and with what kind of technology? What are their situations, capacities, powers and rationalities?

As for the governing system, the assessment focuses on the number and types of institutions established to address fisheries challenges. We ask questions like, which are they? What mandates do they have? How rule intensive are they? Here, we expect an inverse relationship between the diversity of the governing system and governability. As clearly illustrated by Onyango and Jentoft (Chap. 9, this volume), the higher the number of institutions, the more likely the overlap in mandates. In turn, there is a lack of clarity with regard to the responsibility of addressing the emergence of particular problems. Similarly, an increase in rule intensity, as in the case of legal pluralism (Jentoft et al. 2009), results in an increased likelihood of non-compliance. As shown in the case of Lake Malawi fisheries (Song and Chuenpagdee, Chap. 15, this volume), this is due to conflicting norms and principles, and confusion about what rules apply in particular situations.

The second property regards system *complexity*. In Table 18.1, it is operationalized in terms of *relationships*. Complexity refers to the ways in which system components connect to and condition each other. Like species and organisms form communities and trophic chains (Pascual-Fernandez and Chuenpagdee, Chap. 7, this volume), people are organized into networks, groups, communities and institutions for the purpose of managing their interactions and relationships with respect to, for instance, resource uses and conservation (Mahon and McConney, Chap. 16, this volume). These natural and social systems-to-be-governed and the governing system are characterized by mutuality and interdependence that condition what the individual units, as well as the system as a whole, can possibly be, do or become. The hypothesis would be that the more relationships exist in the systems, the more complex they are, and thus the lower the overall governability. One should not assume equilibrium, linearity or complete order in any of these systems. Neither should one expect internal compatibility, coherence or symmetry. Rather, systems may show signs of conflict and disarray, as is the case when stakeholders have diverse interests or when species intrusion occurs. The governing system is then likely to have problems understanding and controlling internal as well as external interconnectedness. Much time and effort thus needs to be concentrated on conflict resolution and the balancing of incongruent demands reactively rather than proactively.

Thirdly, we examine the *dynamics* of the systems. Dynamics relates to the fact that these systems are likely not inert, but they may be unstable and change over time. In particular, the assessment targets *interactions*; how system components actively influence each other, provide inputs and feedback to each other, and how they make the systems change either gradually or in leaps and bounds. In other words, we focus on the systems *in vivo*. The hypothesis is that the more active and dynamic the system is, the lower the governability. We do not assume, however, that fisheries and coastal systems are always proactive or adaptive. In fact, it is often the rigidity and irreversibility that creates governability problems (see Johnson, Chap. 5, this volume). This is the case when fishers defy governing interventions or when governments turn a blind eye to small-scale fisheries issues and concerns (Chuenpagdee 2011). Inertia may stem from the interest and power of some stakeholders that prevents the system from changing. One feature of dynamics is, however, interactive learning, which may over time lead to adaptation and change (Chua et al. 2006; Armitage et al. 2008). The advantage of co-governance and partnership arrangements as governing modes is that they widen the source of knowledge, including local and traditional knowledge, and provide opportunities for interactive learning (Wilson et al. 2006; Chuenpagdee and Mahon, Chap. 14, this volume). It allows stakeholders to learn from each other (see Almerigi et al., Chap. 17, this volume).

Fisheries and aquaculture as natural systems, and the social and governance systems related to them, come in varying spatial and temporal *scales*. Scale is therefore the last system property in this second step of governability assessment. More specifically, the focus is on system *boundaries*, how they confine relationships (like networks) and interactions, and how they define what its components are. System boundaries are actual and analytical, as well as natural and socially constructed. The boundaries determine the scale of the systems, which could be small – like a lake for a natural system or a local community for a social system – or large – like a large marine ecosystem, coastal zone or an entire region (Mahon and McConney, Chap. 16, this volume). At the boundary, relationships and interactions among system components are fewer and less intense. Governability is an issue of scale because large-scale systems, whose diversity, complexity and dynamics are assumed to be more pronounced, are less governable in comparison to smaller scale systems. System boundaries are also a matter of perception and conception (Post et al. 2007), thereby providing the basis on which governing systems are designed. In other words, a governing system is devised for dealing with a system-to-be governed whose boundaries must be determined in order to delineate a mandate. In some instances, multiple governing systems operate within the boundaries, causing problems with coordination and harmonization. This is an issue often referred to as 'legal pluralism' (Jentoft et al. 2009). The governability problem would be enhanced if geographical boundaries of the system-to-be governed and the governing system do not match. This is the case when the jurisdiction of the governing system does not encompass the entire ecosystem or social system. If several governing systems are operative within the same boundaries, governability problems are to be expected. After all, how can overlapping mandates be coordinated? Similarly, what happens at system boundaries is an interesting research issue for governing interactions that generates a host of questions. For example, how are mismatches addressed? Do problems or opportunities fall between the cracks? Boundaries may be more or less permeable. Interactions and relations may cut across boundaries and link systems at the same or different scales. The scale at which a problem is addressed may determine the governability of the system as a whole (Mahon and McConney, Chap. 16, this volume).

Step 3: Evaluating the Governing System

The capability and capacity of the governing system, which are confined by their structural components, play an important role in overall governability (Step 3 in Table 18.1). If, for instance, the governing system is highly capable of dealing with a problem, no matter how wicked it is, and with the systems-to-be-governed, no matter how diverse, complex and dynamics they are, the overall governability will be high. On the other hand, a weak governing system will result in low governability, regardless of the degree of wickedness and irrespective of the system properties. As illustrated in Table 18.1 and also in Bavinck and Kooiman (Chap. 8, this volume), the assessment of the governing system specifically calls attention to three key features: *goodness-of-fit*, *responsiveness* and *performance* of the governing system in addressing challenges. The poorer the institutional fit between the system-to-be-governed and governing system, the greater the governability problem (see for instance Scholtens and Bavinck, Chap. 10 this volume). The appropriateness of the governing system is further determined by the fit of actions to the problems and their images, both of which must be examined (Jentoft et al. 2009). With respect to responsiveness, a highly diverse system-to-be-governed calls for a less centralized governing mode and more involved local actors (as in co-management systems).

On the other hand, a more complex system-to-be-governed would require authority of a central government to facilitate coordination. The performance of a governing system is conditioned upon the images of the problems that are faced, and what and how principles, situated at the meta-order of governance, are established. It also depends on how institutions are arranged at the second order accordingly, and what actions are taken at the first order. Governability at this assessment stage is the outcome of all these considerations. Similar to Step 2, the system-to-be-governed may also act as a governing system for some analytical purposes. This occurs, for example, when fishers set their rules of conduct. The same assessment criteria will then be applied.

At this stage, the assessment therefore turns to the design of the governing system and evaluates its performance. It examines how goals are developed in the first place, and how they are achieved in the second. After all, at the end of the day, governability is about the performance of the governing system relative to the (wicked) problems it tries to address and the specific goals for solving them. More precisely, it looks at the relationship between governing elements, i.e., images, instruments and actions (Kooiman et al. 2005). Images in this context refer to the dominating mental models from which the governing system draws its inspiration, direction and goals. Instruments are the means (steering mechanisms) employed by the governing system in order to realize the goals that are inspired by these images. Actions pertain to the choices made with regard to the implementation of the instruments, such as enforcement, monitoring and surveillance, and the readiness to act on these choices. The expectation is that governability depends on the consistency between images, instruments and actions, and how well they address the wicked problems. When images, instruments, and actions are not clearly defined and formulated, governability is expected to be low. As Chuenpagdee et al. (2008) argue, the effects of instruments can be minimal, and even counter-productive in the long-term, unless the choice of what instruments to employ is based upon images that are accurate and legitimate. Some examples of low compatibility, and hence low governability, can be found where alternative livelihoods are introduced to alleviate poverty in fishing communities without proper attention for the social and cultural identity of the people involved (Onyango 2011; Onyango and Jentoft, Chap. 9, this volume).

The assessment in this step also involves the examination of governing modes; the institutional arrangements for putting goals into action. Interactive governance theory identifies three modes of governance. Hierarchical governance is basically a top-down, command-and-control system that places governments at the apex of the pyramid. Co-governance involves stakeholders working in cooperation with civil society actors and government. The third mode, self-governance, refers to the situation where stakeholders play an autonomous steering role. Here, the expectation is that the more wicked and complex the problem and goal structure are the more difficult it is for government to work alone. The government must instead rely on the involvement of market and civil society. It cannot abstain from assuming responsibility and taking action in addressing problems either. From a governability perspective, co-governance is therefore better suited to improving governability in this context. This is largely due to the broader knowledge base of the multiple stakeholders from which it draws, and the interactivity of collective learning. Co-governance, however, comes with its own challenges. Such is the case when conflicts among stakeholders and power asymmetry in their relations dominate the decision-making agenda. With regards to the governing system, Siry (2006) concludes that the greater the diversity, the greater the need for a decentralized governance mode in order to perceive and deal with details and subtleties. He argues that the huge range of biodiversity, the variation in coastal zones, human populations and regional economies are the main reasons why coastal zone management needs to be decentralized and community-based. Decentralization and self-governance however, entails its own governability problems. These pertain in particular to administrative order and efficiency, rights, equality, and how to deal with mobile users who belong to several systems-to-be-governed.

The last aspect in this step of the assessment is about governing orders and their performance. Interactive governance theory differentiates between three orders of governing activities. The first order pertains to routine decision-making to address problems, as well as to the creation of opportunities for better solutions. The second order is about the design of institutions that frame and facilitate actions at the first order is about. The third order is referred to as the meta-order, and is related to the overall values, norms, and principles that guide institutions and actions. The hypothesis is that the better they perform, relative to each other, the higher the governability. For instance, the principle of justice deliberated among stakeholders must be translated into institutions and the way they operate. If not, the resulting inconsistency and ineffectiveness will likely lead to the abandonment of the governing system. The precautionary principle is also appropriate given the irreversibility and costliness of governance failure.

Step 4: Governing Interactions Analysis

The fourth and final step of governability assessment concerns governing interactions; the institutions and processes through which the system-to-be-governed and the governing system relate to one another (Table 18.1). Several factors concerning interactions have impacts on the governability of the fisheries and aquaculture systems. It is not only the types of interactions or their presence or absence that matter, but also the quality of these interactions and the institutional conditions framing them. Interactive and consultative processes such as those described by Almerigi et al. (Chap. 17, this volume) are therefore useful to facilitate positive governance outcomes.

Governing interactions are about the interconnectedness between the system-to-be-governed and the governing system, and the ways in which these systems reach out to each other. Here we examine both the structure and processes of interactions. The analysis begins with the identification of different types of interactions and the various media through which the system-to-be-governed and the governing system communicate with each other. For instance, Buanes et al. (2005) show that in coastal zone management communication takes many forms, from written memos, oral exchange, public hearings and newspaper announcements to participatory committees, radio, TV and Internet. Given the diversity of involved stakeholders, each with variable resources and opportunities for gaining access to information, we expect that the greater the diversity of interactions, the lesser the governability problem (see Bavinck and Kooiman, Chap. 8, this volume). This will only be the case, however, if the messages convey similar information in a clear and understandable language and form.

At the center of these interactions is the issue of power and power relationships. It is essential to be aware of both the enabling and restricting roles that power plays (Jentoft 2007). Too much or too little power vested in certain stakeholder groups may present a governability challenge. A primary research concern is related to the extent and conditions under which a more balanced distribution of power will facilitate governance processes and lead to more desirable outcomes. One should be open to the possibility that symmetrical power relations may, in some instances, obstruct effectiveness of governance, bringing the entire governing process to a halt.

It must be noted that power is an attribute of all three systems. It is therefore of particular importance that it is examined in the context of governability. Starting from the system-to-be-governed, the assessment of complexity and diversity, in particular, focuses on the existing power relations among stakeholders and examines how some are able to make the system-to-be-governed work to their benefits. Some stakeholders are also capable of influencing the governing system, serving their own interests, and consequently lessening the overall capacity of the governing system to address the basic governance concerns in the fishery. In such a situation, the governability problem becomes entrenched. For instance, in many parts of the world, addressing environmental issues is hampered by a governing system that has been captured by interest groups that oppose such action, leading to governance failure.

Discussion and Conclusion

Applying the governability concept to fisheries and aquaculture problems means acknowledging that these problems share many of the characteristics of those described as wicked by Rittel and Webber (1973). It recognizes that there may be limits to how systematic, effective, rational and socially just a governing system can be in addressing the fundamental concerns of fisheries and aquaculture governance and the wicked problems that are associated with them (Jentoft and Chuenpagdee 2009).

Pullin (Chap. 6, this volume) mentions three essentially non-negotiable limitations to governability: (a) all food is produced in natural or artificial ecosystems; (b) most of these ecosystems are not entirely under human control; and (c) food production is always accompanied with a level of risk, such as natural hazards. In view of these limitations to governability, one must assume that governance outcomes are not always as planned, and that governors are often in situations where they have to accept solutions that are less than ideal. Kooiman et al. (2005) therefore stress that governance involves hard choices. It also means that the governability challenge persists and must be addressed on a continuous basis – just as a wicked problem without a clear end point. Notably, the limits of governability can sometimes be stretched through intervention and innovation. What these limits are and how they can be manipulated is a governability research issue for which the approach has been outlined above.

Interactive governance theory argues that fisheries and aquaculture systems are diverse, complex and dynamic, and that their system characteristics are rich and spatially variable. Consequently, they are not fully controllable, their developments are largely non-linear, and information is never complete. All together, this frequently makes the system outcomes unintended and unpredictable (Degnbol and McCay 2007). The aforementioned reality requires a governance approach that is flexible and adaptive as opposed to one that emphasizes control and stability (Mahon et al. 2008). In governance terms, this is what Scholtens and Bavinck (Chap. 10, this volume) refer to as an institutional match. A lack of compatibility is not only likely to reduce governance capacity and hence effectiveness, but is also prone to conflict that will reduce governing system legitimacy and increase transaction costs. Partly due to institutional inertia, such conflicts tend to persist over time and have to be handled without the prospect of ever being fully resolved. For these reasons, governing goals cannot be assumed to be given *ex ante*. Instead they must be analyzed as outcomes of a continuing negotiation among a diverse group of stakeholders with asymmetric power, differentiated images, incompatible expectations and conflicting demands (Jentoft et al. 2011). This also means that governance is not so much about exercising authority from the top down, as it is about political brokerage from below, where operating goals are at best imperfect and sub-optimal compromises.

Governability issues arise as governing actors attempt to deal with the wicked nature of fisheries and aquaculture problems. More precisely, they appear when definitions of problems are unclear and must be negotiated, and when solutions are pre-conceived and imposed on the negotiators. They also appear when stakeholders have different ideas with respect to when the problem is solved; outcomes do not necessarily fulfill the various demands of stakeholders. Solutions for one group of stakeholders may cause problems for other stakeholders, thereby ensuring the continuation of the governing process without a clear end in sight. For this and for other reasons pertaining to fisheries and aquaculture governance, improving governability is in itself a wicked problem. As a wicked problem, governability is not a quality or capacity that can be achieved once and for all. Rather, it is an outcome of dynamic processes that need to be revisited from time to time. However, governability as a whole, or any of its components, can be influenced by acts of governance. Governance is therefore essentially about promoting governability.

As illustrated in Song and Chuenpagdee (Chap. 15, this volume), stakeholders may share common values. More often than not, however, their interests vary to the point that their activities cause competition, conflicts and resource degradation. In the simplest form, some fishing areas may be overlapped, and tension arising from this may be lessened through proper zoning. Some activities are completely opposed by others, such as when areas used for mariculture are no longer accessible to small-scale fishing, or when a development of an exclusive beach resort prohibits others from enjoying coastal space. A common governing response is to undertake spatial demarcation and boundary control. When, for instance, competing user groups are kept apart, there is less likelihood that they will get entangled in conflict. An alternative response to this problem is to encourage co-operation by forming partnerships (Buanes and Jentoft 2005). After all, the relationship between fisheries, coastal tourism, marine aquaculture, coastal transportation and offshore oil exploration does not need to be antagonistic. How to transform a conflictive relationship into a cooperative and symbiotic one is a wicked problem and a governability challenge in and of itself.

At a general level, governability is a function of processes that occur within and between the system-to-be-governed, governing system and governing interactions. Interactive governance theory thus broadens the perspective of governability to something that is not solely an issue and responsibility of the governing system. Governability further depends on the ability of the totality of these systems to deliver on the challenges and demands brought about by their diversity, complexity, dynamics and scale.

Steps 1 and 2 of the governability assessment described in this chapter help to identify what the system-to-be-governed, governing system and governing interactions are made up of, how they work, and what might possibly explain their problematic nature. The goals of governance and the concerns that they relate to, be they food security, ecosystem health, sustainable livelihoods or social justice, are assumed to be negotiated internally as part of these interactions. They are not preconceived and predetermined as something that the system necessarily tries to achieve. The ways in which goals are actually set with regard to these governance concerns is thus to be assessed empirically. In fact, we anticipate that they are themselves outcomes of the interactions structured by the particular systems under scrutiny. In these systems and interactions, power is assumed to play an important role. Governing system goals and strategies may fall short of what is considered to be ecologically and economically sustainable or socially just. It is for reasons such as these that addressing governability begins by recognizing and reflecting upon the wickedness of the governance problems. Only then can one expect these systems to deliver on the major concerns of fisheries and aquaculture governance, and the goals of governance that follow from them.

As Chuenpagdee and Jentoft (2009) previously argue, assessing governability is part of a reality check that governors must engage in to improve governance effectiveness. Since many fisheries and aquaculture governance problems are essentially wicked, one should not jump to conclusions about solutions. Governability assessments should be part of the governance process and routine, and should be performed prior to, rather than after, major governance reforms are initiated. Several tools and approaches to examine the various characteristics of the fisheries and aquaculture systems for their governability are readily available (see Chuenpagdee and Mahon, Chap. 14, this volume). It is a matter of being mindful and aware of how the various features explained in Table 18.1 may affect governability and how useful these tools are for assessing them. The stakes in fisheries and aquaculture are simply too high for easy solutions and quick fixes. Complex and dynamic social and ecological fisheries and aquaculture systems require systematic and iterative assessments of governability. The great diversity and scale of fisheries and aquaculture situations call for contextualized assessment. It must be proven, for instance, that governance reforms do work to enhance governability, not just for the sector as a whole, but also where fisheries and aquaculture and their sub-sectors actually interact. Detailed analysis of the causal relationship between the system-to-be-governed and the governing system as they are experienced on the ground is therefore needed. The governability assessment framework offers a way to undertake such an analysis.

References

Archeson, J.M. (2006). Institutional failure in resource management. Annual Review of Anthropology, 35, 117–34.
Armitage, D.R., Plummer, R., Berkes, F., Arthur, R.I., Charles, A.T., Davidson-Hunt, I.J., Diduck, I.P., Doubleday, N.C., Johnson, D.S., Marschke, M., McConney, P., Pinkerton, E.W., Wollenberg, E.K. (2008). Adaptive co-management for social–ecological complexity. Frontiers in Ecology and the Environment, 6. doi:10.1890/070089.
Bavinck, M. (2011). The megaengineering of ocean fisheries: A century of expansion and rapidly closing frontiers. In S.D. Brunn (Ed.), Engineering earth: The impacts of megaengineering projects. Dordrecht: Springer.
Buanes, A., & Jentoft, S. (2005). Challenges and myths in Norwegian coastal zone management. Coastal Management, 33(2), 151–165.
Buanes, A., Jentoft, S., Karlsen, G.R., Maurstad, A., Søreng, S. (2005). Stakeholder participation in Norwegian coastal zone management. Ocean & Coastal Management, 48(8–10), 658–669.
Chua T.E., Bonga, D., Bermas-Atrigenio, N. (2006). Dynamic of integrated coastal management: PEMSEA's experience. Coastal Management, 34(3), 303–322.
Chuenpagdee, R. (Ed.). (2011). Contemporary visions for world small-scale fisheries. Delft: Eburon.
Chuenpagdee, R., & Jentoft, S. (2007). Step-zero for fisheries co-management: What precedes implementation. Marine Policy, 31, 657–668.
Chuenpagdee, R., & Jentoft, S. (2009). Governability assessment for fisheries and coastal systems: A reality check. Human Ecology, 37, 109–120.
Chuenpagdee, R., Kooiman, J., & Pullin, R.S.V. (2008). Assessing governability in capture fisheries, aquaculture and coastal zones. Journal of Transdisciplinary Environmental Studies, 7(1), 3.
Degnbol P. (2004). Fisheries science in a developing context. In B. Hersoug, S. Jentoft, P. Degnbol (Eds.), Fisheries development: The institutional challenge. Delft: Eburon.
Degnbol, P., & McCay, B.J. (2007). Unintended consequences of ignoring linkages in fisheries systems. ICES Journal of Marine Science, 64(4), 793–797.
Gilmore, W.S., & Camilius, J.C. (1996). Do your planning processes meet the reality check? Long Range Planning, 29(6), 869–879.
Jentoft, S. (2007). In the power of power: The understated aspect of fisheries management. Human Organization, 66(4), 426–437.
Jentoft, S., & Chuenpagdee, R. (2009). Fisheries and coastal governance as a wicked problem. Marine Policy, 33, 553–560.
Jentoft, S., & Eide, A. (Eds.). (2011). Poverty mosaics: Realities and prospects in small-scale fisheries. Dordrecht: Springer.
Jentoft, S., Bavinck, M., Johnson, D., & Kaleekal, T. (2009). Co-management and legal pluralism: How an analytical problem becomes an institutional one. Human Organization, 68(1), 27–38.

Jentoft, S., Chuenpagdee, R., Bundy, A., & Mahon, R. (2010). Pyramids and roses: Alternative images for the governance of fisheries systems. Marine Policy, 34, 1315–1321.

Jentoft, S., Chuenpagdee, R., Pascual-Fernandez, J. (2011). What are MPAs for? On goal formation and displacement. Ocean & Coastal Management, 54, 75–83.

Jentoft, S., Pascual-Fernandez, J.J., De la Cruz Modino, R., Gonzalez-Ramallal, M., & Chuenpagdee, R. (2012). What stakeholders think about marine protected areas: Case studies from Spain. Human Ecology, 40, 185–197.

Kenyon, J.C., Brainard, R.E., Hoeke, R.K., Parrish, F.A., Wilkinson, C.B. (2006). Towed-diver surveys, a method for mesoscale spatial assessment of benthic reef habitat: A case study at midway atoll in the Hawaiian Archipelago. Coastal Management, 34, 339–349.

Kooiman, J. (2003). Governing as governance. London: Sage.

Kooiman, J., & Jentoft, S. (2009). Meta-governance for natural resources: Choices, values and principles. Public Administration, 87(4), 818–836.

Kooiman, J., Bavinck, M., Jentoft, S., & Pullin, R.S.V. (Eds.). (2005). Fish for life: Interactive governance for fisheries. Amsterdam: Amsterdam University Press.

Mahon, R., McConney, P., Roy, R. (2008). Governing fisheries as complex systems. Marine Policy, 32, 104–112.

Onyango, P.O. (2011). Occupation of last resort? Small-scale fishing in Lake Victoria, Tanzania. In S. Jentoft, A. Eide (Eds.), Poverty mosaics: Realities and prospects in small-scale fisheries. Dordrecht: Springer.

Post, D.M., Doyle, M.W., Sabo, J.L., Finlay, J.C. (2007). The problem of boundaries in defining ecosystems: A potential landmine for uniting geomorphology and ecology. Geomorphology, 89(1–2), 111–126.

Rittel, H.W.J., & Webber, M.M. (1973). Dilemmas in a general theory of planning. Policy Sciences, 4, 155–169.

Sabau, G. (2011). Whose fish is it anyway? Iceland's cod fishery rights. In R. Chuenpagdee (Ed.), World small-scale fisheries: Contemporary visions. Delft: Eburon.

Schön, D.A. (1983). The reflective practitioner. How professionals think in action. London: Temple Smith.

Siry, H.S. (2006). Decentralized coastal zone in Malaysia and Indonesia: A comparative perspective. Coastal Management, 34(3), 267–286.

Wilson, D.C., Raakjær, J.N., Degnbol, P. (2006). Local ecological knowledge and practical fisheries management in the tropics: a policy brief. Marine Policy, 30, 794–801.

Chapter 19
Improving Governability – Reflections for Future Applications

Jan Kooiman

Abstract The purpose of this chapter is twofold. First of all, it presents the major scholarly discussions with regard to the features and systems in interactive governance theory, highlighting the concept of governability. Secondly, it aims to introduce examples from other scholarly work on interactive governance to illustrate where progress has been made and further reinforcement is needed. The author concludes that governability has not only proven to be a rich concept and a source for further theoretical work, but that it can also be fruitfully applied to societal sectors or systems such as fisheries.

Keywords Governability • Diversity • Complexity • Dynamics • System-to-be-governed • Governing system • Governing interaction system

Introduction

The concept of governability has clearly matured since its introduction to the field of fisheries (see Kooiman et al. 2005). Two special issues of journals have since been devoted to the topic, as are a number of papers in a wide range of academic journals, a few doctoral dissertations, and finally the chapters in this book. All these efforts demonstrate that it is a fruitful concept to work with. This volume highlights our endeavor to bring the governability concept to life in a more systematic manner than has been the case thus far. It builds on earlier work by myself

J. Kooiman (✉)
Centre for Maritime Research, University of Amsterdam, Amsterdam, The Netherlands
e-mail: jkooiman@xs4all.nl

M. Bavinck et al. (eds.), *Governability of Fisheries and Aquaculture: Theory and Applications*, MARE Publication Series 7, DOI 10.1007/978-94-007-6107-0_19,

and others in the Fisheries Governance Network and takes some important steps forward, conceptually as well as empirically. In this chapter, I discuss the major elements of the governability framework, looking back and looking forward. My remarks take the form of indicating where progress has been made and where additional work seems appropriate. First I make some suggestions for a strategy to study governability. This is followed by a review of the features characteristic of all societal systems and, thus, a major theme in governability research. Then, I consider the three subsystems – the system-to-be-governed, the governing system and the governing interactions – ending with reflections and suggestions for future assessment work.

Research Strategy

The interactive governance perspective suggests that the research approach to governability should not only be multi-focus, but also multi-level. Moreover, these levels and foci should be related. Although other analytical approaches might be available, I suggest that the classical micro-macro framework serves as a useful guideline. This framework proposes that macro phenomena constitute the frames of reference for the actions and interactions of individuals and groups. At the same time these macro elements are constantly (re)produced, interpreted and modified by individuals at the micro level. This is a two-way macro-micro analysis "with neither direction capable of being grasped without simultaneous reference to the other" (Alexander et al. 1987, 381–382).

Does it make sense to introduce an analytical meso-level in between the other two? Many authors attempt to do just this. The meso-domain approach, designed to explore social organization as recurring patterns of linked contexts (Hall and McGinty 2002), is useful but not required. That is, if a two-way macro-micro and micro-macro analytical approach is systematically applied. The latter is ideal in theory, but hard to do in practice.

For analysing governability, with all of its features and properties, one needs inter-disciplinary insights. Much has been written about the successes and failures, strengths and weaknesses of the collaborations between academic disciplines (Bruce et al. 2004; Maasen et al. 2006). Multi-disciplinarity is the least demanding form because the partners, in effect, remain within the realms of their own bodies of knowledge. Inter-disciplinary work asks for more fine-tuning. Theoretical roadblocks must, at the very least, be partly resolved to be successful. Trans-disciplinary studies, with their involvement of non-scientists, is even more demanding. For the study of governability, no single mode has absolute preference. However, as Brewer (1999) suggests, for real cross-fertilization between disciplines, the risks are high and rewards may come only in the long run. Therefore, simple forms of cooperation may currently be more appropriate for grasping what governability is about.

Governability has been defined as the capacity for governance of the system as a whole. This means that one requires synthesis next to analysis. In the end, whatever we want to say about the system has to be synthetic in one way or the other. At the same time, however, it is useful to divide the system into three sub-systems and start analysis. These two activities – breaking-down into analytical units and building-up into a more holistic whole – should be an iterative process, theoretically, conceptually and empirically. The result is a body of insights into the conditions, factors and forces that form the 'image' we call governability. Additionally, we analyze and synthesize the diversity, complexity, dynamics and scale of the societal system in which we are interested with an inter-level and inter-focus approach.

If scholars take up the challenges for and concerns about fisheries, the need for systematic inter-disciplinary research will become ever clearer. As Pullin argues in Chap. 6, from a food security perspective, and I don't speak of other concerns that may have similar effects, each marine resource has multiple users, each of whom have little or no accountability for the adverse impacts on others. This shortcoming can only be met by an inter-sectoral approach. In practice, the lack of interactions between sectoral studies is a major weakness and drawback in perspectives on the governability of fisheries. The same can be said of inter-level studies. From a feasibility point of view, it is logical that scholars have difficulty in giving micro and macro perspectives research attention simultaneously. However, in micro (case) governability studies, it should become standard practice to at least include a macro oriented positioning of the case. *Vice versa*, a macro study should give indications of issues and consequences on the micro level.

Features

All societal systems are characterized by four features: diversity, complexity and dynamics, each of which is expressed differently at varying scales (see Chap. 2). They form an essential part of the interactive governance approach and figure prominently in this book. For the study of governability, theories of diversity point to the role of differences and similarities in societal systems. Theories of complexity highlight the interrelations within societal systems. Theories of dynamics, in turn, show that linear as well as non-linear patterns of change need to be taken into account. In addition to its analytical role, scale has the important function of ensuring that the three other central features are looked at the same level in any study.

Diversity

Interest in diversity on a general systems level is growing. Diversity is considered a basic concept in explaining the evolutionary, co-evolutionary and self-organizing qualities of natural and societal systems (Kaufmann 1995, 289 ff; also Holland 1995, 27–31). While it is too early to say what these ideas of diversity may mean for the study of governability, they certainly draw attention to its role as a creating force in both natural and human affairs. On a less abstract level, interest in the diversity, heterogeneity and difference of the social and cultural realm has begun to parallel interest in ecological diversity. One sees for example values converge because of media such as television and the Internet. At the same time, regional and local cultures are (re-)discovered all over, often combined with efforts aimed at political autonomy. Although societal systems cannot live without diversity, a certain measure of uniformity in inter-human interactions is needed as well. Unlimited diversity may become counter-productive, as the similarities needed for communication decrease rapidly. The different sides of diversity result in a constant tension between the freedom of action and expression, on the one hand, and the rules and customs to which these actions and expressions need to comply on the other. This tension is a challenge for governance and an important factor influencing governability.

From a governance and governability point of view, there are not many ways to deal with diversity (Kooiman 2003). The method typically used is one that attempts categorize or bring order. However, this not as neutral a governing activity as is often thought: top-down categorization is based on value choices that are frequently hidden. What is needed here instead is inter-action, understood to be a recurrent process in which those ordered or categorized participate.

We must be careful in analysing diversity as a feature of fisheries and an aspect in governability. As far as the natural system is concerned, the observer generally highlights eco-system diversity, where at least two variables play a role, the number of species and the variety of their characteristics (Chap. 7). For socio-economic diversity, the story is somewhat different. In Chap. 5, for example, the diversity of fishing livelihoods is seen as multi-dimensional. Variables such as mobility, gender relations of work, degree of subsistence and technology, to name but a few, play a key role. Each of these variables, in turn, possesses a world of diversity in and of themselves. If we raise the analysis to the level of the diversity of whole systems, the importance of the micro-macro dimension of governability becomes apparent. Detailed case studies of Marine Protected Areas (Chap. 12), demonstrate the potential diversity of a socio-ecological fishery system, its governing system and the interactions between them. The macro study of salmon aquaculture (Chap. 11), on the other hand, indicates that the diversity of these systems varies in a comparative and synthetic perspective. The more these analytical activities are based on theoretical notions of what diversity might mean or stand for and the more they are conducted together with those concerned, the better the role of diversity for the governability in fisheries can be explained.

Complexity

As a concept complexity remains elusive. The distinction between 'simple' and 'complex' is not as sharp as we might intuitively expect. What may appear complex may turn out just to be complicated (a machine), and what appears simple may be immensely complex (the leaf of a tree) (Cilliers 1998). In a systems sense, complexity is defined as the number, intensity and spread of interrelations between its components. This, however, remains rather abstract. Gell-Mann (1994) takes an unusual approach by considering the complexity of a system to be identical to the length of time needed to describe it. One always has to be specific about the level of detail that such a description entails – if finer details are being ignored ('coarse graining'), this adds to the subjectivity of its definition (1994, 28–32). Other authors have made a more rigorous step in the same direction by expressing that complexity is not essentially a material quality of any system, "but only turns on questions posed. It is normative to the degree that complexity arises from the lack of a paradigm" (Zellmer et al. 2006, 171).

Dealing with complexity is thus complex in itself. A two-step procedure can help. First, one decides the level of 'graining' required – the coarser, the less complex the picture of the system. Second, one can use Simon's (1983) discovery that the majority of the systems we deal with are 'nearly-composable'. Simply put, not all relations within a system are equally strong in terms of number and intensity. A combination of decomposition and (re)composition can help reveal the location of the highly related or more autonomous parts of a system (Kickert 1979; Simon 1983).

The complexity of fisheries as natural and social systems (not to speak of the linkages between them) has baffled scholars and practitioners alike. In part II of this book, concerns have been ordered along lines as diverse as food security, social justice, livelihood and employment, and ecosystem health. It is no surprise that these concerns are themselves highly complex. Some of the complexities related to eco-system health (Chap. 7) and livelihoods (Chap. 5), for example, are summarized in the tables of the respective chapters. Complexity becomes even more apparent when one considers the linkages between such concerns and their mutual governability consequences. Simon's (1983) ideas of dealing with complexity can be of help here. If one views the concerns in Part II as a first step in the decomposition of fisheries systems, a next phase might be to recompose them by relating concerns to one another. If Simon (1983) is correct, it should be apparent that not all relations between concerns have an equal impact on each other. By systematically analyzing these interrelations a 'holistic' picture of the complexity of their governability challenges can emerge. It is likely that some of these linkages will be easier to study, while others may remain hidden their governability consequences.

Another procedure is to take a criterion like interdependency as the starting point and analyze it from different angles and intensities. Not all linkages in a fish chain will display the same degree of interdependence or intensity between its parts, e.g. in terms of substitutability. Taking Gell-Mann's (1994) advice that graining is up to the analyst seriously, I suggest that complexity need not always be studied 'till the last detail'.

Dynamics

Dynamics can be seen as a composition of forces that sometimes create gradual developments, but often result in non-linear patterns of change. In an overview of social research, Goldspink (2000) examines the dynamic potential of four recent linear oriented social theories. He argues that functionalists have a strong linear dynamic perspective, while in the eyes of radical humanists the power of ideas forms the main source of societal changes and dynamics. Structuralists (characteristically Marxist) explain concrete social processes in terms of societal dynamics, whereas the self-referring individual tends to self-regulate change patterns in the interpretative view.

According to non-linear theories, dynamic processes are mainly irregular, shockwise, irreversible or chaotic. The theory of non-reversible processes, for example, suggests that open systems, including social ones, are usually unstable or in disequilibrium – they 'fluctuate' constantly. By means of positive feedback processes, these fluctuations can become 'irreversible' (Prigogine and Stengers 1984). Two discoveries underlie 'chaotic' theories of change. One suggests that the initial conditions of a process may be lost in time, while according to the other such processes are critically dependent on initial conditions – the 'butterfly effect'. Although the concept of emergence has very old roots (Aristotle) and is not undisputed (Corning 2002), its leap forward came in the 1980s with scientific work on complex systems. Their emergent properties include radical novelty (features not previously observed), coherence or correlation (integrated wholes), occurrence at the global or macro level, dynamics (arise and evolve over time), and ostensibility (Goldstein 1999).

Each of the theories mentioned thus far has quite a different view, explanation and assumption with respect to the dynamics of modern society, including the fisheries sector. We are therefore left with the following question: which theory or combination of theories should we choose to employ? It is likely that no single theory can explain all instances of societal change. In aquaculture, for example, we may encounter patterns of rapid changes and high dynamics when a new species is brought on the market. In time, however, we can expect stabilizing forces getting the upper hand (Chap. 11). Similarly, a combination of micro-macro analyses can help in understanding the dynamics of phenomena characterized as low, such as patron-client relations (Chap. 5), while macro studies may point to forms of higher dynamics created by the market and, in some cases, state regulation (Chap. 11).

The use of cybernetic principles for handling dynamic qualities of systems, as expressed in terms of positive or negative feedback loops, can be helpful to identify the working of dynamic forces. In positive loops, changes reinforce one another and thus strengthen dynamics. In negative feedback loops, tendencies work to weaken dynamic effects. For example, in the study of the MPA in Chap. 12, where it is shown that internal or external zero-steps in MPA creation are explanatory factors for the dynamics of the way in which they tend to develop and their governability issues, we can discern a 'butterfly' effect. The study of shell-fishing in Galicia (Spain) provides a thorough analysis of the dynamics of the increasing role of gender in which all three sub-systems play a specific role (Chap. 13).

Finally, attention should be paid to the concept of adaptation described in the discussion of fishing livelihoods in Chap. 5. Theories of adaptation, such as complex adaptive systems and resilience and adaptive management thinking, are certainly of interest in the study of fisheries dynamics as they are based on and make use of different notions of dynamics or change.

Scale

Scholarly use of scales has been common in the natural sciences for a long time. In the social sciences, systematic work on the concept is of a more recent date (Gibson et al. 1997). In Chap. 2 of this book we used a simple definition of scale that includes spatial and time dimensions. This definition skirts another analytical dimension; the fact that scale can be used both to measure and to study natural or societal phenomena. The Richter scale, for example, is used to measure the strength of earthquakes, while the term scale, as applied to the distribution of a disease, refers to the scope of study. Scale and level are often used in combination, where level is understood as a unit of analysis located at a position on a scale. The combination of the two meanings can also be seen in socio-political or geographical scales – i.e., international, national, regional, community, household (Grainger 2005).

In interpretations of scale, different types of scales indicate a variety of relations (such as absolute, relative and conceptual) between the units. Generalizations based on theories using scale are, however, not easy to make. Important issues include the transferring of propositions from one scale to another, or the understanding of the workings of nested scales within hierarchical systems. This applies to fisheries as well. Do we, for example, look at a particular a fish farm, a certain type of fish farm or at fish farms as part of the aquaculture sector? In a thorough and critical analysis of the use of the term in small-scale fisheries, Johnson (2006) points out that the contextual meaning of scale is ambiguous, and that the power of its application is of much broader nature. In his view, 'small-scale' is one element in a battlefield between narratives of development, modernization and globalization in fisheries, with serious implications for its governance.

The authors of this volume use scale in different ways. Several chapters contain tables with a column of scale issues in relation to governability. Generally, scale issues in the system-to-be-governed seem to be taken in a spatial sense. By contrast, cross-scale or comparable governability issues are frequently highlighted with respect to governing systems in fisheries. This is in line with the suggestions proposed by Kooiman and Bavinck (Chap. 2), who discuss scale in relation to the boundaries of the system-to-be-governed, on the one hand, and the concept of nestedness for the governing system in particular, on the other.

Three Sub-systems

I regard governability as an integrated quality of any societal system. For analytical purposes, I decompose a societal system into three parts or sub-systems, each with its own specific characteristics. In a synthesis, the three are again combined to create an image of the whole. Many important questions arise around these distinctions, however. One can ask, for example, who or what governs in the end? Is it man or nature? Then, how is one to distinguish operationally between the three sub-systems? Such questions are certainly subjects for future work.

System-to-Be-Governed

In the interactive governance approach, the nature of the system-to-be-governed plays an important role, because we assume that it is a major contributor of governability limits (e.g. Jentoft 2007a). Diversity, complexity, dynamics and scale are variables for analysing and assessing the governability of any system-to-be-governed (also see Chuenpagdee et al. 2008). Other factors such as vulnerability and risk undoubtedly play a role too, but they are of more local importance.

Apart from the possibility of designing a whole new conceptual framework for studying the system-to-be-governed, a more practical choice is to make use of existing frames with sufficient 'governability potential' for our purposes. In our work we have applied two of these to fisheries: fisheries as chains and fisheries as socio-ecosystems (SES). A third frame revolves around concerns or challenges, and is based on our value position that ecosystem health, social justice, livelihood and food security must be taken into consideration when assessing the governability of any fishery. As all three perspectives consist of combinations of worldviews, expert knowledge and principles and values, they might be labelled the three images of fisheries as systems-to-be-governed (see Chap. 2 for an explanation of images).

System-to-Be-Governed as a Chain

From the variety of studies that make use of the chain concept, the Fisheries Governance Network has made most use of the commodity chain approach. The two other approaches mentioned below may be of use in future work on the governability of fish chains, either as alternatives to the commodity chain approach or as supplements.

Global commodity (value) chain (GCC) approaches emphasize the organizational character of the economic integration that takes place around commodities (Gibbon et al. 2008). A GCC consists of "inter-organizational networks clustered around one commodity or product, linking households, enterprises, and states to one another within the world-economy. These networks are situation specific, socially constructed, and locally integrated, underscoring the social embeddedness of economic organization" (Gereffi and Korzeniewicz 1994, 2). Commodity chain studies have contributed much to the knowledge of economic processes. In recent years, their scope has been widened, including more than just economic variables (Bair 2005).

Under the heading of commodity chains one finds studies of world systems, value chains, networks, *filieres*, each with their own special meanings, research communities, ideological and political backgrounds. They deal with textile, automobiles, electronics, food products and cover both the North and South. Of those, I take two that demonstrate a macro and micro approach to a SG, the world-system and the commodity chain.

The *world-systems* approach to commodity chains stresses their role in the structuring and reproduction of capitalist society and the international division of labor at a macro level (Hopkins and Wallerstein 1977, 1986; Kaplinski 2001). They are seen as cyclical contraction and expansion processes over-time (many are historical analyses), with notions of competition at their basis. The theory posits that cyclical patterns have taken place numerous times in the last few centuries. Accordingly, globalization is not understood to be a recent phenomenon, but has instead been in existence for quite some time. This is demonstrated in particular by the emergence of a worldwide commodity system in the sixteenth Century.

The Fisheries Governance Network views fish chains as running from the ecosystem, through harvesting and processing, marketing and distribution, to the end-consumer (Kooiman et al. 2005). Fish chains can be local or regional as they pass from resource to consumer and back. They can also be worldwide as marine life and consumers are sometimes distributed widely across oceans and continents. Various chapters in this volume dedicate sections to fish chains. Chapter 10, for example, pays attention to the different chain roles played by stakeholders. Chapter 13 zeroes in on the (micro) case of mariculture, where shell-fisher women, canning companies, and market dynamics play a role. The authors of the macro analysis of the salmon industry in Chap. 11 use the chain concept to identify a set of governability issues arising from the aquatic environment and the production system, as well as from consumer concerns over food safety. The availability of these examples does not mean that the chain concept has been sufficiently applied to fisheries and aquaculture. It is in many respects a very useful approach. First of all, the chain concept provides a set of images for studying societal field with readily available macro and micro theoretical perspectives. Secondly, the chain concept facilitates the integration of disciplinary insights: natural scientists for the ecological aspects, social scientists on the human side. Thirdly, the concept enables the bridging of geographical scale levels, acknowledging global as well as local processes and events.

System-to-Be-Governed as Socio-ecological System (SES)

There is an immense body of literature on the relation between human and natural systems. Nevertheless, truly interdisciplinary and inter-level perspectives are rare (see e.g. Scoones 1999). Three approaches do, however, come close: Resilience or Panarchy, Long Term Ecological Research (LTER) and Human Ecosystem (HEM). The first of these receives the most attention.

Panarchy, a metaphor of the adaptive (renewal) cycle originally coined by Holling (1973), lies at the basis of resilience studies (Berkes and Folke 1998; Gunderson and Holling 2002; Berkes et al. 2003). It rests on the idea that dynamic natural and societal systems do not tend towards some kind of stable state, but instead pass through phases or cycles. There are four such phases or cycles: exploitation or growth, conservation, release or destruction, and reorganization. The first two are relatively long and marked by slow transformations. The latter two phases, in contrast, are brief periods of time characterized by rapid change. Resilient systems repeat these phases, in which ecological and social memory and novelty play an important role, time and again (Berkes et al. 2003, 16–20).

One of the primary objectives of the *LTER* network is to understand site-specific ecological processes on the basis of systems theory. LTER researchers have recently sought cooperation with social sciences to broaden original ecological approaches (Gragson and Grove 2006). The ultimate aim is to develop a new integrative ecology that explicitly incorporates insights from other disciplines, such as insights on human decisions, cultural institutions, and economic systems.

Although it emphasizes human dimensions, *HEM* tries to integrate both social and biophysical elements into a single approach. A set of critical resources is required to provide the human ecosystem with the necessary services. The social system and these critical resources constantly influence each other, ensuring the adaptation of the social system to new internal and external conditions. It is this interaction that fosters the system's dynamism. Any human ecosystem can be viewed as a nested hierarchy of divergent levels and scales. The model has been devised to account for a large number of aspects. Where necessary, indicators have been developed to adequately capture the approach's features (Force and Machlis 1997).

Members of the Fisheries Governance Network have conducted various studies of coastal zones and fisheries through a socio-ecological systems lens. The tone was set in Chuenpagdee and Jentoft's (2009) paper entitled, "Reality Check". This paper introduced an analytical matrix that, along one axis, listed the three sub-systems and, on the other, the four features. Within the system-to-be-governed, a separate column was reserved to account for both its ecological and socio-economic components (also see Song and Chuenpagdee 2010). This line of inquiry is continued in various chapters of this book. Some chapters focus on research questions regarding the sub-systems (e.g. Chaps. 4 and 5), while others examine, in more descriptive terms, the factors that lead to governability challenges (Chaps. 7, 12, and 13). These are promising lines of inquiry. I believe that it would be worthwhile for follow-up research activities to make use of the conceptual insights generated by the three aforementioned socio-ecological system (SES) research approaches.

System-to-Be-Governed as a Set of Concerns

A third way of conceptualizing the system-to-be-governed is from a concern-driven or a problem-opportunity point of view. This approach is quite common, because most policy, governance and governability studies, in so far as they exist, view systems-to-be-governed from the viewpoint of problems to be solved, opportunities to be made use of, or alternatively, sets of challenges requiring governing attention. Yet, the question of what these problems, opportunities and challenges are remains. This question is easier posed than answered, as the many theories devoted to problem theories show. Even an uncomplicated definition of a problem like, 'a subjectively and negatively experienced difference between an actual and desired situation,' opens the door to a great deal of obscurity (Kooiman 2003). Concern or problem definitions for systems-to-be-governed are social constructions. They are not objective certainties outside the realm of the people (including researchers) who experience them, want to solve them, or study them. The attention span of public arenas largely determines what will filter through and who or what will be the winners and losers in the competition for attention (see section on "Governing Interaction Systems" below).

For many decades, fisheries have received considerable attention. This attention typically concerns the problems fisheries face in terms of natural resource depletion. In part II of this book, which follows-up on *Fish for Life* (Kooiman et al. 2005), the concern approach serves as a foundation for studying the governability of fisheries. This approach culminates in tables that describe fisheries in terms of their diversity, complexity, dynamics and scale. On this basis, increasingly detailed analytical boxes can be composed with appropriate macro and micro components. The four concerns (others may be added) are juxtaposed with one another to form higher-level images of the diversity, complexity, dynamics and scale of a fishery or aquaculture as systems-to-be-governed.

Alternatively, one may use a conceptual methodology I developed earlier (Kooiman 2003). It consists of the following cyclical steps:

Step 1: Stocktaking and ordering of those having a problem or concern – dealing with diversity.
Step 2: Identification of relevant interactions between units and their mutual relations – first reduction of complexity.
Step 3: Bringing the sources of problems or concerns to the surface by locating pockets of tensions in interactions – analysing dynamics.
Step 4: Shift to the solution side in terms of their diversity, complexity, dynamics and scale.
Step 5: Drawing a boundary around the recognized system from the problem and the solution side, focussing on governing options – second reduction of complexity.

The above two routes, a more descriptive one for identifying major characteristics and features of system-to-be-governed, and a conceptually-driven method, are fruitful ways to deal with the features of any fishery.

Governing System

Realms: State, Market, Civil Society, and Hybrids

From the viewpoint of governability, the governing system is the capacity to instigate, organize, implement and evaluate governing activities for societal systems. Individual governors and governing institutions form a governing system together. For many spatially bound fisheries, governing systems can be identified with relative ease, because they are physically situated in, for example, an office. The governing systems for fish chains, however, are frequently more difficult to pinpoint; they are fragmented, fluid and traverse the globe.

An important characteristic of interactive governance is its broad definition of the governing system. Although it is a major governing agency, the state is by no means the sole actor. The market, civil society and the assorted hybrids that exist between them are recognized as partaking in governing action as well. Like the state, they too organize, implement and evaluate governing activities, albeit in a different way than those sanctioned by the state.

Chain Governance

Chain theorists view the market, operating at the firm level, as the main governing agency (Gereffi et al. 2005). They, for example, consider the (agro) food industry to be buyer-driven. Supermarkets, as major lead-firms, use quality requirements in addition to pricing mechanisms as a form of private or market governance (e.g. Konefal et al. 2005). Other types of structures, namely in the form of producer-driven (usually industrial products) and the recently developed processor-driven chains (e.g. coffee) have been added as yet another way of envisioning chain governance (Gibbon 2001). The verb 'to drive' is utilized to indicate that lead firms should be understood as core actors setting parameters for others in a commodity chain. Next to the governance of chains by the market, chain theorists also distinguish external actors: governments and non-governmental organizations (NGOs) that, for example, interfere in food chains for public health and environmental reasons. Network and regime concepts too are used as ways to capture arrangements (e.g. Raynolds 2004; Bush and Oosterveer 2007; Bair 2008), in which private and public actors govern together (Ponte 2002).

Governance of Socio-ecological Systems

There is an abundance of literature that sets out to create a perspective of the adaptive management or governance of socio-ecological systems. Folke et al. (2005) therefore argue that new insights about the nature of change processes in SES ask for the transformation of managing and control capacities from ones that assume stability to those with the capacity to reorganize. Adaptive governance theories extend this approach by including social contexts, creating conditions for ordered rule-making and collective action, establishing institutions for social coordination, and allowing for broad-based decision-making and power-sharing (Gregory et al. 2006; Gunderson and Light 2006; Pahl-Wostl 2007; Rammel et al. 2007).

In applying these observations to the role of the governing system in fisheries, as discussed in this volume, it is clear that the state is predominantly seen as the principal governor. Almost every chapter in this book alludes, in one way or another, to this role. However, it should also be abundantly clear that such governance is often sub-standard and faces severe limitations (Jentoft 2007a). It is too easy in this connection to blame politics for not taking its responsibilities seriously, or to accuse authorities of inefficiency and not being up to their tasks. Entry point is therefore the idea that expectations of what governance is or can do are simply too high or tasks are too difficult in relation to governance capacities. This is not a plea to let things go. Instead, it is an appeal for a better understanding of what might be behind these limitations. This applies not only to the state as governor, but to the market and civil society as well. Future work on the governing systems of fisheries at the micro and macro level should pay closer attention to the role of the market and civil society.

Capacity of the Governing System

State, market, civil society and hybrid governors use a variety of means to carry out their governing activities. I call this composite set of variables of the governing system its governance capacity. One can find several approaches to state capacity and capacity-building in the literature. One line of work focuses on task attributes with lists as political, regulatory, administrative, technological, extraction and delegational (Olowu 2002). Cummings and Nørgaard (2004) develop a model of a more conceptual nature in which four dimensions for state capacity are elaborated: ideational, political, technical and implementational. Another conceptual framework aims at assessing capacity gaps where tasks are not listed, but seen as those defined by necessity, history, or the situation in specific contexts (Grindle 1997).

The capacity of the governing system to govern, or the lack thereof, is probably one of the most studied in fisheries research. In *Fish for Life*, which is essentially a study of this capacity, we have made our own contribution to the debate. However, this is not to say that we know enough. In the research strategy proposed at the start of this chapter, the study of the system-to-be-governed and governing interactions in fisheries should in my opinion have higher priority than the study of the governing system itself – maybe with the exception of the governing role of the market and civil society.

I will not discuss the three elements (images, instruments and action) and the three orders (problem-solving/opportunity creation, institutional care and meta governance) as attributes of capacity separately. It is clear that each of these contribute to governing capacity in and of themselves. What is of special importance, however, is the way in which they are linked. That is, whether they are consistent or fit with each other. For example, it might be hypothesized that the dominance of eco-economic policies in fisheries management in the last few decades can be explained by the coherence of a set of images matched by a set of top-down instruments as well as ideology-based political support. To address this imbalance, counter images, instruments and action potentials should be developed as well. The usage of the 'rose' image of fisheries governance as opposed to the conventional 'pyramid' (Jentoft et al. 2010) is a good attempt in this direction. Several chapters of this book, such as the one on MPAs (Chap. 12) and another on mariculture (Chap. 11), demonstrate similar tendencies.

The same kind of reasoning may be used for coherence between the three orders of governing. In Chap. 4 of this volume, for example, Jentoft develops a coherent set of ideas that show how the three governance orders can sustain a model for the role of social justice in the governance and governability of fisheries. In my opinion, institutions play a key two-way, connecting role between problem-solving/opportunity creation, on the one hand, and the application of meta principles, on the other. Research efforts should be directed to the study of the role of the value base and the promotion of a principle based vision of the governance of fisheries. Chapter 15 provides a good example of efforts to this end.

Governing Interaction Systems

I consider governing interaction systems as the whole of all interactions between governing entities, public as well as private (Kooiman 2003). They do not exist without their two companion sub-systems, the system-to-be-governed and the governing system, and constitute as it were the embodiment of the traffic between them. In societal interaction systems, interactions have two main directions: one emanating from the system-to-be-governed to the governing system and the other in the opposite direction. In a full-scale interaction system, the two directions may become so intertwined that they assume a circular form. The more an interaction system is carefully fashioned, the more it will display its own system features. Horizontal as well as vertical aspects of governing interactions can be distinguished, and the actor and structural dimensions can be spelled out as well.

The Actor Level

As argued in Chap. 2, many aspects of governing interactions are relevant for governability. However, some have yet to be addressed and deserve attention here. Governance interactions, in whatever form, can be seen as mixes of cooperation and conflict. After all, societal governance usually deals with highly contentious sociopolitical issues. Especially long-lasting governance interactions will show more cooperative periods alternating with more conflictive ones. Some interactions can exhibit both phenomena at the same time; in certain areas there will be cooperation, while in others interests can clash and interactions may be competitive. Governance interactions can be arranged on a scale from (highly) conflictive to (highly) cooperative with all kinds of shades in between. The TWINS matrix developed in the field of trans-boundary water governance is a useful example (Zeitoun and Mirumachi 2008).

Not all governing interactions have the same intensity, however. Some are deliberative, while others play a role in decision-making. In fact, one could devise a scale of intensity varying from 'light' to 'heavy'. The following are examples that illustrate points on such a scale.

Deliberation has gradually become a central concept in thinking about democracy, where it is seen as a necessary supplement to classical 'electoral' forms. Deliberative democracy is advocated as a means of bridging differences, yielding policies that are more legitimate, rational and just (Rosenberg 2007, 335–336). I adhere to a somewhat narrower interpretation by which deliberation might lead to a better understanding of the diversity, complexity and dynamics (and other aspects) of governability issues. However, we also need to keep an open eye for hidden agendas, strategic behaviour, motives other than to convince, power games and the possibilities of skewed participation and exclusion (see e.g. Bächtiger and Steiner 2005). The higher the expectation of the role of deliberation in democratic governance, the more one has to be aware of the possibility of hidden agendas.

Interactive learning is a second important form of governing interaction. Typically, governance addresses many concerns that are difficult to reconcile and the capacity to learn is a major part of such processes (Jentoft et al. 2009; Mahon et al. 2005). The question is not only one of substance, i.e. what should be learned, but also one of process, i.e. how to learn how to learn interactively. There are several ways to improve learning capacities in governing interactions. For example, there is what is called social learning, which incorporates strong notions of learning-by-doing (see Armitage et al. 2007), and inter-organizational or network learning (Carlton and Payne 2003), with a methodology for multi-party learning processes that stress the interdependence of parties (Bouwen and Tallieu 2004).

Interactive decision-making is the most intensive type of governing interactions. Without decisions, interactions remain somewhat of a half empty glass; the pudding has to be eaten to become real. What decisions are, how decisions are made and how to analyze them has been the subject of much discussion (Parsons 1995). There are schools that view decisions as rational, ordered and comprehensive. They are formally organized and based on expert knowledge. Others consider decisions to be incremental, step-by-step and less organized. Here, common sense, practical experience and indigenous knowledge dominate. Questions about (interactive) decision processes are especially useful (Hogwood and Gunn 1984, 49 ff). It is important to ask, for example, whether or not there are time constraints or political overtones; whether positions among participants have been fixed; how central the decision is for those concerned; and, how value-laden the decisionto be taken is.

With interactions central to the interactive governance perspective, it is logical that members of the Fisheries Governance Network have regarded them as an essential part of their governability studies. For example, when comparing two MPA cases, one in Canada and the other in Mexico, Chuenpagdee (2011) gave interactions a central place, analysing them as a 'new lens' in their governance. Almost all chapters in this book pay attention to interactions in some form. Some chapters do this in a descriptive sense. Chapter 16, for example, presents a systematic analysis of interactions as networks. Chapters in part III explore their role in the actual governing of fisheries, while chapters in part II zero in on interactions as a source for asking questions about fisheries.

The actual picture of interactions taking place in fisheries governance is rather bleak. The chapters in this book highlight their ad-hoc and haphazard nature, a lack of mutual understanding and the dominance of state authorities. In this context, it is important to raise concerns about the need for participation and interaction with other groups of societal stakeholders. In the chapter on food security (Chap. 6) for example, Pullin argues for a broader definition of governance interactions. Instead of the usual mono-sectoral management definition, he espouses a cross-sectoral approach that seriously considers the nature of a broader set of governance interactions.

The Structural Level

We can also focus on the last term in the concept of governing interaction systems – the notion of 'system'. 'System' suggests that we look at phenomena as wholes. Diversity, complexity and dynamics play a considerable role here and must be considered. Similarly, scale, boundaries, nestedness and many other system attributes warrant investigation. I will focus on three aspects that are of prime importance when we looking at governing interactions in relation to governability. I will firstly mention the features, diversity, complexity, dynamics and scale, as they are at the center of our governability approach. Secondly, I want to call attention to what Jentoft considers one of the most neglected issues in fisheries governance, power (2007b). Thirdly, I will discuss the mix of the three governance modes, self, co- and hierarchical, as a special subject for future research.

Features

The *diversity* of the interaction system can be expressed by the ways in which such systems differ from each other. One might even go so far as to suggest that, in relation to its governing system, every system-to-be-governed has its own interaction system. For investigating the *complexity* of a governing interaction system one has to realize that governance interactions and modes have multiple interrelations and often overlap. Thus, 'corporate actors' or 'policy entrepreneurs', for example, derive their role and influence from their participation in many governing interactions, and use these strategically for special purposes. With respect to the *dynamics* of the governing interaction systems, one may hypothesize that shock changes will appear more frequently at the actor level, while evolutionary shifts between modes of governance can be expected to occur at the structural level. Finally, it will be quite apparent that *scale* is an important variable in the study of governing interaction systems. For example, a small-scale fishery in India will show quite different interaction patterns than those characteristic of the European Union.

Systematic analyses of interaction patterns at the actor and structural levels are relatively scarce. This book is a step in that direction, and an introduction to such analyses can certainly be found in its chapters. The study of the role of Galician women in mariculture activities (Chap. 13) makes it quite clear that interactions at all governing levels played an important role in increasing the governance position of these women, as well as in contributing to the governability of the resource. Looking at the discussions in the chapters of this book, however, this is more of a promise for the future than a governance reality. What many of the authors in this book show (e.g. in the summary matrices of Part II) are of the types of (research) questions that should be asked about interactions at actor and structural levels.

Power

Applying influence and bringing power into play in governance are facts of life that remain difficult to pinpoint and study. In addition to the general literature on the nature of power (see for a recent overview and discussion Jentoft 2007b), all kinds of useful distinctions between types of power can be found. Numerous typologies of power are potentially relevant for governance studies: episodic, dispositional, and facilitative power (Clegg 1989); relational, dispositional and structural power (Arts and van Tatenhove 2005); and, strategic, institutionalized and structural power. Jentoft (2007b) points to the fact that, within governance, power can be viewed in a productive sense. It makes governance possible and supports it. It may also be construed as destructive force when, for example, it blocks useful governing efforts. It is because the interaction system reflects the attributes of the system-to-be-governed and the governing system in many ways that it is a suitable conceptual and empirical location to study power.

Although the difficulties of studying power in a satisfying manner remains one of the reasons why it is, as Jentoft (2007b) claims, an under-researched subject with respect to fisheries and coastal zones, there are good examples of methodologies for analysing power and influence in decision processes. The six sources of power – authority, status, information, expertise, solidarity and finances – distinguished by Lasswell and Kaplan (1950) and others provide a starting point (Kooiman 1970). These sources can be combined with the types of power distinguished above to form a framework for analysis in interaction systems that may potentially be expanded to cover both the governing system and the system-to-be-governed.

Mixes of Governing Modes

Chapter 2 introduced the three modes of self, co- and hierarchical governance. Although much more can be said about them individually, the most important governability challenge resides in their combination, mix or fit. This is unchartered theoretical territory and the scope of this chapter is insufficient to even scout this field. It is certain that each societal sector has its own mix that reflects the particular relations that exist between the system-to-be-governed and the governing system. There are sectors where the weight in the balance of the different modes is on the hierarchical side (police, education), with minor contributions of co- and self- governance. In other sectors and countries where, for example, a (neo) corporatist political culture exists, co-governance modes are well known and of great importance.

The application of these ideas on mixes of governance modes to fisheries is, in and of itself, mixed. Much of the existing work has been done by members of Fisheries Governance Network and refers largely to the co-governance mode of governance. Less research energy has clearly been spent investigating the two other modes, at least in a systematic manner.

There are interesting initiatives in some areas of study regarding governance modes (e.g. Chap. 10). In *Fish for Life* (Kooiman et al. 2005, 221), we argued that self-governance often results in some degree of de-stabilization. On the contrary, self-governance may also have a stabilizing effect in situations of rapid change. We also mention that co-governance frequently requires a stick based on hierarchical governance in the background if not for other reasons than to counter free-rider behaviour. These earlier observations still hold true, and it is certainly worthwhile to pursue and bring them into a more generalized context.

Fit: The Relation Between Sub-systems

One of the central propositions in our approach to the governability of any societal system is that governability largely depends on the relationship between the system-to-be-governed, the governing system, and the governing interaction system. For example, one might hypothesize that the more the features of a system-to-be-governed are represented in or matched by a governing system, the more a system is governable. One might also expect to find that the better an interaction system represents attributes of a system-to-be-governed and a governing system, the better it can play its role as 'in-between'. Several concepts are available to try to express such relations. Chief among them are terms like representation, fit, match and correspondence. But what does it mean for the features of a system-to-be-governed like a (fish) chain to be represented, corresponding or reflected in its governing system? Is this a case of parallelism, whereby a particular pattern of diversity, complexity and dynamics in the fish chain is, for example, mirrored in the governing system? Or is it a matter of adjustment, whereby a governing system makes maximum use of the opportunities that arise?

Members of the Fisheries Governance Network have formulated ideas that tackle these questions in various studies. In an earlier publication, I phrased a set of propositions on governability using representation as central concept. I posited, for example, that the representation of societal features, in particular diversity, complexity and dynamics, can be differentiated according to the elements and modes of governance (Kooiman 2003, 206). Additionally, Mahon et al. (2005) argue that the dynamics affecting fish chains should be matched by the dynamics of the 'learning organization'. Here, dynamics are met with dynamics – a clear instance of mirroring. According to the same authors, the diversity and complexity of fish chains should be matched by 'partnership' between governing actors. This is not a matter of one-to-one reflection, but a case of positive attunement (Mahon et al. 2005). Jentoft argues that the diversity of the system-to-be-governed asks for contextual sensitivity on the part of the governing system, and thus must become structurally diverse itself (2007b). The question as to whether the term representation cannot be better replaced by reflection and attunement is further raised in a study of the fisheries in the Bay of Bengal (Bavinck and Salagrama 2008). The issue is taken up again in Chap. 8, when the concept of compatibility is introduced next to attunement in the study of matching between the sub-systems of India's Palk Bay. Compatibility is seen as a match in spatial scale and the answer for mismatches in structured interactions. Attunement is used for (mis-) matches in diversity and dynamics, while responses in modes of governance demands are seen as responses to governing demands formulated in the system-to-be-governed.

Now this might all look as though it is merely a question of terminology, all 'representing' in different terms but basically being the same central idea. There is, however, more to it than that. We have progressed far enough to propose further steps in the development of this basic idea of representation. Pitkin's (1967) distinction between a more factual or descriptive type and a more indefinite or symbolic version of representation might help us further (cf. Cilliers 2005). I suggest that we can treat the relation between the system-to-be-governed, governing system, and interaction system, and particularly the idea of the representation of diversity, complexity and dynamics at different scales, in these two ways. The symbolic version is a two-way interpretation process from the system-to-be-governed to the governing system in terms of reflecting these features, while the factual version encourages us to think in terms of attuning. Both are researchable processes, but each is of a different character: the first is more synthetic and interpretative, while the second is more analytic and descriptive. In both of these processes, governing interactions play a key role.

Conclusion

In this chapter I have gone through some of the major components of the conceptual framework developed around governability. I have paid attention to both theoretical issues as well as applications in fisheries. Anyone who has followed the work of the Fisheries Governance Network can see that this is only a minor selection of the topics raised around governance in the last few years. Our efforts have made it abundantly clear that governability is not only a rich concept and a source for further theoretical work, but also that it can be fruitfully applied to societal sectors or systems like fisheries, aquaculture and coastal zones. Furthermore, there is no reason to assume that these systems represent the concept's limits. Governability studies may be useful for other societal fields as well.

As is to be expected when working with a concept like governability, more questions emerge as new insights are gained. Its potential richness has only just been skimmed. It is my strong belief that it can stand its own in the field of (fisheries) governance studies; hopefully this chapter will have contributed to the credibility of this statement.

References

Alexander, J.C., Giesen, B, Münch, R., Smelser, N.J. (Eds.). (1987). The micro-macro link. Berkeley: University of California press.
Armitage, D., Berkes, F., Doubleday, N. (Eds.). (2007). Adaptive co-management: Collaboration, learning, and multi-level governance. Vancouver: University of British Columbia Press.
Arts, B., & van Tatenhove, J. (2005). Policy and power: A conceptual framework between the 'old' and 'new' policy idioms. Policy Sciences, 37, 339–356.
Baechtiger, A., & Steiner, J. (2005). Introduction to special issue on political deliberation. Acta Politica, 40, 153–168.
Bair, J. (2005). Global capitalism and commodity chains: Looking back, going forward. Competition and Change, 9(2), 153–180.
Bair, J. (2008). Analying economic organization: embedded networks and global chains compared. Economy and Society, 37/3, 339–364.
Bavinck, M., & Salagrama, V. (2008). Assessing the governability of capture fisheries in the Bay of Bengal – A conceptual enquiry. The Journal of Transdisciplinary Environmental Studies, 7(1), 13 pp.
Berkes, F., & Folke, C. (Eds.). (1998). Linking social systems and ecological systems. Cambridge: Cambridge University Press.
Berkes, F., Colding, J., Folke, C. (Eds.). (2003). Navigating social-ecological systems. Cambridge: Cambridge University Press.
Bouwen, R., & Tallieu, T. (2004). Multi-party learning as social learning for interdependence. Journal of Community & Applied Social Psychology, 14, 137–152.
Brewer, G.D. (1999). The challenges of interdisciplinarity. Policy Sciences, 32, 327–337.
Bruce, A., Lyall, C., Tait, J., & Williams, R. (2004). Interdisciplinary integration in Europe. Futures, 36, 457–470.
Bush, R.S., & Oosterveer, P. (2007). The missing link: Intersecting governance and trade in the space of place and the space of flows. Sociologia Ruralis, 47/4, 384–399.
Cilliers, P. (1998). Complexity and postmodernism. London: Routledge.
Cilliers, P. (2005). Knowledge, limits and boundaries. Futures, 37, 605–613.
Carlton, J.M., & Payne, S.L. (2003). Coping with paradox: Multi stakeholder learning dialogue as a pluralist sense making process for addressing messy problems. Business and Society. 42/1, 7–42.
Chuenpagdee, R. (2011). Interactive governance for marine conservation: An illustration. Bulletin of Marine Science, 87(2), 197–211.
Chuenpagdee, R., & Jentoft, S. (2009). Governability assessment for fisheries and coastal systems: A reality check. Human Ecology, 37, 109–120.
Chuenpagdee, R., Kooiman, J., Pullin, R. (2008). Assessing governability in capture fisheries, aquaculture and coastal zones. The Journal of Transdisciplinary Environmental Studies, 7(1), 20 pp.
Clegg, S.R. (1989). Frameworks of power. London: Sage.
Corning, P.A. (2002). The re-emergence of "Emergence": A venerable concept in search of a theory. Complexity, 7(6), 18–30.
Cummings, S., & Norgaard, O. (2004). Conceptualising state capacity. Political Studies, 52, 685–708.
Folke, C., Hahn, T., Olsson, P., Noerberg, J. (2005). Adaptive governance of social-ecological systems. Annual Review of Environmental Resources, 30, 441–473.
Force, J.E., & Machlis, G.E. (1997). The human ecosystem part II: Social indicators in ecosystem management. Society and Natural Resources, 10(4), 369–387.
Gell-Mann, M. (1994). The quark and the jaguar. London: Little, Brown and Company.
Gereffi, G., & Korzeniewicz, M. (Eds.). (1994). Commodity chains and global capitalism. Westport: Praeger.
Gereffi, G., Humphrey, J., Sturgeon, T. (2005). The governance of global value chains. Review of International Political Economy, 12(1), 78–104.

Gibbon, P. (2001). Upgrading primary production: A global commodity chain approach. World Development, 29(2), 345–363.
Gibbon, P., Bair, J., Ponte, S. (2008). Governing global value chains: An introduction. Economy and Society, 37(3), 315–318.
Gibson, C., Ostrom, E., Ahn, T.K. (1997). Scaling issues in the social sciences (IHDP Working Paper No. 1). Berlin: IHDP.
Gragson, T.L., & Grove, M. (2006). Social science in the context of the long-term ecological research program. Society and Natural Resources, 19, 93–100.
Grainger, A. (2005). The role of spatial scale and spatial interactions in sustainable development. In M. Purvis & A. Grainger (Eds.), Exploring sustainable development (pp. 51–84). London: Earthscan.
Gregory, R., Ohlson, D., Arvai, J. (2006). Deconstructing adaptive management. Ecological Application, 16(6), 2411–2425.
Goldspink, C. (2000). Contrasting linear and nonlinear perspectives in contemporary social research. Emergence, 2(2), 72–101.
Goldstein, J. (1999). Emergence as a construct: History and issues. Emergence, 1(1), 49–72.
Grindle, M.S. (Ed.). (1997). Getting good government: Capacity building in the pubic sectors of developing countries. Cambridge: Harvard University Press.
Gunderson, L.H., & Holling, C.S. (Eds.). (2002). Panarchy. Washington, DC: Island Press.
Gunderson, L., & Light, S.S. (2006). Adaptive management and adaptive governance in the everglades ecosystem. Policy Sciences, 39, 323–334.
Hall, P., & McGinty, P.J.W. (2002). Social organization across space and time: The policy process, mesodomain, and breadth of perspective. In S.C. Chew & J.D. Knottnerus (Eds.), Structure, culture, and history (pp. 303–323). Lanham: Rowman and Littlefield.
Hogwood, B.W., & Gunn, L.A. (1984). Policy analysis for the real world. Oxford: Oxford University Press.
Holling, C.S. (1973). Resilience and stability of ecological systems. Annual Review of Ecology and Systematics, 4, 1–23.
Hopkins, T., & Wallerstein, I. (1977). Patterns of development of the modern world-system. Review, 1(2), 11–45.
Hopkins, T., & Wallerstein, I. (1986). Commodity chains in the world prior to 1800. Review, 10(1), 157–170.
Holland, J.H. (1995). Hidden order. Reading: Perseus Books.
Jentoft, S. (2007a). Limits of governability? Institutional implications for fisheries and coastal governance. Marine Policy, 4, 360–370.
Jentoft, S. (2007b). In the power of power: The understated aspect of fisheries and coastal management. Human Organization, 66(4), 426–437.
Jentoft, S., Friis, P., Kooiman, J., & van der Schans, J.W. (2009). Knowledge-based fisheries: Opportunities for learning. In J. Kooiman, M. van Vliet, S. Jentoft (Eds.), Creative governance: Opportunities for fisheries in Europe (pp. 239–259). Aldershot: Ashgate.
Jentoft, S., Chuenpagdee, R., Bundy, A., & Mahon, R. (2010). Pyramids and roses: Alternative images for the governance of fisheries systems. Marine Policy, 34, 1315–1321.
Johnson, D.S. (2006). Category, narrative and value in the governance of small-scale fisheries. Marine Policy, 30, 747–756.
Kaplinski, R. (2001). Globalisation and unequalisation: What can be learned from value chain analysis. Journal of Development Studies, 327(2), 117–146.
Kaufmann, S. (1995). At home in the universe. London: Viking.
Kickert, W.J.M. (1979). The organization of decision-making. Amsterdam: North Holland
Konefal, J., Mascarenhas, M., & Hatanaka, H. (2005). Governance in the global agro-food system: Backlighting the role of transnational supermarket chains. Agriculture and Human Values, 22, 291–302.
Kooiman, J. (1970). Besturen is beslissen. Meppel: Boom.
Kooiman, J. (2003). Governing as governance. London: Sage.

Kooiman, J., Bavinck, M., Jentoft, S., Pullin, R. (Eds.). (2005). Fish for life: Interactive governance for fisheries. Amsterdam: Amsterdam University Press.

Lasswell, H.D., & Kaplan, A. (1950). Power and society. New Haven: Yale University Press.

Maasen, S., Lengwiller, M., Guggenheim, M. (2006). Practices of transdisciplinary research. Science and Public Policy, 33(6), 394–398.

Mahon, R., Bavinck, M., Roy, R. (2005). Governance in action. In J. Kooiman, M. Bavinck, S. Jentoft, R. Pullin (Eds.), Fish for life: Interactive governance for fisheries (pp. 351–374). Amsterdam: Amsterdam University Press.

Olowu, D. (2002). Capacity building for policy management through twinning: Lessons from a Dutch-Namibian case. Public Administration and Development, 22, 275–288.

Pahl-Wostl, C. (2007). Transition towards adaptive management of water facing climate change and global change. Water Resource Management, 21, 49–62.

Parsons, W. (1995). Public policy. Cheltenham: Edward Elgar.

Pitkin, H. (1967). The concept of representation. Berkeley: University of California Press.

Ponte, S. (2002). The 'Latte revolution'? Regulation, markets and consumption in the global coffee trade. World Development, 7, 1099–1122.

Prigogine, I., & Stengers, I. (1984). Order out of chaos. New York: Bantam.

Rammel, C., Stagl, S., & Wifling, H. (2007). Managing complex adaptive systems. Ecological Economics, 63, 9–21.

Raynolds, L.T. (2004). The globalization of organic agro-food networks. World Development, 32(5), 725–743.

Rosenberg, S.W. (2007). Rethinking democratic deliberation: The limits and potential of citizen participation. Polity, 39(3), 335–360.

Scoones, I. (1999). New ecology and the social sciences. Annual Review of Anthropology, 28, 479–507.

Simon, H.A. (1983). The architecture of complexity. Proceedings of the American Philosophical Society, 106, 467–482.

Song, A.M., & Chuenpagdee, R. (2010). Operationalizing governability: Case study of a Lake Malawi fishery. Fish and Fisheries, 11, 235–249.

Zellmer, A.J., Allen, T.F.H., Kesseboehmer, A. (2006). The nature of ecological complexity. Ecological Complexity, 3, 171–182.

Zeitoun, M., & Mirumachi, N. (2008). Transboundary water interaction I: Reconsidering conflict and cooperation. International Environmental Agreements: Politics, Law and Economics, 8(4), 297–316.

Author Bios

Sharon Almerigi is Principal Associate of People Dynamics Associates in Barbados and a Certified Professional Facilitator. She provides facilitation services and capacity-building training that includes the Technology of Participation (ToP) facilitation methods, teambuilding and leadership skills. A certified mediator she teaches conflict resolution, negotiation and mediation.

Maarten Bavinck is associate professor at the Department of Human Geography, Planning and International Development Studies of the University of Amsterdam, The Netherlands. He is co-founder of the social science Centre for Maritime Research (MARE). Maarten specializes in coastal and marine issues, with special relevance to India and Sri Lanka.

Ratana Chuenpagdee is Canada Research Chair in Natural Resource Sustainability and Community Development, Memorial University. Her research emphasizes interdisciplinary, innovative and interactive approaches to coastal, fisheries and ocean governance. She is currently leading a major initiative called Too Big to Ignore, which examines issues and concerns related to small-scale fisheries around the world.

Raquel De la Cruz Modino is a researcher at the Institute of Social and Political Sciences, University of La Laguna (ULL, Tenerife, Spain). She obtained her PhD in 2008 at ULL, combining fisheries, tourism and MPA analysis in the Canary Islands and Spain's Mediterranean coast. The research was awarded the national prize, Marqués de Lozoya, in 2009. She has since undertaken post-doc training at the University of Tromsø, Norway and Memorial University, St. John's, Canada.

Lucia Fanning is Associate Professor and Director of the Marine Affairs Program at Dalhousie University, Halifax, Canada. Her research focuses on the roles of policy networks in influencing marine policy decisions; the assessment and effectiveness of evolving oceans governance regimes in managing marine uses; ecosystem-based approaches to the management of coastal and marine space and use.

M. Bavinck et al. (eds.), *Governability of Fisheries and Aquaculture: Theory and Applications*, MARE Publication Series 7, DOI 10.1007/978-94-007-6107-0,

Katia Frangoudes is a senior researcher at UMR AMURE, joint research Unit, of University of Brest (France). She specializes in women in fisheries and aquaculture in Europe and institutional issues in fisheries management.

Svein Jentoft is a sociologist and a professor at Norwegian College of Fishery Science, University of Tromsø, Norway. He has published a great number of books and journal articles on fisheries and coastal zone management. He is the co-editor of the recent book, *Poverty Mosaics: Realities and Prospects in Small-Scale Fisheries* (Springer 2011).

Derek Johnson is an Associate Professor in the Department of Anthropology at the University of Manitoba. In addition to interests in governance and governability, he also works on the political ecology of development in fisheries and other natural resource sectors. His major research focus is the fisheries of Gujarat State, India.

Jan Kooiman is Em. Professor of Public Organization and Management at the Erasmus University Rotterdam and Guest Scholar at the University of Amsterdam. He studied Dutch and comparative Parliaments, public organization and management, and has recently concentrated on conceptual studies of governance and governability and its application to fisheries.

Yajie Liu has a PhD in Resource and Environmental Economics from University of British Columbia, Canada, and currently a research scientist with SINTEF Fisheries and Aquaculture, Norway. Her research interests include economics and management of marine resources with a particular expertise in applied bio-economic modeling, productivity and efficiency analysis, and valuation of environmental issues.

James R. McGoodwin is author of *Crisis in the World's Fisheries: People, Problems, and Policies* (Stanford University Press, 1990). A former Fellow in the Marine Policy and Ocean Management Program of the Woods Hold Oceanographic Institution, he is *Professor Emeritus* in the Department of Anthropology, and Research Affiliate in the Institute of Arctic and Alpine Research at the University of Colorado, Boulder, Colorado.

Robin Mahon is Professor of Marine Affairs and Director of the Centre for Resource Management and Environmental Studies (CERMES), University of the West Indies, Barbados. His research is on marine resource governance, in particular, institutional arrangements for governance in complex systems and assessment of governance arrangements for transboundary systems.

Begoña Marugán Pintos is a lecturer at the University Carlos III (Madrid, Spain). She obtained her PhD in Sociology at the University Complutense of Madrid. Her research focuses on applying gender perspectives in Spanish fisheries, specifically to Galician shell-fishers (mariscadoras), and has been published nationally and internationally. She has also worked for the Spanish Government's Social Marine Institute.

Patrick McConney is Senior Lecturer in Marine Resource Management Planning at the Centre for Resource Management and Environmental Studies (CERMES), The University of the West Indies (UWI) Cave Hill Campus in Barbados. His work covers many aspects of small-scale fisheries and marine protected areas, but especially governance, livelihoods, socio-economics and developing capacity.

Paul Onyango is lecturer at the Department of Aquatic Sciences and Fisheries University of Dar es Salaam, Tanzania. His research focuses on fisheries management, governance challenges of poverty alleviation among small-scale fishers, optimal fishing and socio-economic aspects of aquaculture. Paul has 18+ year's research experience in small-scale fisheries in East Africa.

José J. Pascual-Fernández is Professor Titular of Social Anthropology at the University of La Laguna (Tenerife) since 1993. Since the 1980s, his research has been related to fisheries, focusing most recently on governability and MPAs in Spain. He is currently the Director of the Institute of Social and Political Sciences at the same University.

Roger Pullin lectured in Marine Biology for the University of Liverpool and was Senior Scientist at the International Center for Living Aquatic Resources Management. Based in Manila, Philippines, he is currently a Consultant for responsible aquaculture development, conservation and sustainable use of aquatic biodiversity, and oil spill damage assessments.

Joeri Scholtens holds an MSc in Environment and Natural Resource Management from the VU University Amsterdam. He currently conducts his PhD at the Centre for Maritime Research, University of Amsterdam. His research revolves around governance arrangements for fisher and resource conflicts, with a regional focus on South Asia.

Andrew Song is a doctoral candidate in the Department of Geography, Memorial University of Newfoundland, Canada. His research interest lies in governance and institution with regard to natural resources, with particular attention to fisheries. He holds some familiarity with several fishery settings including the Philippines, Malawi, and most recently South Korea.

U. Rashid Sumaila, Professor and Director, Fisheries Centre, the University of British Columbia, has authored over 125 journal articles; won awards including the Leopold Leadership Fellowship and the Pew Fellowship for Marine Conservation; and given talks at the UN, the White House, the Canadian Parliament and the British House of Lords.

Index

M. Bavinck et al. (eds.), *Governability of Fisheries and Aquaculture: Theory and Applications*, MARE Publication Series 7, DOI 10.1007/978-94-007-6107-0,

Printed by Printforce, the Netherlands